DOS
UTILITIES
SORT/MERGE
MULTIPROGRAMMING

UTILITIES
SORT/MERGE
MULTIPROGRAMMING

By:

Gary B. Shelly
Educational Consultant
Instructor, Long Beach City College

&

Thomas J. Cashman, CDP, B.A., M.A.
Long Beach City College
Long Beach, California

ANAHEIM PUBLISHING COMPANY
1120 E. Ash, Fullerton, Ca. 92631

Seventh Printing

May 1979

ISBN 0-88236-275-5

PREFACE

The widespread use of the Disk Operating System includes not only the use of the operating system for executing user-written programs but also the use of the service programs which are a part of DOS. In the past, this use of the service programs provided with the Disk Operating System has, for the most part, not been included in the curriculum for beginning data processing students. It is the intent of this textbook to introduce the student to the Utility and Sort/Merge programs which are a part of the Disk Operating System and to provide practical examples of these programs as they are commonly used in many business data processing installations. In addition, the concept of DOS multiprogramming and examples of the practical use of multiprogramming are presented to familiarize the student with this important DOS capability.

The student is first given an overview of the Disk Operating System including the Utility Programs, the Sort/Merge program, the use of Multiprogramming, and the concept of Batched-Job Processing. The Unit-Record Utility programs, card-to-card and card-to-print, are then discussed including a simple 80-80 listing, a simple card reproducing example, and the use of the Field Select options available with the card-to-card and card-to-print utility programs. These programs are explained through the use of complete job streams which have been tested and run on a System/360 operating under DOS.

The utility programs which process direct-access files are explained next and include the card-to-disk, the disk-to-print, the disk-to-card, and the disk-to-disk utility programs. All of the possible variations of these utility programs are illustrated through the use of fully tested job streams. They include copying, reblocking, and field selecting for all of the programs. Field selection used with the direct-access utility programs includes omitting fields, packing and unpacking numeric fields, and formatting printed reports. All of the required DOS job control statements, including the ASSGN, DLBL, and EXTENT job control cards, are illustrated together with the Utility control statements. Both the list and the display formats of the print utility programs are illustrated as well as the use of an indexed sequential file as input to the disk-to-print utility program. The magnetic tape utilities are completely covered also including labelled and unlabelled tapes, multi-file volumes, and multi-volume files. All of the options available to the programmer are illustrated by showing complete and ready-to-be-executed job streams.

The Disk Operating System Tape and Disk Sort/Merge Program (360N-SM-483) is explained in detail including tape input and output, disk input and output, disk and tape work files, multiple control fields, multiple input files, fixed and variable length records, unlabelled tape files, I/O device pooling, and the use of Exits from the sort program. All of the options which are available to business programmers are explained again through the use of fully-tested job streams.

The concept and use of DOS Multiprogramming is covered by presenting examples of the use of an interrupt and then building on this concept to illustrate how multiprogramming is possible. The student is then introduced to the required job control and attention routine commands to be used for both Single Program Initiation and Batched-Job processing in a multiprogramming environment. The rules and assignment statements used for placing System Logical units on magnetic tape are covered in detail together with the methods of loading a "SYSIN" input tape and printing the "SYSLST" output tape.

This text assumes that the student is familiar with Assembler Language and COBOL although no examples in the text require a detailed knowledge of any programming language. Student assignments are included at the end of the chapters on utilities and the sort/merge program to enable the student to have experience in writing job streams utilizing these programs. In addition, test data is included in the appendices which may be used with these assignments so that the job streams may be processed on a System/360 operating under DOS.

After the student has completed the study of the material in this text, he should be familiar with all of the capabilities of the DOS Utility programs and the DOS Sort/Merge Program. In addition, the job streams presented in the text provide a good reference for practicing programmers who have need of using the utility or sort/merge programs. The understanding of the concept of multiprogramming will aid the student and the practicing programmer in fully utilizing the capabilities of the Disk Operating System.

This text may be effectively used in a course dealing with DOS Concepts and Facilities. This text, together with DOS JOB CONTROL FOR COBOL PROGRAMMERS or DOS JOB CONTROL FOR ASSEMBLER LANGUAGE PROGRAMMERS, Anaheim Publishing Company, forms a complete introduction to and the use of the facilities available to the business programmer using the Disk Operating System.

The authors would like to thank International Business Machines for permission to reproduce sections of copyrighted material used in the appendices of this text.

Gary B. Shelly

Thomas J. Cashman, CDP

TABLE OF CONTENTS

CHAPTER 1

DISK OPERATING SYSTEM

INTRODUCTION

To effectively use the capabilities of the System/360 a programmer should have not only a knowledge of one or more of the programming languages available with the system, such as Assembler Language, COBOL, RPG, PL/I, and Fortran, but he should also have an understanding and usable knowledge of the additional types of software provided as a part of the operating system. Two important software components of the disk operating system include utility programs and the sort/merge programs. These programs have been incorporated into the operating system to enable the programmer to perform basic data processing operations common to all installations, such as listing cards on the printer, reproducing cards, sorting, etc. without the necessity writing a program in one of the programming languages such as COBOL. For example, with utility programs, operations such as producing an 80/80 listing from punched card input, can be performed by developing the required entries on a single card!

DISK OPERATING SYSTEM

The DISK OPERATING SYSTEM is a disk resident system that provides operating system capabilities for 16K and larger System/360 computers. The Disk Operating System consists of a comprehensive set of language translators and service programs operating under the supervisory coordination of an integrated control program to provide the service of an operating system.

The following diagram illustrates the various components of the Disk Operating System.

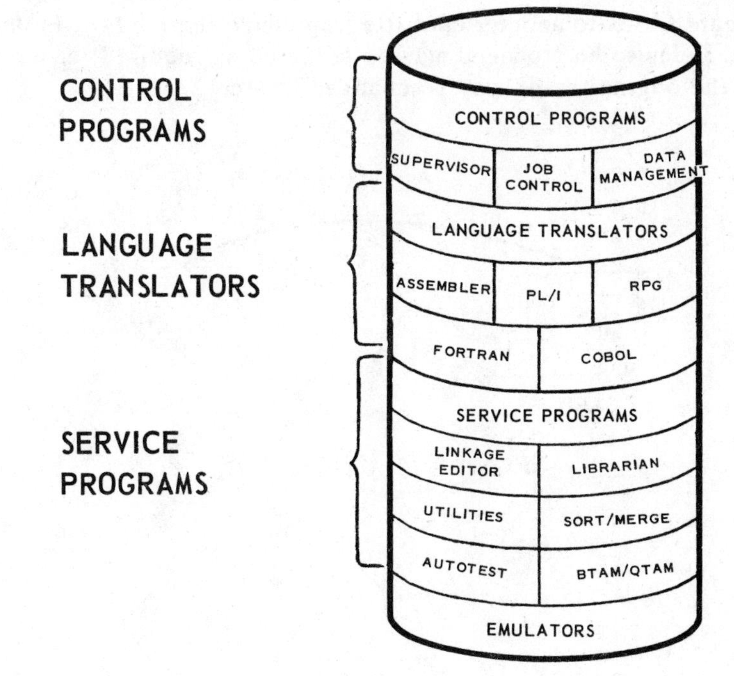

Figure 1-1 Components of the Disk Operating System

As can be seen in Figure 1-1, the Disk Operating System is segregated into three sections: Control Programs, Language Translators, and Service Programs. The Control Programs consist of a Supervisor, the set of programs which constitute Job Control, and Data Management routines which coordinate the Input/Output operations. The Language Translators, which are used to translate source programs into machine-language code, include the Assembler Language translator, the COBOL compiler, the RPG compiler, the FORTRAN compiler, and the PL/I compiler.

SERVICE PROGRAMS

The Service Programs which are provided by IBM with the Disk Operating System are used to perform functions on the computer which are required by a majority of users in the normal course of data processing. The function of file copying, that is, copying data from one file to another, is accomplished through the use of the DOS Utility Programs. The function of sorting or merging data is performed by the DOS Sort/Merge Program. Special functions, such as teleprocessing or computer emulation, are accomplished through the use of specialized programs which enable the user to specify his particular needs and then accomplish them using the generalized service programs.

DOS Utility Programs

The DOS Utility Programs consist of generalized programs for file copying and special purpose programs which are used for specific functions such as initializing a disk pack or assigning alternate tracks on a disk pack. Each of the general file copying utility programs are "tailored" to the needs of the user through the use of control cards in which the user states the specific file specifications and options desired. The following is a list of the general file copying programs which are a part of the DOS Utilities:

1. Card to Card (Punch): This utility program enables the user to copy a card file onto another card file (reproduce the cards). In addition, the fields to be produced may be selected as required, or the format of the original cards may be changed if necessary.

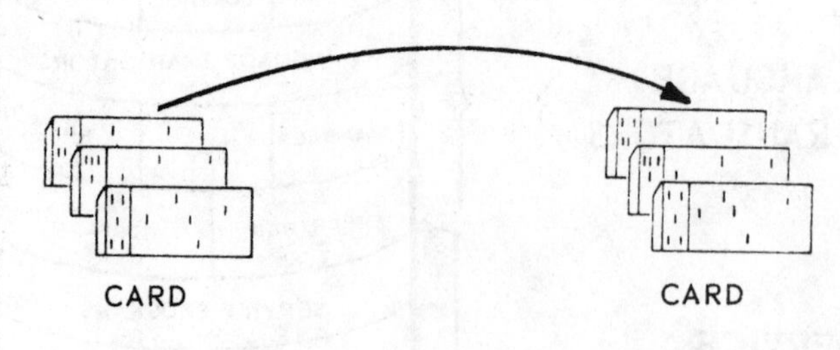

Figure 1-2 Card to Card

2. Card to Printer: This utility program enables the user to copy a card file using the printer (80-80 list). In addition, the input may be field selected and the printed report may be formatted with headings in order to make the report more readable.

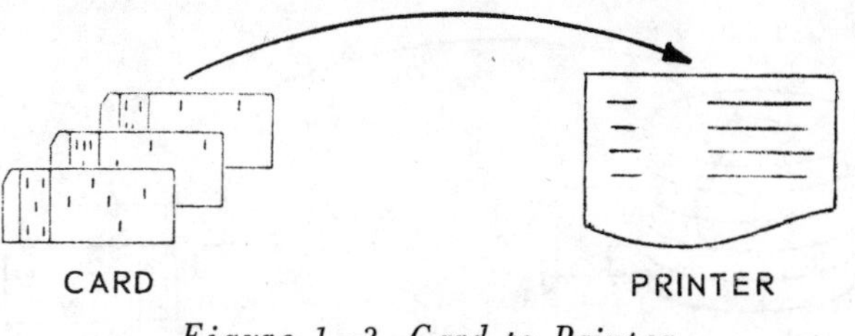

Figure 1-3 Card to Printer

3. Card to Disk: This utility program allows the user to copy a card file, that is, a group of data cards, to a disk file. In addition to copying, the utility program allows reblocking of the file so that the disk file may be written as a blocked file and also allows field selecting, that is, selecting fields from the card record to be written on the disk record or writing the entire card on the disk. The card-to-disk program also enables the conversion of data from zoned-decimal to packed-decimal format.

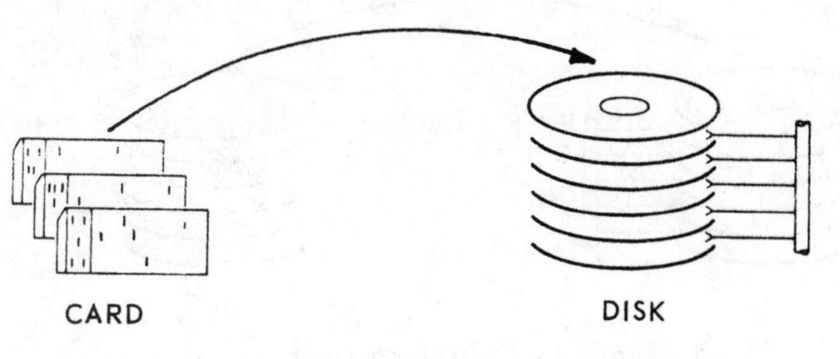

Figure 1-4 Card to Disk

4. Card to Tape: This utility program provides the capability of copying a card file to a magnetic tape file. The options of field selection and reblocking are included in the card-to-tape program. Tape labelling may be with standard DOS labels, no labels, or user labels.

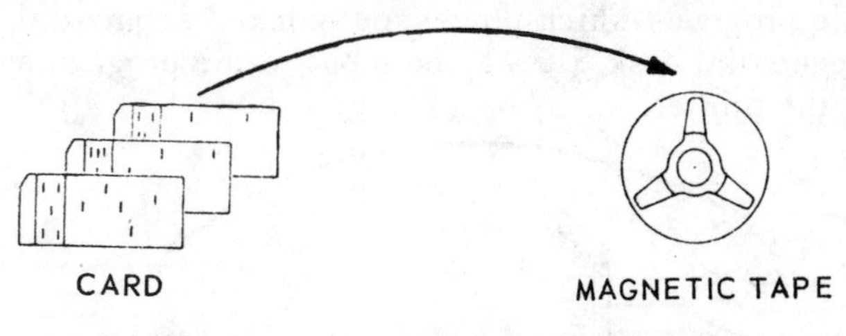

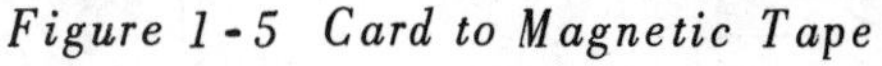

Figure 1-5 Card to Magnetic Tape

5. Disk to Card: The disk-to-card utility program allows a disk file to be punched on cards. The disk file can be reblocked to single block records when it is punched on the cards and field selecting may take place. In addition, conversion from packed decimal format in the disk record to zoned decimal on the card may be accomplished.

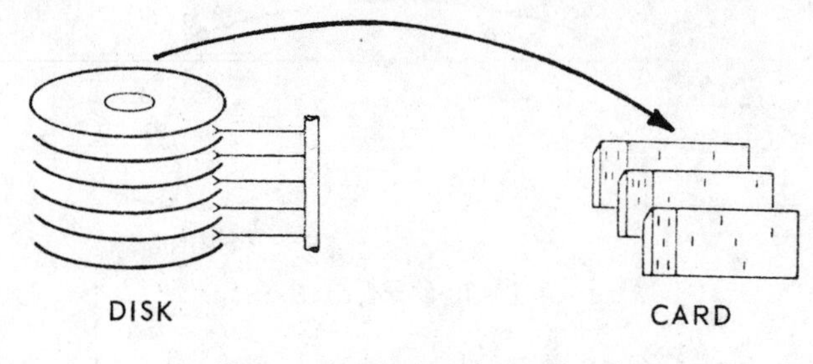

Figure 1-6 Disk to Card

6. Disk to Disk: This utility program allows one disk file to be copied to another disk file. The files may be reblocked and fields may be selected from the input disk file to be written in the output file. Data conversion from packed-decimal to zoned-decimal or vice versa may also take place.

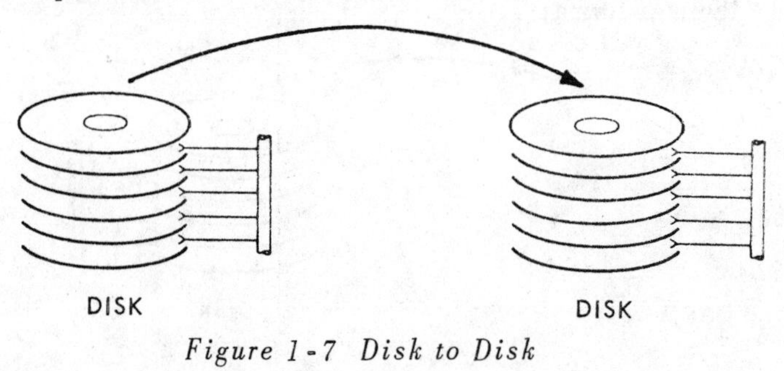

Figure 1-7 Disk to Disk

7. Disk to Printer: A file stored on disk may be printed on the printer using this utility program. The report may be formatted through the use of the field select feature and data may be printed in either the zoned-decimal format or a hexadecimal format. In addition, headings may be on the report. This utility program is the only program in the file-to-file programs which processes indexed sequential files. An indexed sequential disk file may be input to the program and be written on the printer.

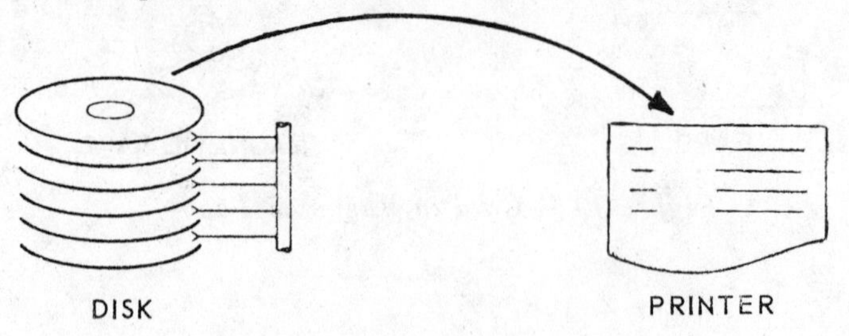

Figure 1-8 Disk to Printer

8. Disk to Tape: A disk file may be copied to a magnetic tape file through the use of this utility program. The file may be reblocked and selected fields may be copied to the tape file. Data conversion from packed-decimal to zoned-decimal or zoned-decimal to packed-decimal is also possible.

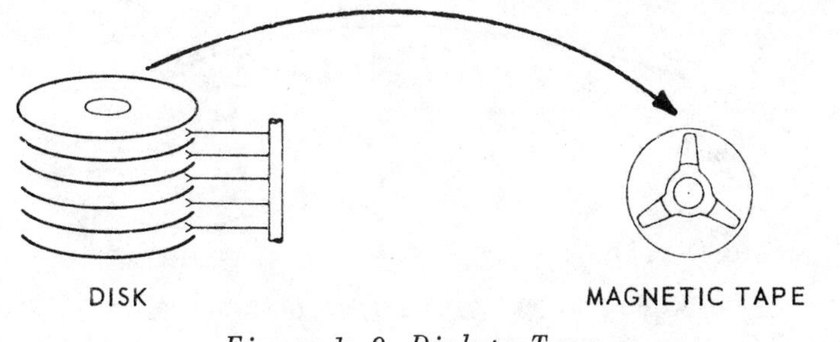

Figure 1-9 Disk to Tape

9. Tape to Card: The records in a tape file may be punched on cards through the use of this utility program. If the records on the tape are blocked, they can be unblocked and certain fields within the tape record may be selected for punching. Data conversion may also be requested.

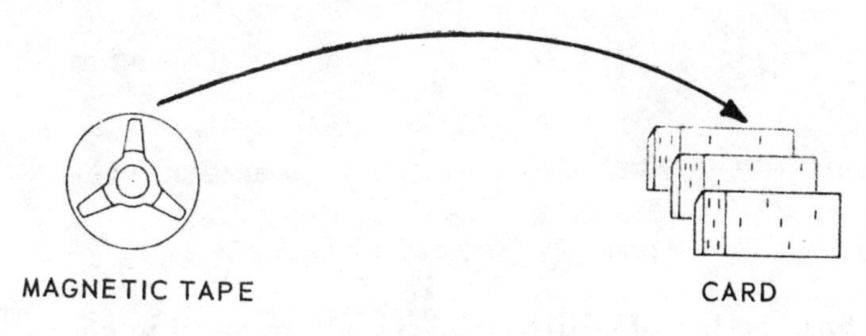

Figure 1-10 Tape to Card

10. Tape to Disk: The tape-to-disk utility program is used to copy a file from magnetic tape to a disk. The record stored on the tape may be reblocked and selected fields from the tape record may be written or omitted on the disk record.

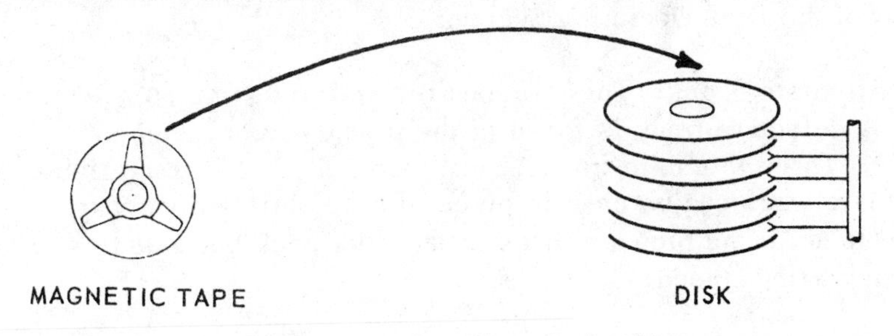

Figure 1-11 Tape to Disk

11. Tape to Printer: The records on the tape file are printed on the printer by this utility program. Field selection may occur and the data in the tape record may be printed in a character format or a hexadecimal format. The report may contain headings in order to identify fields.

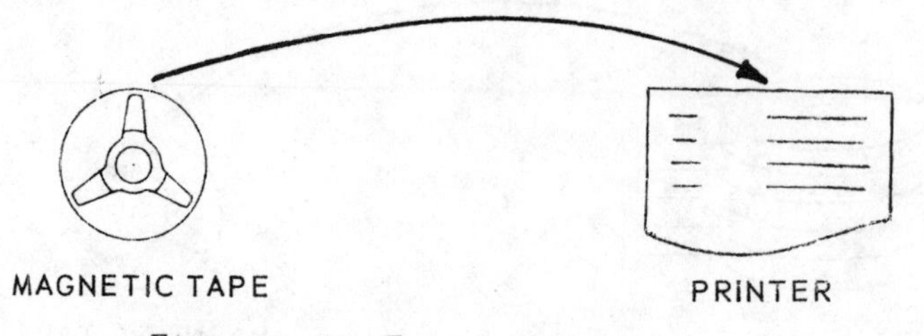

Figure 1-12 Tape to Printer

12. Tape to Tape: This utility program is used to copy one file on tape to another tape. The options of field selection and data conversion are available with the tape-to-tape utility.

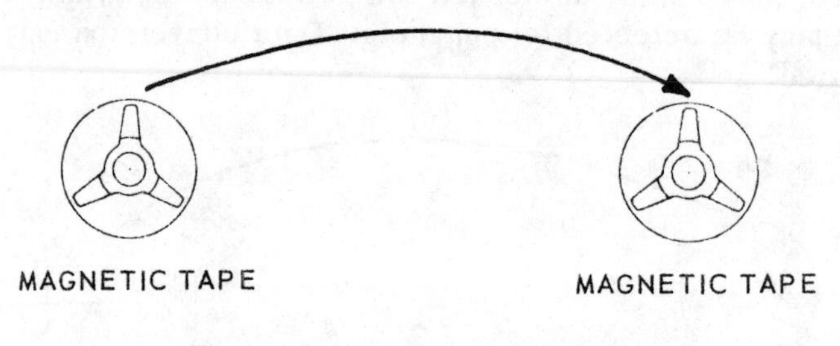

Figure 1-13 Tape to Tape

As can be seen from the list of utility programs, all are used to copy files from one storage medium to another. A detailed description of the use of these file-to-file utility programs is contained in Chapter 2.

The special purpose utility programs are used to perform tasks which are commonly required in an installation but which are not considered "file-oriented", that is, they do not deal with data files. The following is a list of the special purpose utility programs which are a part of the Disk Operating System:

1. Assign Alternate Track: This program is used to assign an alternate track on a disk pack when a defective track is found in the primary data areas on a pack, that is, cylinders 0-199. Three alternate tracks, numbered 200-202, are available on disk packs so that an alternate track can be used in place of a defective primary track. This utility program is used to make the proper entries in the disk pack home address record so that the alternate track will be used.

2. Clear Disk: This utility program is used to clear data from areas on a disk pack and to format a pack for use with application programs. It is commonly used to format a disk pack when the pack is to be used for a direct-access file.

3. Copy and Restore Disk: The copy and restore disk utility program is used to both copy an entire disk or file onto tape and to "restore" the tape back onto the disk. This program is normally used when the entire disk pack is to be copied regardless of the format of the logical files stored on the pack. It is quite useful in order to create backup for indexed sequential files which cannot be processed by the disk-to-disk program.

4. Initialize Disk: This special purpose utility program is used to "initialize", or format, a disk pack for use with logical IOCS and all programs processed under DOS. In addition, it performs a surface analysis which tests each of the tracks on the disk pack in order to ensure that there are no defective tracks.

5. Initialize Tape: This program is used to write a volume label on a tape volume. It can be used when the tape is initially brought into an installation or it can be used after the standard volume label has been destroyed by a program not using standard labels.

6. Tape Compare: The tape compare utility program is used to compare two tapes to ensure that they contain the exact same data. If values in the records stored on the tapes are found to be not equal, a message is issued on SYSLST identifying the bytes not equal and the record number on the tape. This program may be used to determine that back-up files contain the same information as master tape files or for any other application in which two identical tapes should be produced but which must be checked.

The special purpose utilities as described above are part of the Disk Operating System and may be used whenever the need arises.

DOS Sort/Merge Programs

A very common function which must be performed in the data processing environment is sorting, or placing records in a prescribed sequence based upon a controlling field or key. In addition, many applications require that sorted records from two or more files be "merged" into a single sorted file. These functions of sorting and merging are performed through the use of one of three sort/merge programs which are provided by IBM with the Disk Operating System.* The general features of all three of the sorts are explained below.

1. Each of the sort/merge programs processes standard DOS tape and disk labels and, in addition, provide linkage to user-written subroutines to process user labels.

2. Support 9-track or 7-track drives.

3. Allow multivolume input and output files to be processed.

** Due to the changing nature of IBM support for the DOS sort/merge programs, the sort/merge programs documented in SRL GC24-5030-9 CONCEPTS AND FACILITIES FOR DOS AND TOS, Release 25, will be explained.*

4. Provides linkages to user-written routines at various points in a sort or merge operation.

5. Provide checkpoint, interruption, and restart procedures for the sort operation. This allows a great deal of flexibility in the execution of the sort and merge operations.

6. Allows input and output files to be spread over multiple I/O devices.

7. Provides for specification of alternate input and output tape drives for either a sort or a merge operation.

8. Provides the ability to bypass unreadable data blocks, or to indicate the need for operator intervention. Provides a message to the operator indicating that a block has been bypassed.

9. Sequence checks the records during the final pass of a sort or merge run.

In addition to the above features, each of the sort/merge programs provides the capability of sorting 9 different files and merging up to 8 input files. The sort/merge programs are general programs which are tailored to the user's needs through the use of sort/merge control cards. The use of the control cards and the related job control statements which must be used with the sort/merge programs is discussed in Chapter 5, Chapter 6 and Chapter 7.

DOS OPERATING CAPABILITIES

The Disk Operating System, like most operating systems, is designed to provide continuous operation of the computer system. It does this through the use of a core resident control program called a SUPERVISOR and a set of programs called JOB CONTROL which provide the link between the execution of user and system programs. In addition, DOS allows programs to be executed in a MULTIPROGRAMMING environment, that is, more than one program may be stored in core storage and be processing data at the same time. Programs executed under DOS may be executed as BATCHED JOBS or as single jobs in a multiprogramming environment.

MULTIPROGRAMMING

Multiprogramming is a technique whereby two or more programs may concurrently share the resources of a computer system in such a way that the operation of any one program is independent of the operation of any other program. Thus, under DOS, more than one program may be operating at any one time. In order to store the programs in core storage for concurrent operation, core storage is divided into PARTITIONS. A partition is a contiguous portion of core storage which is capable of holding a distinct problem program. Under DOS, a maximum of three partitions may be allocated for operation at any one time. These partitions are called BACKGROUND, FOREGRCUND 2, and FOREGROUND 1. The diagram below illustrates the allocation of core storage when all three partitions are being used.

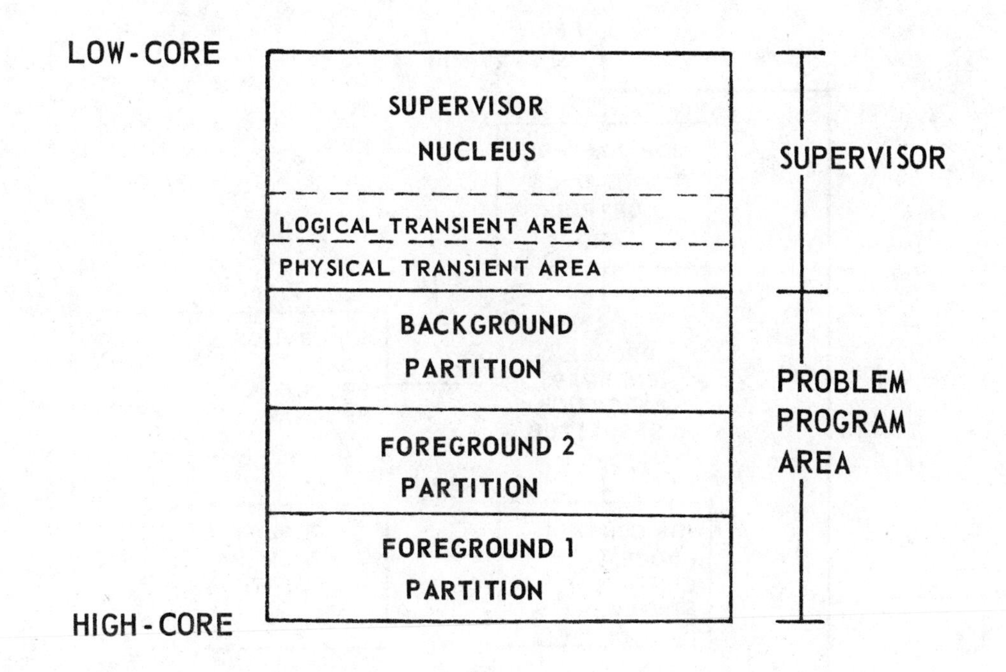

Figure 1-14 Allocation of Core Storage for MULTIPROGRAMMING

In Figure 1-14 it can be seen that core storage is "partitioned" into four separate areas. The Supervisor, which must always be in core storage, is placed in core storage beginning with the first available position (Hex Address '000000'). Immediately following the Supervisor is the BACKGROUND PARTITION. If only one partition is to be operating, it is the background partition. It always immediately follows the supervisor in core storage. The FOREGROUND 2 PARTITION, if it is used as in Figure 1-14, follows the Background Partition in core storage. It is followed by the FOREGROUND 1 PARTITION, provided that the foreground 1 partition is to be used. It should be noted that the Foreground 1 partition can be allocated without the Foreground 2 being allocated but that the Background partition must always be allocated.

The Supervisor normally is allocated 10K – 16K, depending upon the options specified for the system. The background partition must be allocated at least 10K and the background partition is required. The foreground partitions are optional and need not be allocated or used. If a foreground partition is allocated, it must be a minimum size of 2K and must be allocated in 2K increments, that is, it must be 2K, 4K, 6K, etc. If batched job processing is to take place in a foreground partition, the partition must be a minimum of 10K in length. The allocation and use of the Foreground partitions and the use of DOS Multiprogramming is discussed in Chapter 8 and Chapter 9.

BATCH JOBS

Batch Jobs are jobs which are executed one after another without any required operator intervention. A general flowchart of batched job processing is illustrated below.

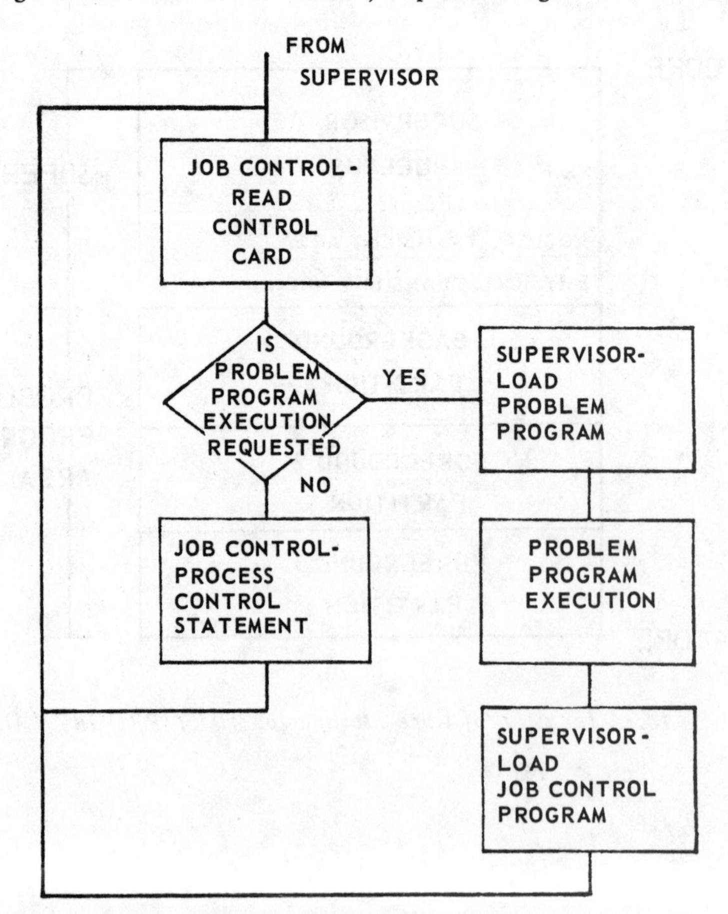

Figure 1-15 Flowchart of BATCH JOB Operation

Note in the flowchart illustrated in Figure 1-15 that the Job Control program is in core storage to process control statements when a problem program is not being executed. When an Execute card is found in the job stream, the problem program is loaded into core storage by the Supervisor and it is then executed. When the problem program terminates, the Supervisor again loads the job control program which then reads more control statements. Thus, in a batched job environment, problem programs may be executed with no intervention from the operator because the job control programs provide the interface between the executions of the problem programs.

It should be noted also that all programs executed in the Background partition are executed as Batched Jobs. Jobs executed in the Foreground partitions may be processed as batched jobs or may be executed as single programs. If they are executed as single programs, the execution is initiated by the operator, not through the use of job control.

The problem programs which are executed in a batched job environment may be any user-written program or one of the service programs provided with DOS, such as a Utility program. If the programs are to be executed as batched jobs in a Foreground partition, it may be necessary to link-edit the program to be executed in the foreground partition rather than the background partition.

SYSTEM LOGICAL UNITS

The control statements which are read by job control must be read from specific devices in order to be processed by job control. In addition, when job control writes on an output device, the device must use a specific symbolic unit name. The symbolic unit names used by Job Control are called SYSTEM LOGICAL UNITS. System Logical Units are used by the DOS system programs for input and output operations.

The job control programs always read the job control statements from the device assigned to SYSRDR. In many instances, SYSRDR is assigned to the card reader and the job control statements are punched on cards and read from the card reader. When batched jobs are to be processed in a multiprogramming environment, however, the card reader cannot be used for both partitions because it can only be assigned to one partition, that is, the card reader cannot be used in two partitions at the same time. It is possible, therefore, to assign SYSRDR to a magnetic tape drive or a disk drive in order that the "job stream", that is, the job control statements, can be read from a device other than the card reader.

The device assigned to SYSLST is used by job control for writing control statements and outputting other messages. In addition, many of the system programs, such as the language translators, write the output on SYSLST. As with the card reader, the printer cannot be assigned to more than one partition at one time. Thus, SYSLST must be assigned to a magnetic tape drive or to a disk drive if it is to be used by more than one partition. The use of System Logical Units on devices other than unit-record devices is explained in detail in Chapter 9.

SUMMARY

A knowledge of utility programs by one or more of the programmers within an installation can often times greatly contribute to the efficient operation of the data processing installation. For example, the utilization of card to print utility programs when required rather than writing a similar program in COBOL, RPG, or Assembler Language can result in the savings of many hours programming time throughout the year. Similarly rather than sorting cards using a unit record sorter, through the use of the DOS sorting programs sorting time can be reduced from hours or hundreds of hours to minutes during a given period of time! Thus, a thorough understanding of the functions and capabilities of the utility programs available under the Disk Operating Systems should be an integral part of the training of all System/360 programmers.

CHAPTER 2

DOS FILE-TO-FILE UTILITIES

INTRODUCTION

As noted in Chapter 1, the file-to-file utility programs are used to copy data from one file to another. The files may be on the same storage medium, for example, cards, or may be stored on different mediums, such as disk and tape. Each of the utility programs which are used to copy files are stored in the core image library with unique phase names so that they may be identified and executed through the use of the job control Execute card (//EXEC phasename).

In addition to having unique phase names, the utility programs must read control cards which identify the characteristics of the file to be processed and also request special processing by the utility program such as field selecting or reblocking the file.

CARD-TO-CARD UTILITY

The Card-To-Card Utility program is used to produce a punched card output file from a punched card input file. The punched card illustrated in Figure 2-1 is to be reproduced.

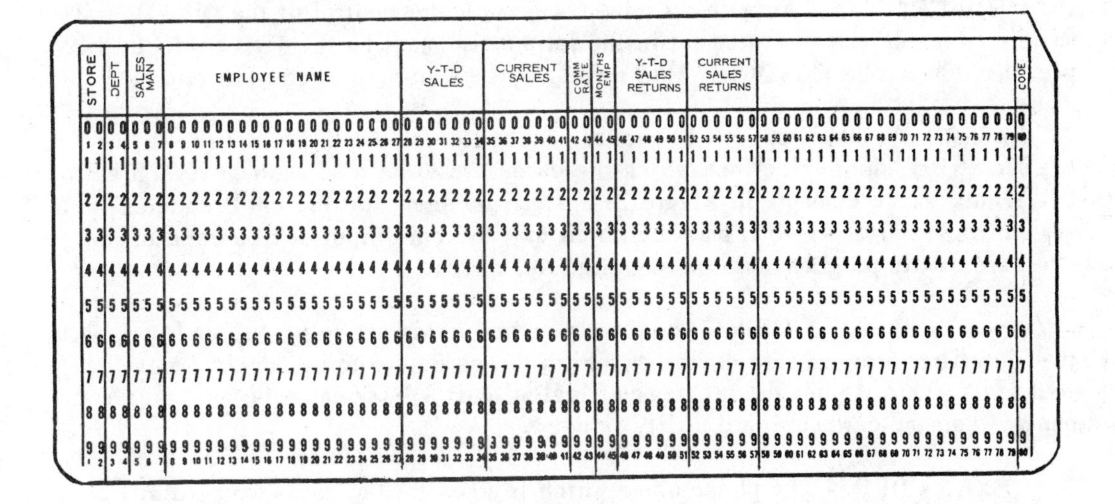

Figure 2-1 Example of card which is to be reproduced

In the following example, the entire card as illustrated in Figure 2-1 is to be reproduced. The job stream to accomplish this using the DOS Card-To-Card Utility program is illustrated in Figure 2-2.

EXAMPLE

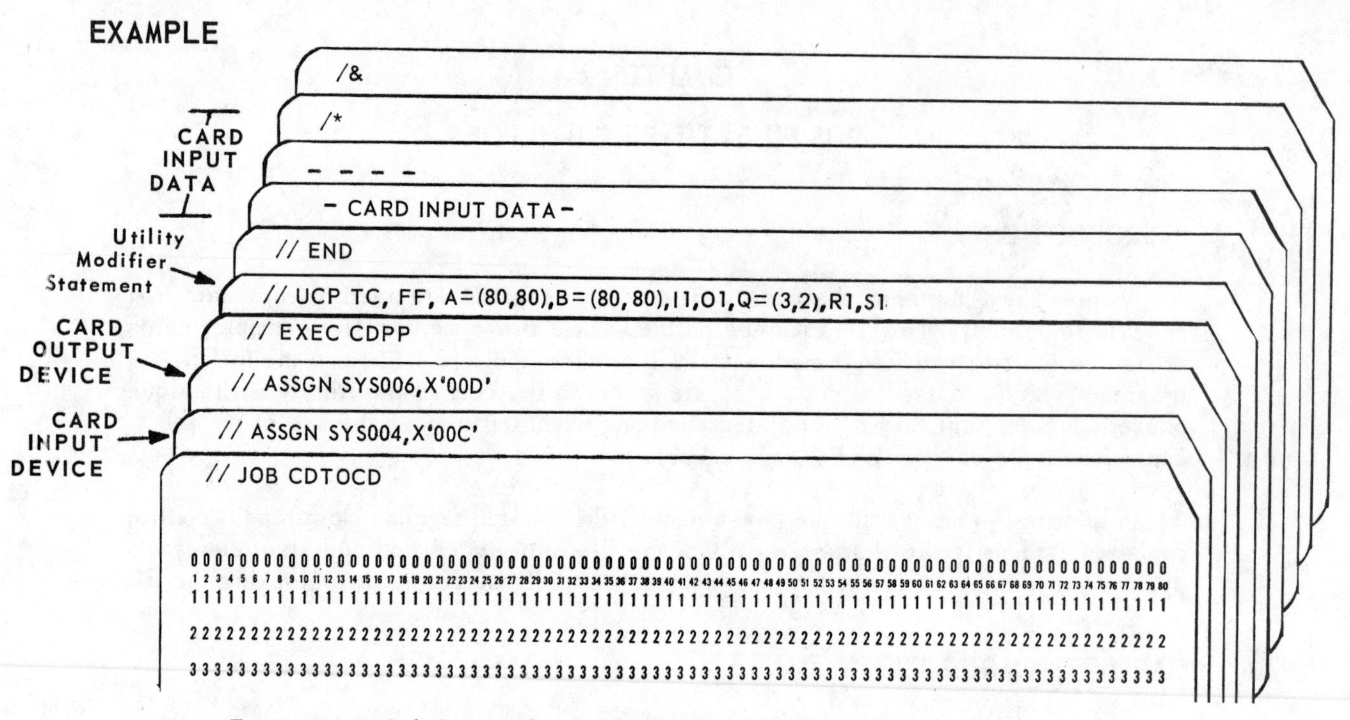

Figure 2-2 Job Stream for Card-To-Card Utility Program

The job stream illustrated in Figure 2-2 is explained below.

1. // JOB CDTOCD – As with all jobs processed under control of the Disk Operating System, the first job control statement in the job stream must be the JOB card. In the example, the jobname is CDTOCD. This name may be 1 to 8 characters in length.

2. // ASSGN SYS004,X'00C' – Whenever punched cards are to be input to a DOS file-to-file utility, the device from which the cards are to be read must be assigned to SYS004. Thus, in the example in Figure 2-2 it can be seen that the card reader, X'00C', is assigned to SYS004. Again it must be noted that the card input device used in the card-to-card utility program must be assigned to SYS004.

3. // ASSGN SYS006,X'00D' – Whenever punched cards are to be output from a DOS file-to-file utility program, the device on which the cards are to be punched must be assigned to SYS006. Again, the programmer logical unit SYS006 must be assigned to the card punch when the card-to-card utility is used.

4. // EXEC CDPP – The phase name which is used for the card-to-card utility program is CDPP. This program phase is normally stored in the Core Image Library. Thus, in order to execute the program, the Execute card is used with the phasename as the operand. As can be seen, when this Execute card is processed, the program phase CDPP will be read into core storage and it will gain control.

5. UTILITY MODIFIER STATEMENT – The Utility Modifier Statement is used to "tailor" the general file-to-file program to the specific needs of the user. The general format of the Utility Modifier Statement for a card-to-card operation is illustrated below.

```
// UCP TC,FF,A=(80,80),B=(80,80) [,Ix][,Ox][,Q=(x,y)][,Rx][,Sx]
```

Figure 2-3 General Format of Card-To-Card Utility Modifier Statement

As can be seen from the general format of the Utility Modifier Statement in Figure 2-3, it resembles the general job control cards which are used with DOS Job Control. The first two columns of the card must contain the characters //. These slashes in card columns 1 and 2 indicate to the utility program which reads the card that it is a modifier card. A single blank must follow the two slashes and this blank is followed by the Utility Identifying code "UCP". The UCP is a required entry for the card-to-card utility program and must be written as shown. All values which are written in capital letters in the general format above and in the remainder of the text must be included and must be written as shown.

The letters UCP are followed by a single blank and then the entry "TC" is included. The "T" indicates that the type of processing to be performed is specified next. The "C" indicates that a straight copy is to be done. This parameter is required as shown when an 80-80 copy is to be performed. The "TC" entry is followed by a comma with no blanks. All of the entries in the Utility Modifier Statement must be separated by a comma with no blanks included.

The "F" entry is used to indicate the type of record which is to be processed. The value "FF" indicates that the records are fixed-length. All card records are fixed-length. Note again that the entry is followed by a comma with no blanks.

The next entry is used to specify the record length and block size of the input record. The general format of the entry is A=(r,b) where r is the record length and b is the block length. When processing punched cards, the record length and the block length are always the same because card files cannot be blocked. Thus, the entry A=(80,80) as shown in Figure 2-3 is required for card input. Note that the record length and the block length must be enclosed within parentheses and the entire entry is followed by a comma.

The next entry is used to specify the record length and the block length of the output records. The general format is the same as the format for the input records, that is, B=(r,b), where r is the record length and b is the block length. As illustrated in Figure 2-3, a card output record always contains 80 characters and the file is unblocked. Thus, the record length and the block length are the same.

The remaining entries in the Utility Modifier Statement are optional entries. They may be included but if they are not, default values are used. Note that each of the optional entries is enclosed within brackets ([ ]), which indicates that the entries are optional. This convention of enclosing optional entries within brackets will be followed throughout this text.

The "I" identifier is used to indicate the type of format in which the data in the card is stored, that is, whether it is in the EBCDIC format or the Binary format. In the example in Figure 2-2 it can be seen that the value included on the Utility Modifier Statement is "I1", which indicates that the data is in the EBCDIC format. The other possible entry is "I2", which indicates that the data is stored in a binary fomat. The default value, which is used if this optional entry is not included, is "I1".

The "O" identifier is used to specify the type of format which is to be used for the output data. The value "O1" indicates that the punched output is to be in the EBCDIC format and the value "O2" specifies that the output should be in the binary format. The default value is "O1".

In many applications it is desirable to sequence-check the card input, that is, ensure that the card input is in a specified sequence from a particular field in the card. In the example, it is desired that the cards be in sequence by department number (card columns 3-4). The "Q" entry when used in the Utility Modifier Statement for the card-to-card utility, specifies which columns, if any, should be checked for an ascending sequence. The general format of the "Q" entry is Q=(x,y), where x is the beginning column for checking and y is the number of columns to check. The maximum value which can be specified for y is 10. In the example in Figure 2-2 it can be seen that the entry is Q=(3,2) which indicates that two columns in the card input record are to be checked beginning with column 3. Note from the card format illustrated in Figure 2-1 that this corresponds to the department number field in the card. If this field is not included in the Utility Modifier Statement, no sequence checking is performed on the input file.

The "R" entry in the Utility Modifier Statement indicates the first logical record which is to be copied onto the output file. Thus, through the use of this parameter, records at the beginning of a file may be bypassed in the file copy processing. The x value which follows the "R" specifies the beginning record number. In the example in Figure 2-2, the value is R1, which indicates that processing is to begin with the first logical record. An entry of R5 would cause the first four logical records to be bypassed and the first card to be copied would be the fifth card in the input data. If this entry is omitted, copying begins with the first input record.

The "S" entry is used with the card-to-card utility program to specify which output stacker is to be used for the punched output cards. Although the values which may be specified for this entry vary with different devices, the values which may be specified for a 2540 card punch are S1, S2, or S3, indicating that the output cards are to be placed in the first, second, or third stacker respectively. If this optional entry is omitted, all output cards will be placed in Stacker 2.

6. // END – The last statement which must be read by the card-to-card utility program is the END Statement. This statement indicates to the utility program that all control cards have been read and that the next card to be read is a data card which is to be copied. This END Statement must be included in every file-to-file utility.

7. Data Cards – The card input data is placed immediately following the // END card. These cards will be read by the utility program and reproduced on the card punch. The data cards may contain any information desired; there is no restriction on the data which may be punched in them.

8. /* – The end of the input data is indicated through the use of the /* card. The first two columns must contain the characters /* and card columns 3-80 must be blank in order that this card indicate end of data. If values other than blanks are contained in column 3 through column 80, this card is treated as data and will not indicate end of data to the utility program. A /* card which indicates end-of-data is required for all card input utility programs.

9. /& – This card indicates end of job and is read by job control after the utility program has completed processing and returned control to job control.

As can be seen from the previous example, the general utility program CDPP is modified by the Utility Modifier Statement to perform the task desired by the user. The following example illustrates the detail processing which takes place when the utility program is executed.

EXAMPLE

Step 1: The job control cards are read by the job control program. When the Execute card is read, the card-to-card program (CDPP) is loaded into core storage by the Supervisor.

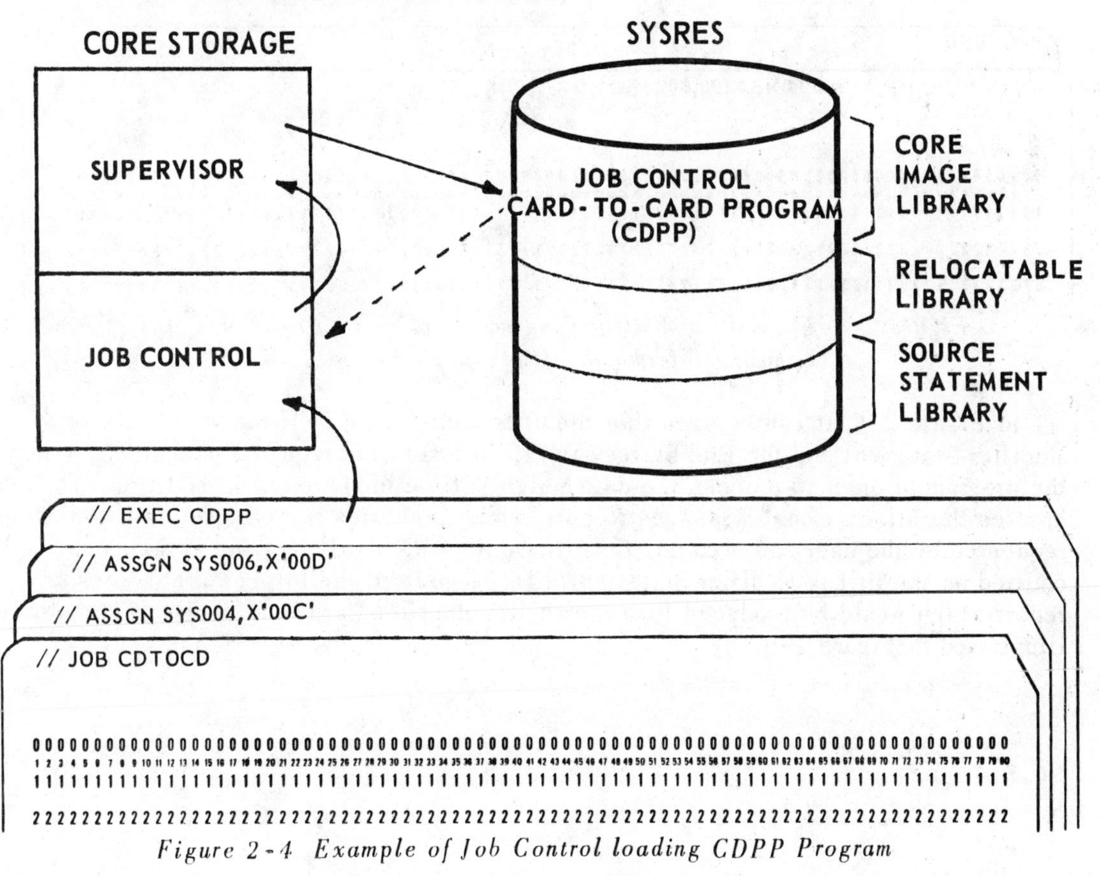

Figure 2-4 Example of Job Control loading CDPP Program

Note from the example in Figure 2-4 that when job control reads the // EXEC CDPP card, it gives control to the Supervisor which in turn loads the card-to-card utility program into core storage.

Step 2: The CDPP Utility program reads the utility control cards and creates a report containing informational messages which indicate the processing which is to be accomplished by the Utility.

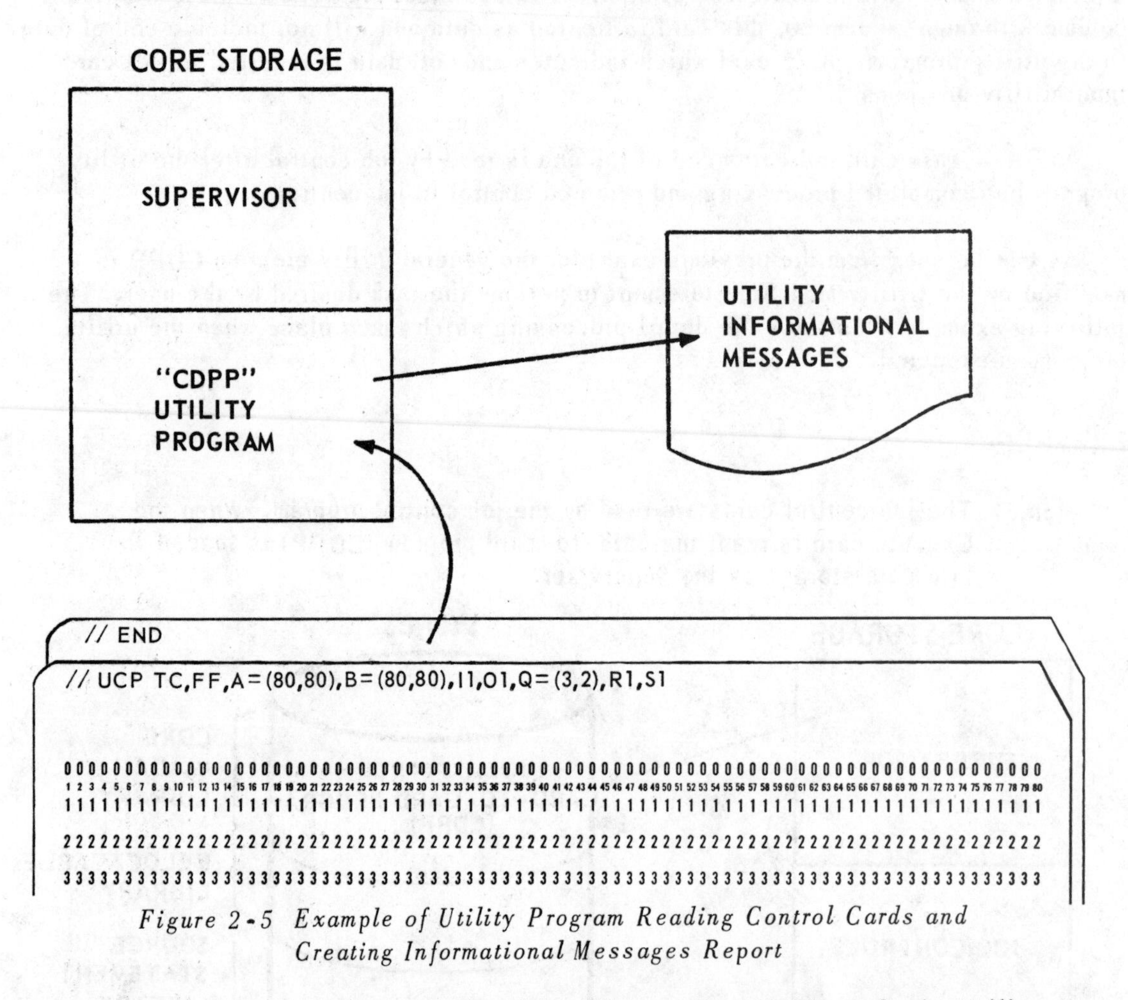

Figure 2-5 Example of Utility Program Reading Control Cards and Creating Informational Messages Report

In Figure 2-5, it can be seen that the utility program CDPP reads the Utility Modifier Statement and the End Statement. It then sets the required parameters within the program in order to process the data which will be input to it. In addition, it creates the Informational Messages Report, which indicates the parameters which were requested by the user and also any default values which will be used if they were omitted on the Utility Modifier Statement. An example of the Informational Messages report which would be produced from the Utility Modifier Card read in Figure 2-5 is illustrated in Figure 2-6.

INFORMATIONAL MESSAGES

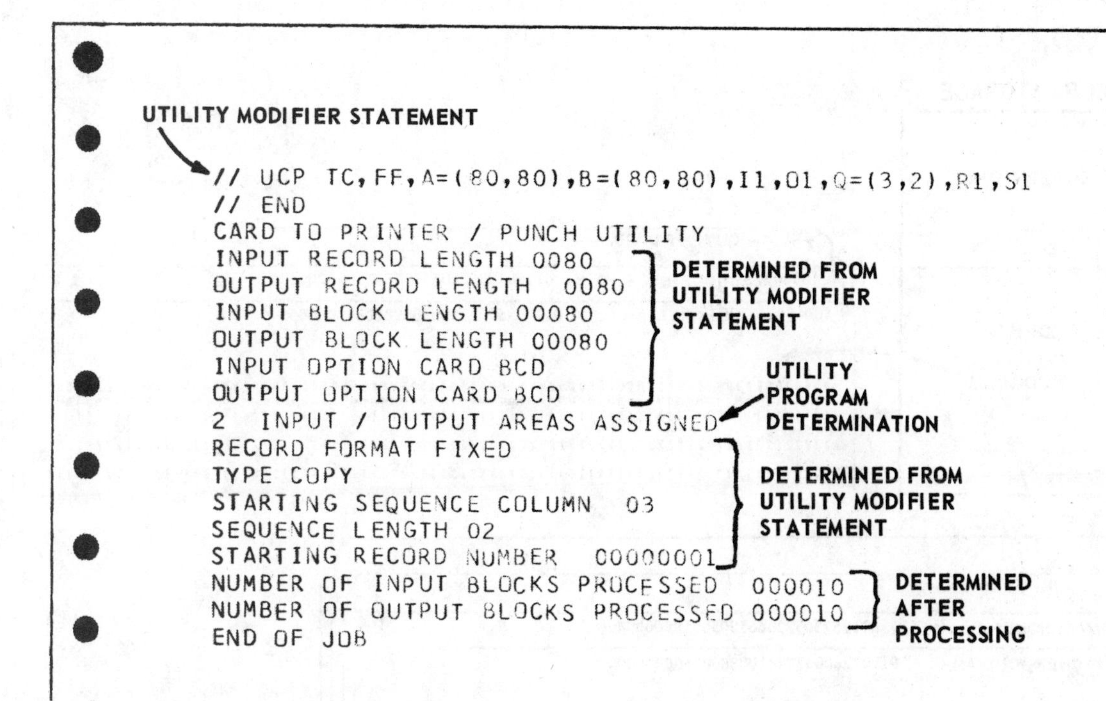
UTILITY MODIFIER STATEMENT

```
// UCP TC,FF,A=(80,80),B=(80,80),I1,O1,Q=(3,2),R1,S1
// END
CARD TO PRINTER / PUNCH UTILITY
INPUT RECORD LENGTH 0080
OUTPUT RECORD LENGTH  0080
INPUT BLOCK LENGTH 00080
OUTPUT BLOCK LENGTH 00080
INPUT OPTION CARD BCD
OUTPUT OPTION CARD BCD
2  INPUT / OUTPUT AREAS ASSIGNED
RECORD FORMAT FIXED
TYPE COPY
STARTING SEQUENCE COLUMN   03
SEQUENCE LENGTH 02
STARTING RECORD NUMBER    00000001
NUMBER OF INPUT BLOCKS PROCESSED  000010
NUMBER OF OUTPUT BLOCKS PROCESSED 000010
END OF JOB
```


Figure 2-6 Example of Utility Informational Messages

As can be seen from Figure 2-6, the report produced by the Utility program specifies the parameters which will be used when the data is processed. The Utility Modifier Statement and the End Statement are reproduced and then the various factors determined from the Utility Modifier Statement are reported. Thus, the input record length, the output record length, the input block length, the output block length, the format of the input card (BCD is the same as EBCDIC), and the format of the output card are reported as determined from the Utility Modifier Statement. The number of input/output areas, or buffers, is determined by the Utility program at the time of execution. The block size of the input file, the block size of the output file, and the size of the core storage area in which the utility program is to be processed are the factors which the program uses to determine the number of buffers to be used.

The record format, type of processing to be done, the sequence check data, and the starting record number are also determined when the Utility Modifier Statement is read and analyzed by the Utility program. The number of input blocks processed and the number of output blocks processed are determined by the program after all of the data has been copied from the input file to the output file. Thus, from Figure 2-6, it can be seen that ten cards were read as input and ten cards were punched as output. The end of job message is printed to indicate that the utility program has gone to a successful end of job. It should be noted that the number of blocks processed and the end of job message are not printed on the report until after Step 3 has been completed, that is, until after the data has been read and processed.

Step 3: The card input data is read by the Utility program and the output cards are punched on the card punch device.

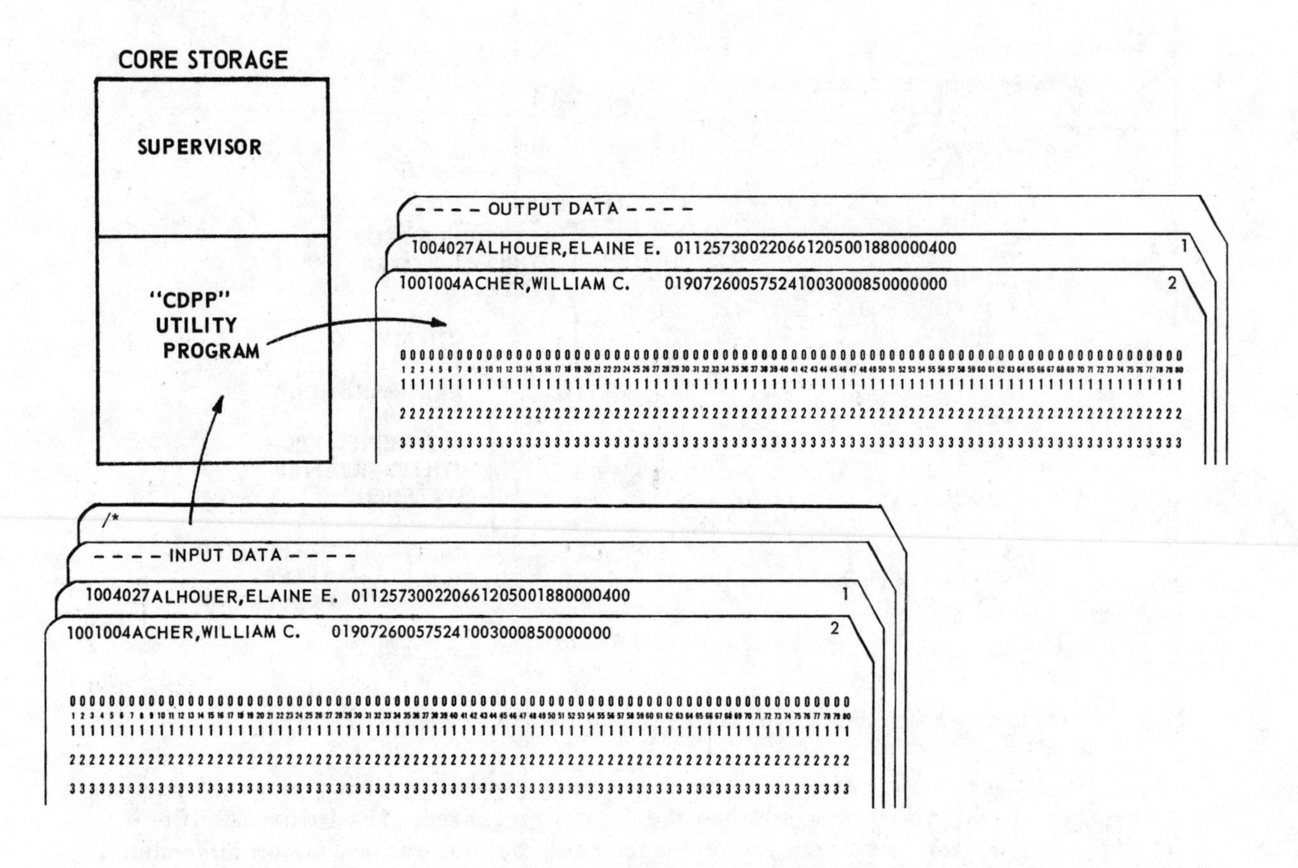

Figure 2-7 Example of Utility program reading and punching cards

Note from the example above that after the Utility program has established all of the options requested by the user, the processing of the data begins. The input cards, which must be on a card reader assigned to SYS004, are read and the output cards are punched on the device assigned to SYS006. Note again that these assignments must always be used for the card-to-card utility program.

The /* card indicates to the program that the last data card has been read. It must be noted that columns 3-80 must contain blanks in order for the /* card to indicate end of data. If these columns do not contain blanks, the card is treated as a data card.

Step 4: The Utility program returns control to the Supervisor which loads the job control program. Job control reads the last card in the job stream.

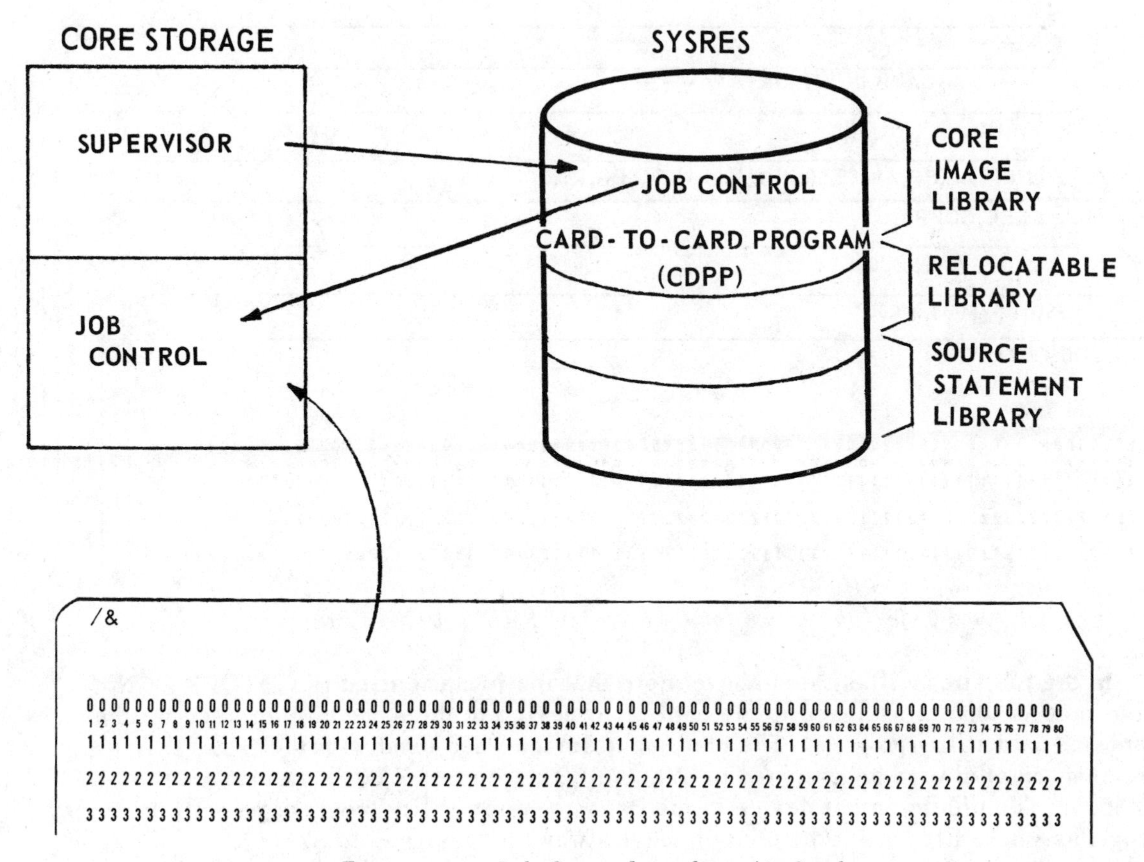

Figure 2-8 Job Control reading /& Card

Note in Step 4 that the job CDTOCD would be completed because the /& card is read. If more than one job step were to be performed in the job CDTOCD, the additional job control cards would follow the /* card instead of the /& card.

CARD-TO-PRINT UTILITY

As noted previously, the file-to-file utilities are used to copy data from one storage medium to the same or different medium. The card-to-card utility is an example of copying a file to the same medium as the input medium. The card-to-print utility, however, is used to copy punched cards to the printer. The job stream illustrated in Figure 2-9 could be used to provide an "80-80 list", that is, provide a report which merely has the punched card input file listed on the printer.

EXAMPLE

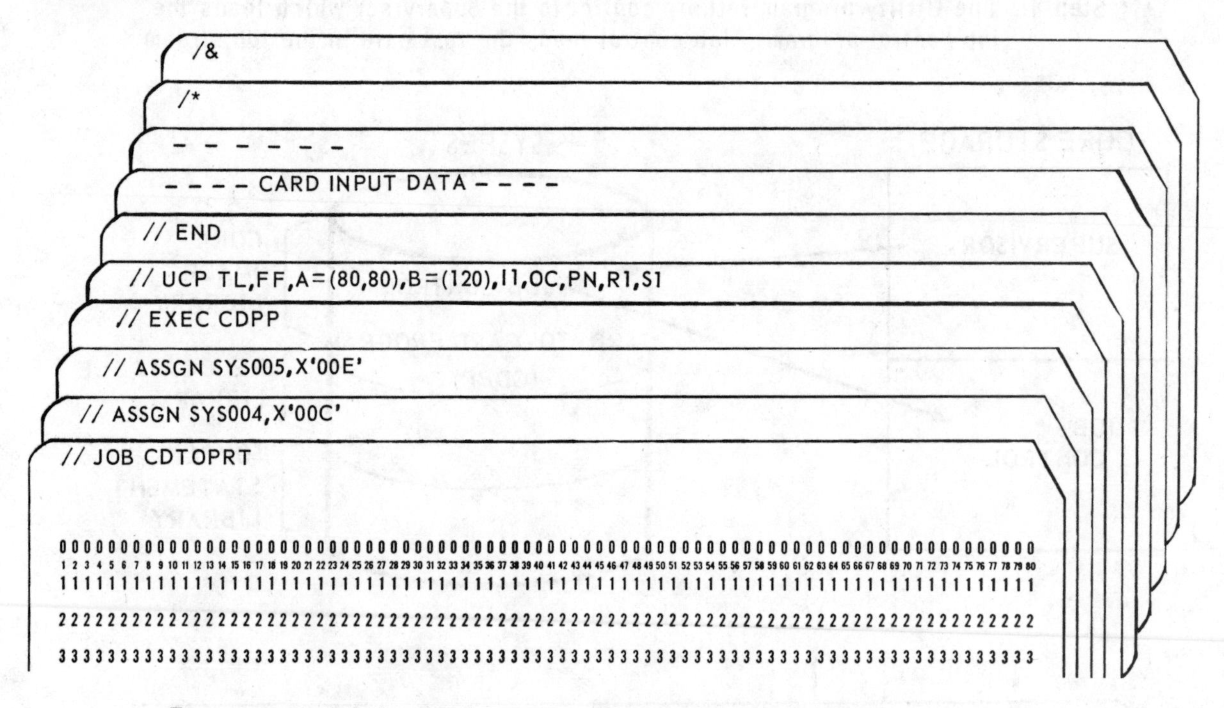

Figure 2-9 Job Stream for CARD-TO-PRINT Utility Program

In the job stream illustrated above, note that the jobname used is CDTOPRT. Note also that SYS004 is assigned to the card input device in the same manner used for the card-to-card utility program. The card input device when used with the DOS utility programs must always be assigned to SYS004. The printer, X'00E', is assigned to SYS005. The printer output device in the card-to-print utility program, as well as other programs where files are to be printed, must always be assigned to SYS005.

The program which is used for the card-to-print operation is the same as the program used for the card-to-card program, that is, the program with the phase name CDPP. Thus, the execute card illustrated in Figure 2-9 specifies that the phase CDPP should be loaded into core storage for execution.

The utility modifier statement used for the card-to-print utility program is used to submit user parameters so that the program will be tailored to the application. The general format of the Utility Modifier Statement used to list cards on the printer is illustrated in Figure 2-10.

// UCP TL,FF,A=(80,80),B=(xxx) [,Ix] [,Ox] [,Px] [,Rx] [,Sx] [,Q=(x,y)]

Figure 2-10 General Format of Utility Modifier Statement for Card-To-Print

In the general format of the Utility Modifier Statement illustrated in Figure 2-10, it can be seen that the format is almost identical to that used for the card-to-card utility program. This commonality between Utility Modifier Statements is true for all of the file-to-file utility programs. Certain statements may have additional parameters or the optional parameters may have different meanings, but the general format and the rules for punctuation, etc. remain the same for all Utility Modifier Statements.

The Utility Modifier statement for the card-to-printer utility begins with the characters // in card columns 1 and 2 and these slashes are followed by a single blank. The identifier UCP is entered to indicate the type of modifier statement which is being used. A single blank must follow the identifier. The function indicator "T" follows with the letter "L", which indicates that the cards are to be Listed on the printer in the same format as they are read. The "FF" value indicates that the records being processed are fixed-length records. The "A" parameter is again used to specify the record length and the block length of the input file. Since cards are 80 bytes in length and are unblocked, the parameter A=(80,80) is used for card input files.

The "B" entry is used with the card-to-print program to indicate the length of the print line on the printer. It is in the format B=(xxx), where xxx may have the values 120, 132, or 144. In the job stream illustrated in Figure 2-9, note that the value 120 is entered, thus indicating that the print line is to be 120 bytes long on the printer.

The parameters described above are required when the card-to-print program is to be executed. The remaining parameters are optional and if they are not included, the card-to-print utility program assumes default values. The "Ix" parameter is used to specify the format of the data in the input card. If x is equal to 1, the data is in an EBCDIC format and if x is equal to 2, the data is stored in the binary format. In the job stream in Figure 2-9 it can be seen that the value I1 is used indicating that the data is in the EBCDIC format. If the optional "I" parameter is not used, the utility program assumes that the data is in the EBCDIC format.

The "O" parameter is used to specify the format in which the data is to be printed on the printer. The general format, "Oy", is used. If "y" is equal to "C", the output will be printed in a character format. If "y" is made equal to an "X", the data will be printed in a hexadecimal format. Thus, the user has the option of displaying the data as character or hexadecimal output through the use of the "O" parameter. If this optional parameter is omitted, the utility program assumes character output.

The "P" parameter is used to specify whether page numbering is to be done by the utility. If the value PN is used, it indicates that no page numbering is to be done. If PY is used, each page will contain an ascending page number as the last line printed on each page. In the job stream in Figure 2-9, it is requested that the pages not be numbered. If this parameter is omitted, the utility assumes that page numbering is requested.

The "R" parameter is used for the same function in the card-to-print utility program as it is in the card-to-card program, that is, it specifies the logical record which will be the first to be copied. If this parameter is omitted, the value R1 is assumed.

The "Sx" entry in the Utility Modifier card specifies the spacing which is to take place on the printed output. The x may have the values 1, 2, or 3 specifying single, double, or triple spacing respectively. In the job stream in Figure 2-9, single spacing is requested on the report. If this optional parameter is not included in the Utility Modifier Statement, the Utility program will double space the report.

The "Q" entry is used for the same purpose as in the card-to-card program, that is, it is used to specify the desired sequence-checking on the input records. The x entry specifies the beginning column in the card and the y entry indicates the number of bytes in the field to be checked. No sequence checking takes place if this parameter is omitted.

As with all utility programs, when the program is executed, it reads the Utility Modifier Statement and produces the utility informational messages. The following messages were produced when the job stream in Figure 2-9 was executed.

EXAMPLE

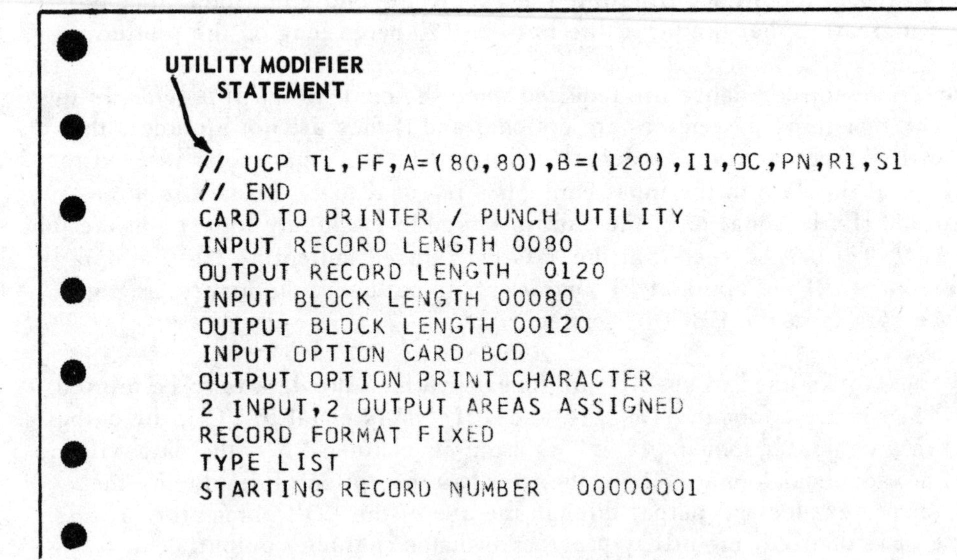


```
// UCP TL,FF,A=(80,80),B=(120),I1,OC,PN,R1,S1
// END
CARD TO PRINTER / PUNCH UTILITY
INPUT RECORD LENGTH 0080
OUTPUT RECORD LENGTH  0120
INPUT BLOCK LENGTH 00080
OUTPUT BLOCK LENGTH 00120
INPUT OPTION CARD BCD
OUTPUT OPTION PRINT CHARACTER
2 INPUT,2 OUTPUT AREAS ASSIGNED
RECORD FORMAT FIXED
TYPE LIST
STARTING RECORD NUMBER  00000001
```

Figure 2-11 Utility Informations Messages

Note from the print-out in Figure 2-11 that the Utility Modifier Statement is printed together with the factors which will be used to process the card input data and create a report. Note also that 2 buffer areas will be used for the input data and two will be used for the printer output area because of the block sizes of the files and the area alloted to the utility program.

The report which is generated by the card-to-print utility program is an "80-80 listing" of the card input. The report is illustrated in Figure 2-12.

LISTING

```
1001004ACHER, WILLIAM C.    01907260057524100300085000000O ←—— CARD IMAGE ——→ 2
1004027ALHOUER, ELAINE E.   011257300220661205001880000400                    1
1003030ALLOREN, RUTH W.     213202000000001510114000020933                    2
2006100BATES, TONY F.       081906602076450504088340005170                    1
2008102BELLSLEY, ARTHUR A.  088300000990001509019033002762                    2
1009105BOYLE, RALPH P.      087804401430551206016070007787                    1
1002111CARTOLER, VIOLET B.  028099800750061204003020001140                    2
2005122CENNA, DICK L.       035577700440011008003270000000                    1
1004171COSTA, NAN S.        058035600560021210013020000903                    2
2006179DAMSON, ERIC C.      035250201808880502000223000223                    2
1003181DELBERT,EDWARD D.    125065901305541509037540000000                    2
1001185DONNEMAN, THOMAS M.  065042300900191007006700001020                    1
2005207EBERHARDT, RON G.    056400700940091006006300000980                    1
1007214EDMONSON, RICK T.    007906700330571003000290000000                    1
1009215EDSON, WILBUR S.     060705000820051207011885001575                    2
2008282ESTABAN, JUAN L.     198405500000001510040505005050                    2
1006292EVERLEY, DONNA M.    033200001775000502000322000322                    2
2001300FELDMAN, MIKE R.     025000000300001009000600000000                    2
1002304FROMM, STEVE V.      065000001200001205006090001823                    2
1005308GLEASON, JAMES E.    014500000390001004000925000000                    3
1007310GORMALLY, MARIE N.   038902200640551006007320000866                    1
2003311GROLER, GRACE B.     230064302054201510260480048369                    1
2009315HALE, ALAN A.        127400001680001208025954003829                    2
2004317HANBEE, ALETA O.     038500000395001210007530001180                    2
2003318HANEY, CAROL S.      097500001450001505012020006019                    3
1008322HARLETON, JEAN H.    078089901200891506010180000000                    2
2001325HATFIELD, MARK I.    015112200205391007001840000220                    3
2007332HELD, ANNA J.        024400000295001009005115000926                    3
1006409ICK, MICK W.         041012201950800502000590000590                    1
1003487KING, MILDRED J.     185089601804291510038260005322                    1
1008505LAMBERT, JERRY O.    015400100407451504002840000000                    5
1005568LYNNE, GERALD H.     092448701333111007016530001428                    1
1009574MELTZ, FRANK K.      075436600590301210015760000000                    1
2002590NEIL,CLARENCE N.     075006500950231208014000002324                    2
1006607ODELLE, NICHOLAS P.  058250701875070503006220006220                    3
2007689OWNEY, REED M.       043778800530661009008635001132                    1
2004721RASSMUSEN, JOHN J.   081000001000001208017680002346                    1
1001730REEDE, OWEN W.       090055501051441008010520002310                    1
2004739RIDEL, ROBERT R.     055750400825711207033303007224                    3
2009740RIDGEFIELD, SUZY S.  180417801905061210033024002310                    1
2002801SCHEIBER, HARRY T.   009523700325081203000520000520                    3
2007802SHEA, MICHAEL H.     064203300820091008012920004680                    1
2004806STOCKTON, NORMAN Q.  067250700725881209013380013380                    1
1009820TELLER, STEPHEN U.   195704401770091210051920032222                    3
1006825TILLMAN, DON M.      123044401995090505019230004665                    1
2003834TRAWLEY, HARRIS T.   057732600550001510011000001329                    2
1005909UDSON, DORIS M.      015449900303251004000780000000                    1
1008921ULL, GEORGE A.       180200002000001509034090012080                    3
2002956WANGLEY, THEO. A.    012000000150001207001720000275                    2
1001960WINGLAND, KEITH E.   005000000350001002000000000000                    1
 END OF DATA
```

INFORMATION MESSAGE

```
NUMBER OF INPUT BLOCKS PROCESSED   000050
NUMBER OF OUTPUT BLOCKS PROCESSED  000050
END OF JOB
```

Figure 2-12 Report and Information Messages from Card-To-Print Utility

In the listings illustrated above, it can be seen that the report contains the card images of the cards which were read as input. Each line on the report represents one card which was read. The data is printed exactly as it is contained on the card. The information messages, which are printed on a separate page after all of the input data has been processed, contain a count of the number of input records read and the number of output records written. In the example above, 50 cards were read as input and 50 lines were written on the report. The "End Of Job" message is printed to indicate that the utility program processed to a successful conclusion.

CARD-TO-PRINT FIELD SELECTION

In the previous example of the card-to-print utility program, the report was an "80-80 listing" of the card input, that is, the cards which were read were reproduced exactly on the printed report. In many applications, it is desirable to select certain fields from an input record and to also format the report so that the selected fields may be easily identified. It is also desirable to be able to label the fields on the report with headings. This capability is provided with the card-to-print utility program through the use of the field select option and the heading control card. An example of a report created when the cards illustrated in Figure 2-1 are read to print the store, department number, salesman number, and employee name is illustrated in Figure 2-13.

EXAMPLE

```
     STORE   DEPT    S/M      EMPLOYEE NAME  <---- REPORT HEADINGS

      20      01     325   HATFIELD, MARK I.

      20      07     332   HELD, ANNA J.          } REPORT IS
                                                  } DOUBLE-SPACED
      10      06     409   ICK, MICK W.

      10      03     487   KING, MILDRED J.

      10      08     505   LAMBERT, JERRY O.

      10      05     568   LYNNE, GERALD H.

      10      09     574   MELTZ, FRANK K.

      20      02     590   NEIL,CLARENCE N.

      10      06     607   ODELLE, NICHOLAS P.

      20      07     689   OWNEY, REED M.

      20      04     721   RASSMUSEN, JOHN J.

      10      01     730   REEDE, OWEN W.

      20      04     739   RIDEL, ROBERT R.

      20      09     740   RIDGEFIELD, SUZY S.

      20      02     801   SCHEIBER, HARRY T.

      20      07     802   SHEA, MICHAEL H.

      20      04     806   STOCKTON, NORMAN Q.

      10      09     820   TELLER, STEPHEN U.

      10      06     825   TILLMAN, DON M.

      20      03     834   TRAWLEY, HARRIS T.

      10      05     909   UDSON, DORIS M.

      10      08     921   ULL, GEORGE A.

      20      02     956   WANGLEY, THEO. A.

      10      01     960   WINGLAND, KEITH E.

      END OF DATA

      PAGE   2  <---- PAGE NUMBER
```

Figure 2-13 Example of Field-Selected Report

Note from the report illustrated in Figure 2-13 that the fields are formatted in such a way that they are easily readable instead of being in the exact card image. Note also that only the store number, the department number, the salesman number, and the employee name are printed on the report. The remaining fields on the card are omitted from the report. This formatting of a report is accomplished through the use of the field select option available with the DOS utility programs.

In order to field select input and output files, the Utility Modifier Statement must contain the proper entries and a Field Select Control Card must also be used. The job stream to create the report illustrated in Figure 2-13 is illustrated below.

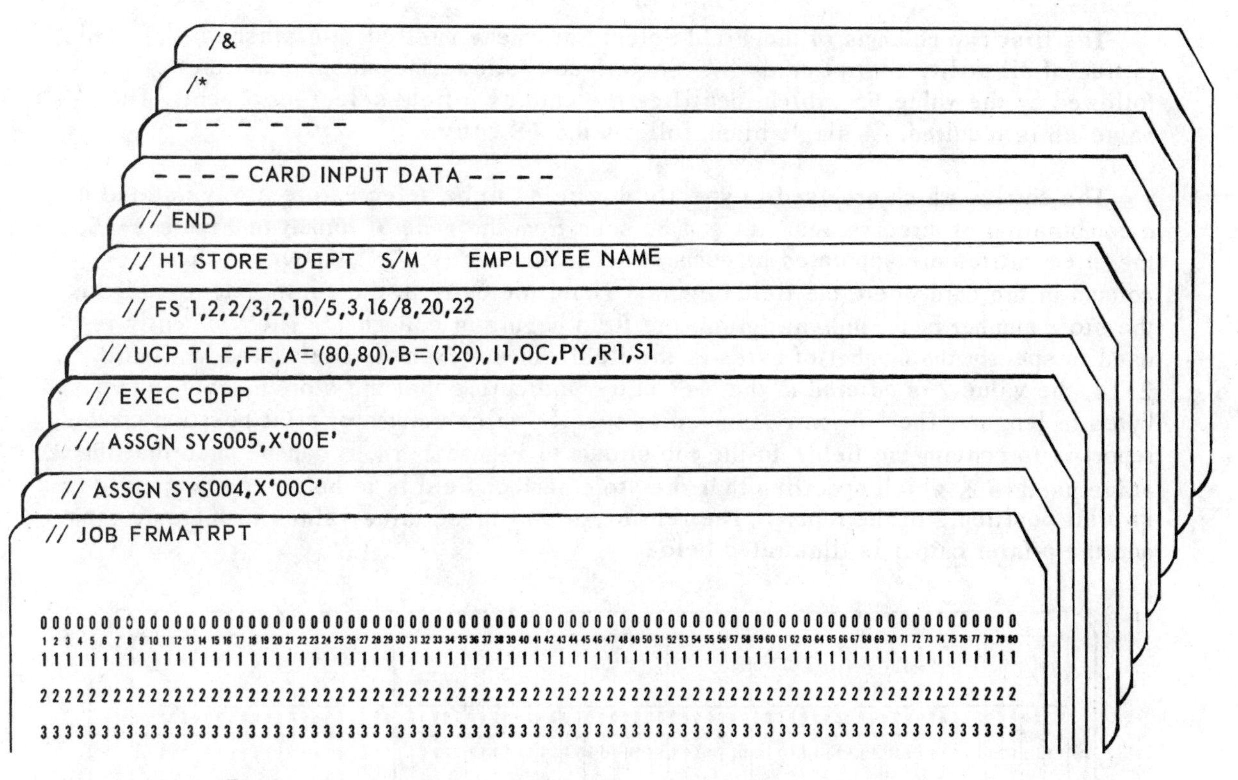

Figure 2-14 Job Stream for Field-Selected Report

Note from the job stream above that the card reader, X'00C', must be assigned to SYS004 and the printer, X'00E', must be assigned to SYS005. The program name is CDPP, which is the same name used for the 80-80 listing. The Utility Modifier Statement and the Field Select Statement are used to indicate to the program that field selection should take place.

The entries in the Utility Modifier Statement for field selection are the same as the entries for the straight list except for the function field which contains the value "TLF", instead of TL. The entry "TLF" indicates that the card input is to be field selected. When "TLF" is used as the function to be performed, a field select card must be contained in the Utility control cards to specify which fields are to be selected. Note also from the utility modifier statement that the pages are to be numbered (PY) and that the report is to be double-spaced (S2).

The field select card is used to specify which fields are to be selected from the input record and also where in the output the field is to be written. Note from Figure 2-14 that the field select card must immediately follow the Utility Modifier Statement in the job stream. The general format of the field select card is illustrated in Figure 2-15.

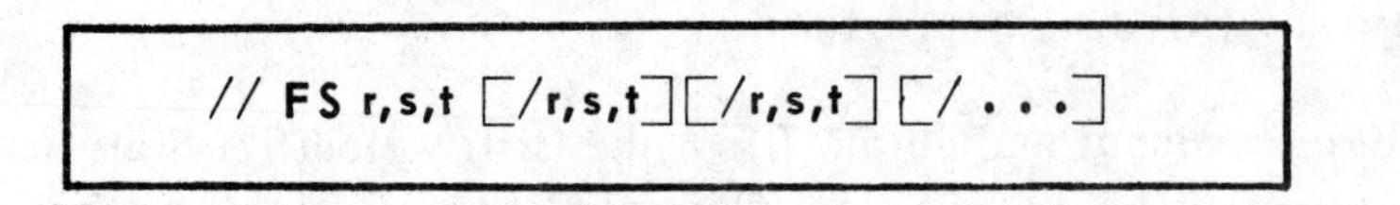

Figure 2-15 General Format of FIELD SELECT STATEMENT

The first two columns of the Field Select Statement must contain slashes (//). This is true of all utility control cards. A single blank follows the slashes and this is followed by the value FS, which identifies the card as a field select statement. The value FS is required. A single blank follows the FS entry.

The entries which are used to specify the fields to be selected are always stated in a combination of three values. As can be seen from the general format in Figure 2-15, the three entries are separated by commas. The first entry, "r", specifies the column in the card where the field begins. Thus, the entry in the Field Select card for the store number is 1, indicating that the field begins in column 1. The "s" entry is used to specify the number of bytes in the field. In the example illustrated in Figure 2-14, the value 2 is entered in the "s" entry, indicating that the store number is two bytes in length. The "t" entry is used to specify which column or print position on the report is to contain the field. In the job stream in Figure 2-14, it can be seen that the value used is 2, which specifies that the store number field is to be placed beginning in print position 2 of the report. The relationship of these three values to the card input and the printer output is illustrated below.

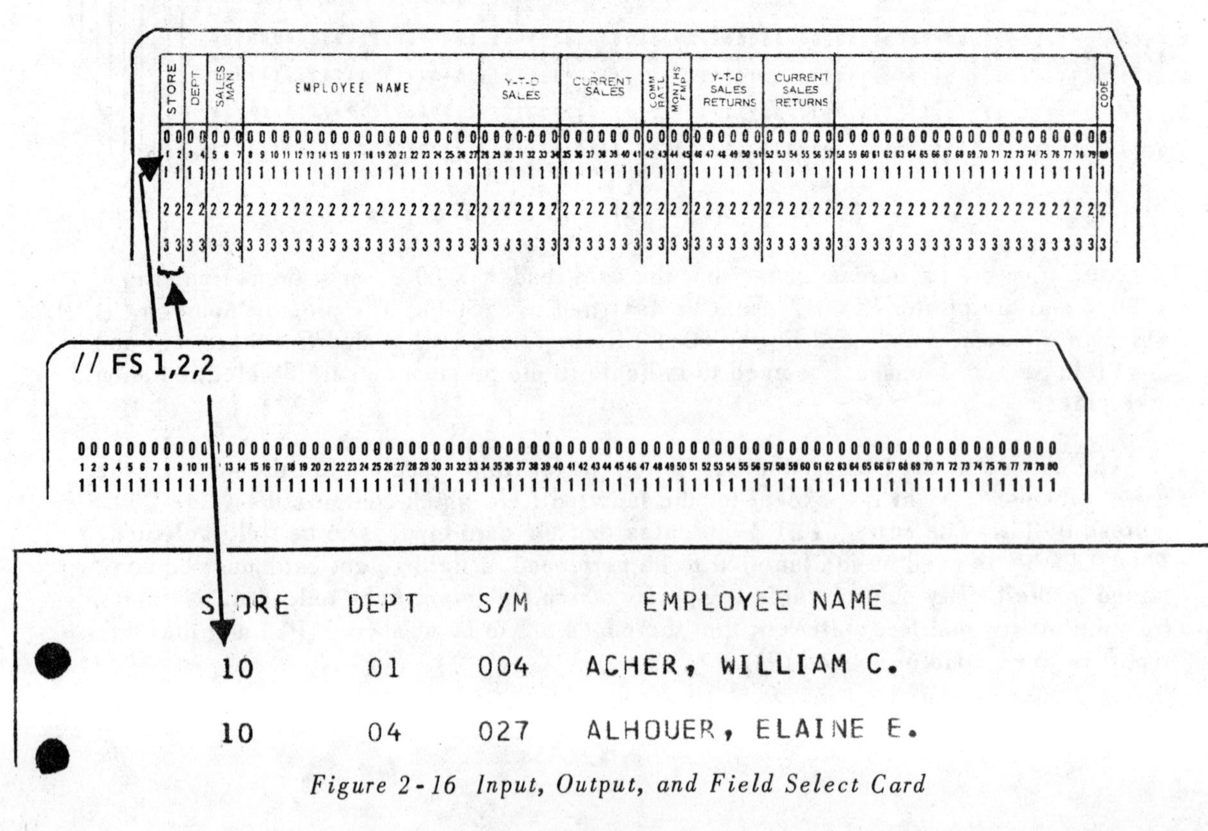

Figure 2-16 Input, Output, and Field Select Card

In the drawing in Figure 2-16 it can be seen that the "r" entry in the Field Select card references the column where the field to be selected begins, the "s" entry specifies the number of bytes in the field, and the "t" entry indicates the beginning print position on the report where the field is to be printed. Thus, the store field begins in column 1 of the input card and is 2 bytes in length. The value in the two-byte store field is to be printed on the report beginning in position 2 (note that the heading "STORE" begins in position 1).

If more than one field is to be selected from an input record, another combination of "r,s,t" entries must be made in the field select card. Each of the three entries must be separated from one another by a slash (/) with no blanks included. The field select card used for the report : again illustrated in Figure 2-17.

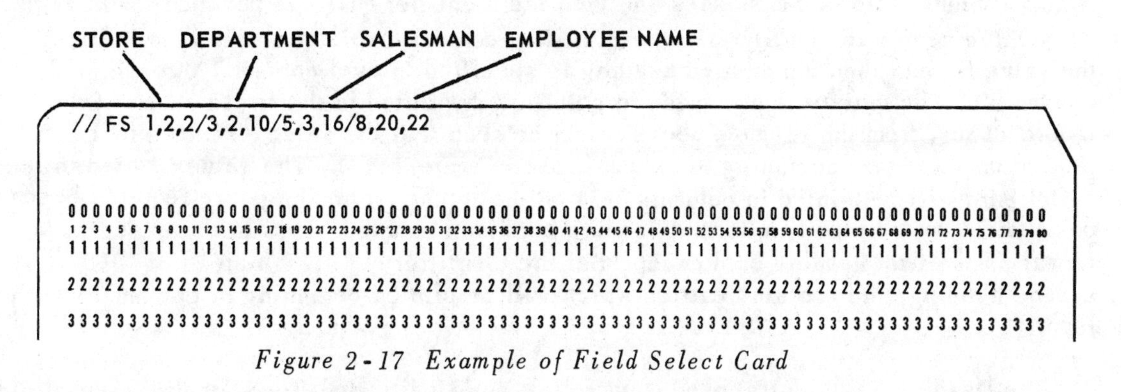

Figure 2-17 Example of Field Select Card

In the Field Select card illustrated above, note that four fields are to be selected from the input card. These fields are the store number, the department number, the salesman number, and the employee name. The store number begins in column 1 of the card, is two bytes long, and is to be printed beginning in position 2 on the report. Thus, positions 2 and 3 of the report will contain the store number. The department number begins in column 3 of the card and will be printed in positions 10 and 11 on the report. The three byte salesman number, which begins in column 5 of the input card, will be printed in positions 16, 17, and 18 on the report. The employee name, which is 20 bytes in length, will be moved from the card beginning in column 8 and be printed on the report beginning in position 22.

Note from the above example that the remaining fields in the input card are not printed on the report. When the field select feature is being used in the card to print utility, every field which is to be printed must be specified in the field select statement. If a field is not specified, it will not be copied to the printed report.

Headings are produced on the printed report through the use of the Heading Statement. The Heading Statement used for the report illustrated in Figure 2-13 is shown in Figure 2-18.

Heading Statement

```
// H1 STORE  DEPT  S/M      EMPLOYEE NAME

00000000000000000000000000000000000000000000000000000000000000000000000000000000
1 2 3 4 5 6 7 8 9 10 11 12 13 14 15 16 17 18 19 20 21 22 23 24 25 26 27 28 29 30 31 32 33 34 35 36 37 38 39 40 41 42 43 44 45 46 47 48 49 50 51 52 53 54 55 56 57 58 59 60 61 62 63 64 65 66 67 68 69 70 71 72 73 74 75 76 77 78 79 80
11111111111111111111111111111111111111111111111111111111111111111111111111111111

22222222222222222222222222222222222222222222222222222222222222222222222222222222

33333333333333333333333333333333333333333333333333333333333333333333333333333333
```

Figure 2-18 Example of HEADING Statement

Note from the above example that the Heading Statement begins with the characters // in card columns 1 and 2. Every utility control card must begin with these characters. A single blank follows the slashes and then the identifier "H1" is punched in the card. This value serves to identify the type of utility control card. A single blank follows the value H1 and then the desired heading is specified in card column 7 through card column 80. The entry which is made in column 7 is printed in the first column of the report. Thus, from the example above it can be seen that the value STORE will be printed on the report beginning in column 1 (See Figure 2-13). The values stated in the "H1" card will be printed in columns 1-74 on the report. If headings are required beyond position 74 on the report, a second heading card may be used. It is in the exact same format as the first heading card except that the identifier is "H2" instead of "H1". The second heading card contains entries which will be printed beginning in column 75 of the report.

The Heading Statement(s) used must follow any Utility Modifier Statements and Field Select Statements which appear for the card-to-print utility program. As with the card-to-print program which did not use field select, the last control card must be the // END card which indicates the end of the utility control cards. The card input data immediately follows the // END card and the end of the data is indicated through the use of the /* card (see Figure 2-14).

When the card-to-print utility program is executed, an information messages listing is always produced. The listing produced from the job stream in Figure 2-14 is illustrated below.

```
// UCP TLF,FF,A=(80,80),B=(120),I1,CC,PY,R1,S2
// FS 1,2,2/3,2,10/5,3,16/8,20,22
// H1 STORE    DEPT    S/M         EMPLOYEE NAME
// END
CARD TO PRINTER / PUNCH UTILITY
INPUT RECORD LENGTH 0080
OUTPUT RECORD LENGTH  0120
INPUT BLOCK LENGTH 00080
OUTPUT BLOCK LENGTH 00120
INPUT OPTION CARD BCD
OUTPUT OPTION PRINT CHARACTER
2 INPUT,2 OUTPUT AREAS ASSIGNED
RECORD FORMAT FIXED
TYPE LIST,FIELD SELECT
STARTING RECORD NUMBER  00000001
```

Figure 2-19 Information Message for Card-To-Print Program

Note that in addition to the Utility Modifier Statement, the Field Select Statement and the Header card are also printed in the listing. The remainder of the listing indicates the values which are to be used by the Utility program. Note that the "type" entry specifies both list and select as the type of processing which will be accomplished.

CARD-TO-CARD FIELD SELECT

As with the card-to-print utility program, the field select option may be used when the card-to-card utility program is executed. The job stream to select the salesman number and the employee name from the card illustrated in Figure 2-1 and punch the salesman number in card columns 1, 2, and 3 and the employee name beginning in card column 10 is illustrated in Figure 2-20.

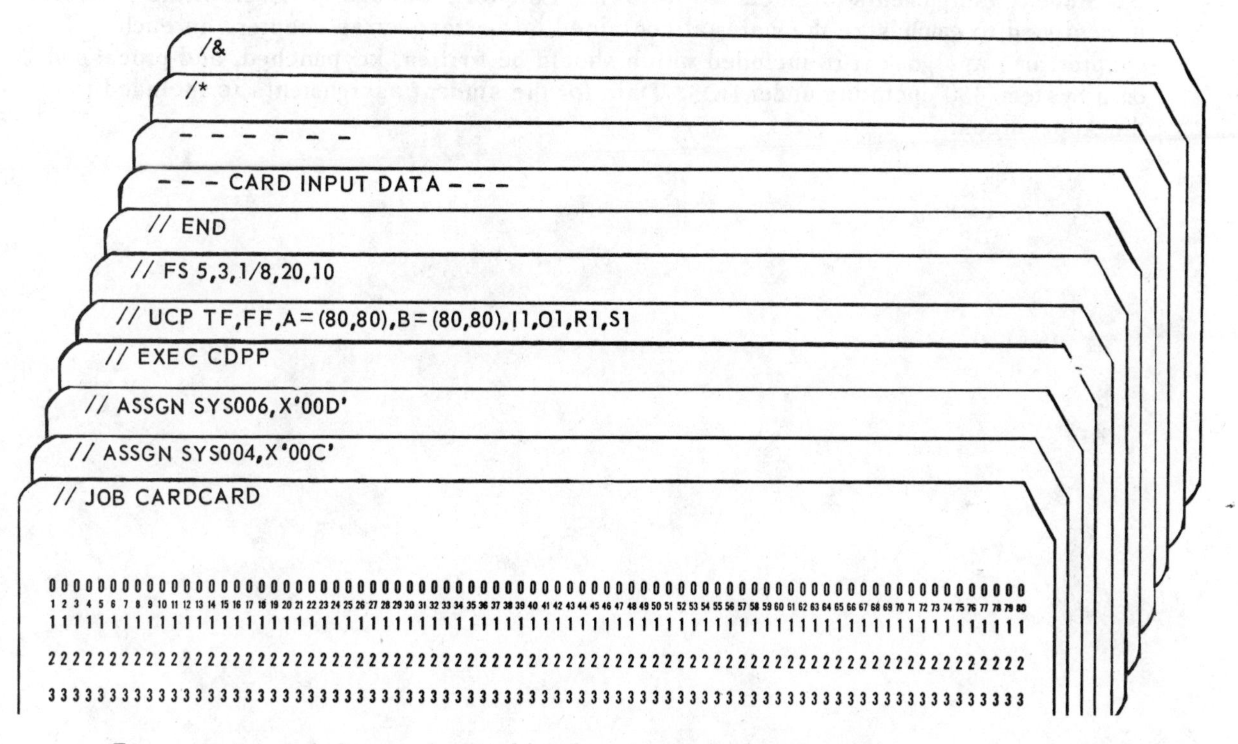

Figure 2-20 Job Stream for Field Select in Card-To-Card Program

Note from the above job stream that the card reader, X'00C', is assigned to SYS004 and the card punch, X'00D', is assigned to SYS006. These assignments are required for the card-to-card program. The Utility Modifier Statement contains the type indicator "TF", which specifies that field select is to take place for the card-to-card program. Note that the type indicator for the card-to-card program which did not utilize the field select feature was TC.

Whenever the card-to-card program is to field select the data, a Field Select Statement must be submitted in the control cards read by the Utility program immediately following the Utility Modifier Statement. The format of the Field Select Statement is identical to that used for the card-to-print program, that is, the combination "r,s,t" must be used to specify the beginning column in the input card (r), the length of the field (s), and the beginning column in the output card (t). The sets are separated from one another by a slash (/). Note in the example in Figure 2-20 that the salesman number field, which is 3 bytes long, is to be punched in the first column of the output card and the name is to be punched beginning in column 10.

STUDENT ASSIGNMENTS

Student assignments are included following Chapter 2 through 7. Each of the problems is designed to emphasize the material contained in the respective chapter. In each chapter, one assignment is included which should be written, keypunched, and processed on a System/360 operating under DOS. Data for the student assignments is included in the Appendices.

STUDENT ASSIGNMENTS

1. Write the job stream to reproduce a deck of cards using the card-to-card utility program. The card input device is X'00C' and the card output device is X'00D'. Both the input and the output cards are in the EBCDIC format and the output cards should be placed in Stacker 2 of the 2540 Card Read Punch.

1	2	3	4	5	6	7	8	9	10	11	12	13	14	15	16	17	18	19	20	21	22	23	24	25	26	27	28	29	30	31	32	33	34	35	36	37	38	39	40	41	42	43	44	45	46	47	48	49	50	51	52	53	54	55

2. Write the job stream to print a deck of cards on the printer using the card-to-print utility program. The input device is X'00C' and the output device is X'00E'. The report, which is to be double-spaced, should contain page numbers. Sequence check the input cards on columns 6-10.

1	2	3	4	5	6	7	8	9	10	11	12	13	14	15	16	17	18	19	20	21	22	23	24	25	26	27	28	29	30	31	32	33	34	35	36	37	38	39	40	41	42	43	44	45	46	47	48	49	50	51	52	53	54	55

3. Write the job stream to process a card file, stored on X'00C', and print it on device X'00E'. The single-spaced report should contain page numbers and the following fields from the card:

Col 1 - 5: Salesman Number
Col 10 - 30: Salesman Name
Col 40 - 42: Commission Rate
Col 50 - 51: Months Employed

The report should be in the format Salesman Name, Salesman Number, Months Employed, and Commission Rate. Headings should be included on the report. The columns used on the report are to be determined by the programmer.

1	2	3	4	5	6	7	8	9	10	11	12	13	14	15	16	17	18	19	20	21	22	23	24	25	26	27	28	29	30	31	32	33	34	35	36	37	38	39	40	41	42	43	44	45	46	47	48	49	50	51	52	53	54	55

4. Write the job stream to reproduce cards from the input cards in Problem #3 in the same format as they are written on the report.

1	2	3	4	5	6	7	8	9	10	11	12	13	14	15	16	17	18	19	20	21	22	23	24	25	26	27	28	29	30	31	32	33	34	35	36	37	38	39	40	41	42	43	44	45	46	47	48	49	50	51	52	53	54	55

5. Write the job stream to reproduce the data cards produced in Problem #4 and rearrange them in the following order:

Col 1 - 2: Months Employed
Col 3 - 5: Commission Rate
Col 6 - 10: Salesman Number

The salesman name is not to be included in the cards.

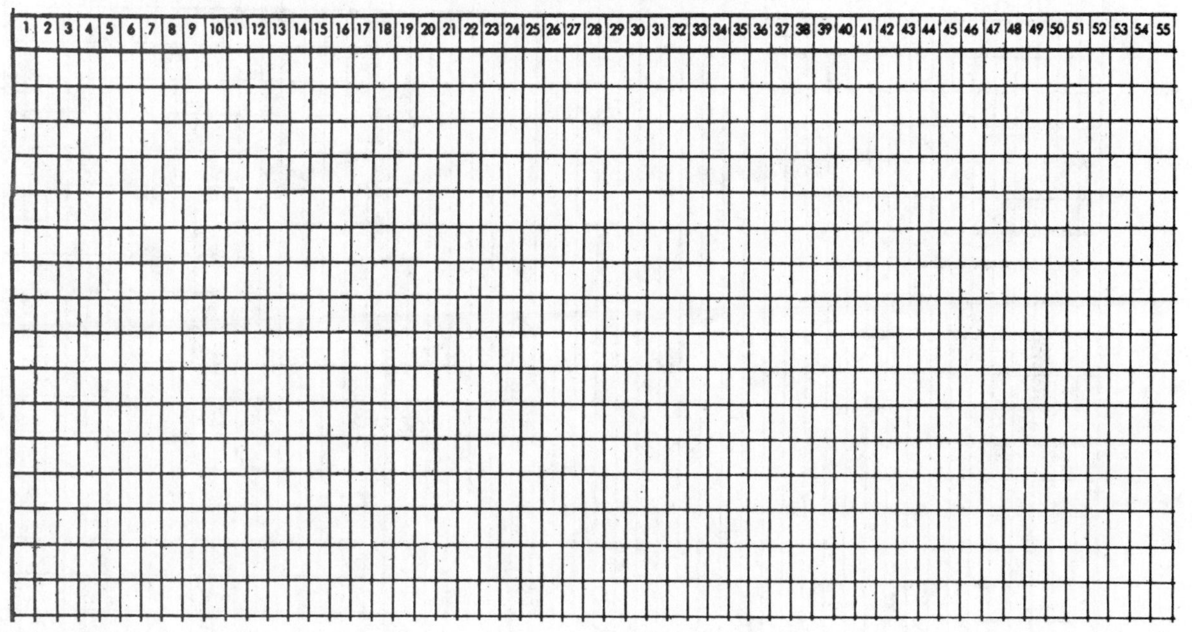

6. Create a report using the card - to - print program of the cards produced in Problem #5. The report format and the other options available in the program are to be determined by the programmer.

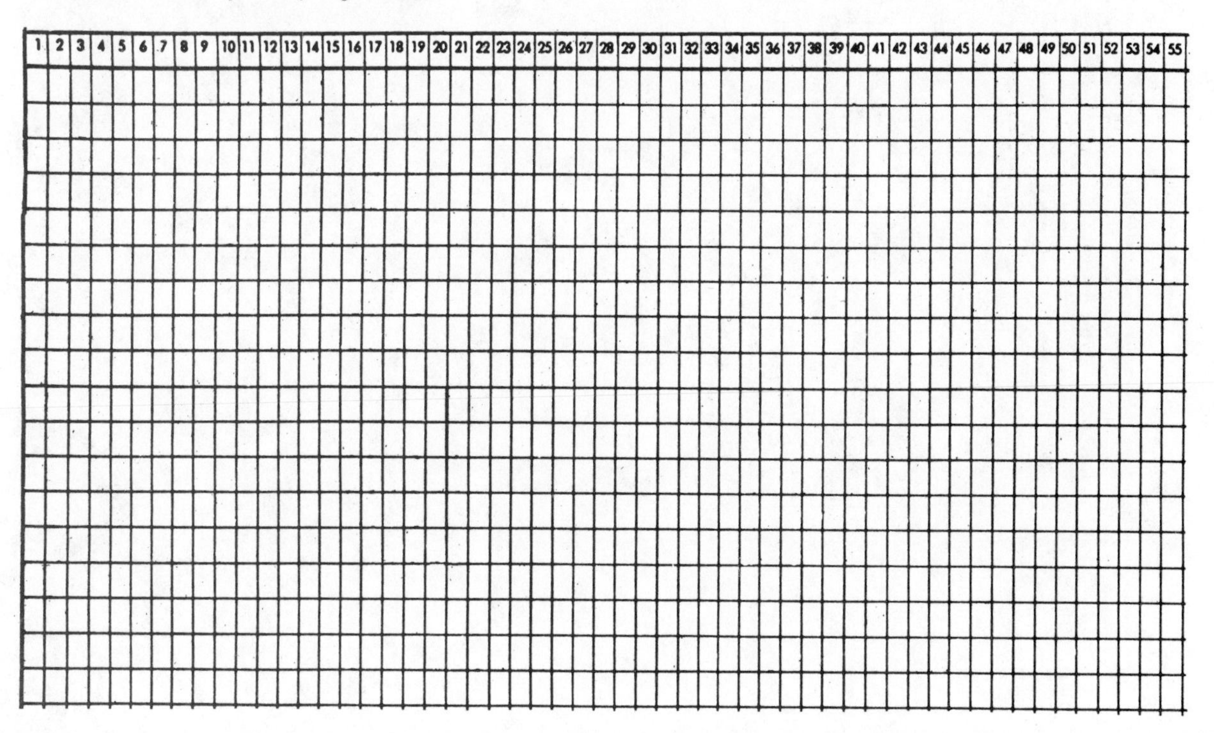

7. A report is to be produced from the data cards contained in Appendix E. The report should include the invoice number, the branch, salesman, city, state, customer number, quantity and description. The report format, including headings, should be determined by the programmer. The job stream should be written and punched and the job processed on a System/360 operating under DOS using the data cards in Appendix E.

1	2	3	4	5	6	7	8	9	10	11	12	13	14	15	16	17	18	19	20	21	22	23	24	25	26	27	28	29	30	31	32	33	34	35	36	37	38	39	40	41	42	43	44	45	46	47	48	49	50	51	52	53	54	55

CHAPTER 3

DIRECT-ACCESS UTILITY PROGRAMS

INTRODUCTION

As noted previously, the DOS Utility Programs are used to perform functions which are normally required within a data processing environment. The direct-access devices within an installation are many times used for files and data which play an important part in the systems which are executed on the computer. Thus, one of the more important sets of utility programs are the ones which process direct-access files. Included in these are the card-to-disk, the disk-to-print, the disk-to-card, the disk-to-disk, the tape-to-disk, and the disk-to-tape utility programs. This chapter will cover the card-to-disk, disk-to-print, disk-to-card, and disk-to-disk programs. Chapter 4 contains the information concerning the magnetic tape utility programs.

CARD-TO-DISK Utility Program

A card-to-disk utility program may be used to load cards onto a disk file. The job stream presented below illustrates the required entries.

EXAMPLE

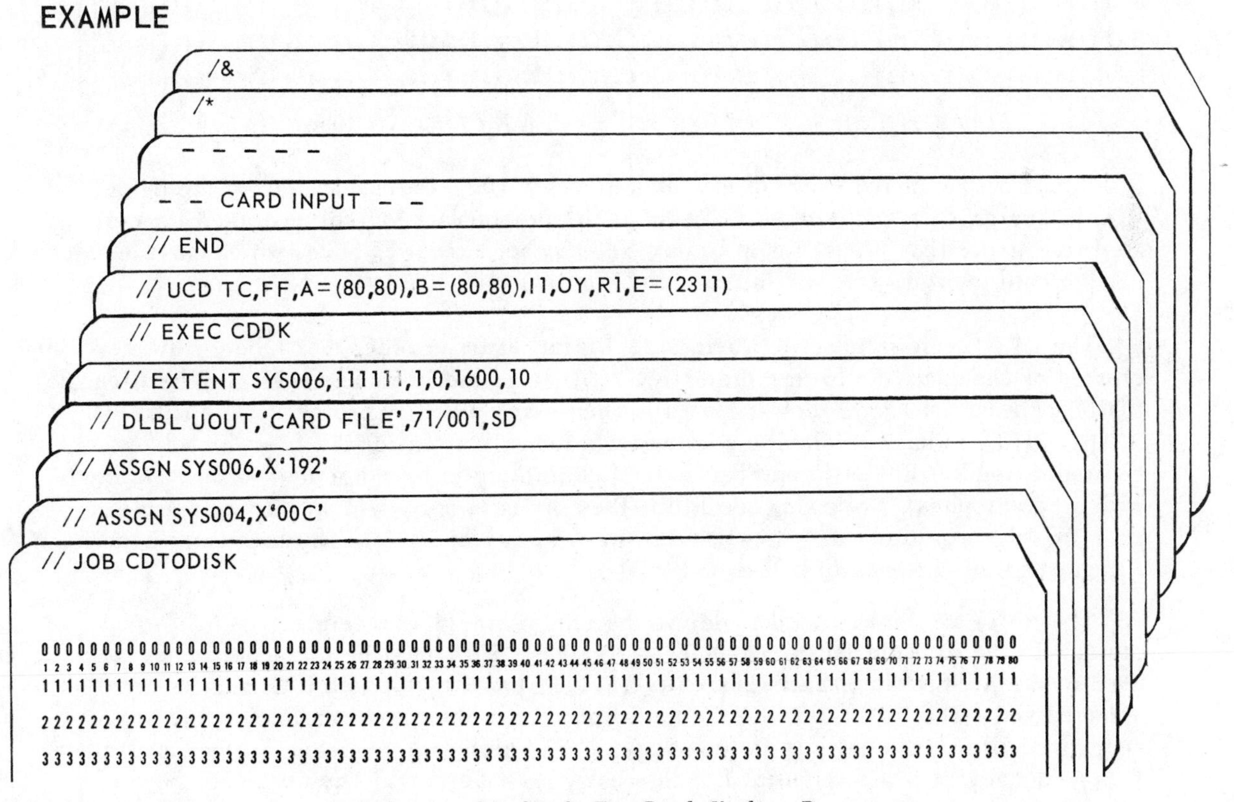

Figure 3-1 Job Stream for Card-To-Disk Utility Program

Note from Figure 3-1 that the sequence of job control and utility control cards is the same as those used with the card-to-print and card-to-card programs, that is, the ASSGN cards and, in the case of disk files, the DLBL and EXTENT cards, precede the execute card. The phase name of the card-to-disk program is CDDK. Thus, the job control card // EXEC CDDK will cause the card-to-disk utility program to be loaded into core storage and control will be given to it. The Utility Modifier Statement and the End Statement follow the execute card in the same manner as used for the utility programs discussed previously.

Whenever a Utility program is used to reference a direct-access device, standard direct-access labels must be used and the area on the disk which is to contain the data must be defined. This is accomplished by the DLBL and the EXTENT card. The ASSGN, DLBL, and EXTENT cards used for the output file in Figure 3-1 are again illustrated below.

EXAMPLE

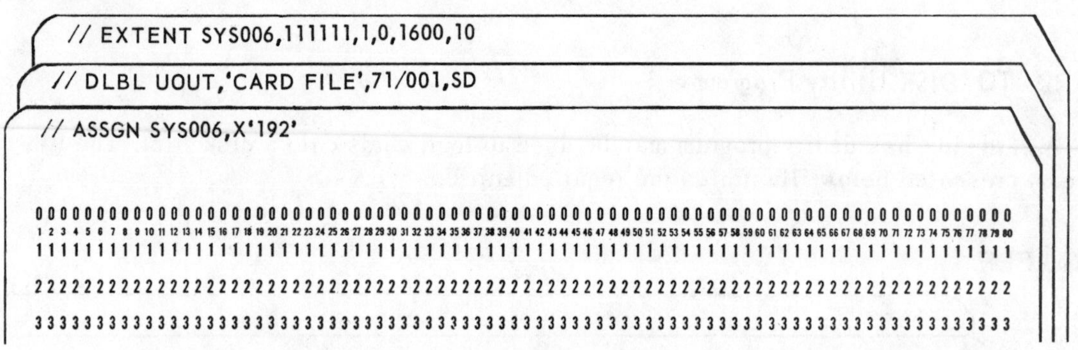

Figure 3-2 ASSGN, DLBL, and EXTENT Cards used with Utility Program

Note from the above example that disk drive, X'192', which is used for the output file, is assigned to symbolic unit SYS006. The programmer logical unit used for the card-to-disk utility program may be any SYS number except SYS004, which must be used for the card input device as illustrated in Figure 3-1.

The DLBL card is used to provide label information in order that labels may be created or checked. Following the entry // DLBL is the file name UOUT. The filename UOUT must be used for a disk output file when using a utility program. The file I-D, 'CARD FILE' is used to provide a unique file identification for the file which is being created or read. This entry can be up to 44 characters in length and must be contained within apostrophes. Following the file I-D is the date entry which, in the example, specifies the expiration date for the file (the first day of 1971). The last entry specifies a sequential disk file is to be created (SD).

The EXTENT card is used to define the area on the disk where the data is to be stored. In the example the SYS number is SYS006. This must be the same number as used in the ASSGN statement. The entry 111111 specifies the serial number of the disk volume used. The "1" following the serial number is the entry required to specify a sequential file. This entry is followed by a "0" which is a sequence number for the EXTENT card. For a sequential file this entry must be zero. The next entry specifies the relative track number which indicates where the file is to be stored. In the example, the relative track is 1600, that is, cylinder 160, track 0. The final entry, the "10", indicates the number of tracks which are allocated for the file.

For a more detailed explanation of job control see DOS JOB CONTROL FOR COBOL PROGRAMMERS, Shelly & Cashman, Anaheim Publishing Company.

Again, it should be noted that the phase name of the card-to-disk program is CDDK. The execute card used to load the program and cause its execution is illustrated below.

EXAMPLE

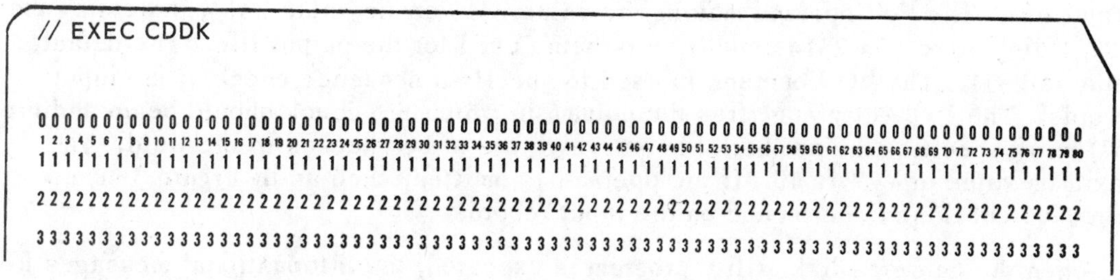

```
// EXEC CDDK
```

Figure 3-3 Execute Card for Card-To-Disk Utility Program

The Utility Modifier Statement which is used for the card-to-disk utility program is identical in format to that used for the other file-to-file utility programs and it serves the same purpose, that is, to define for the program the parameters which the user wishes to use for the processing. The general format of the Utility Modifier Statement used to copy a card file to a disk file is illustrated below.

```
// UCD TC,FF,A=(80,80),B=(80,80)[,Ix][,Ox][,Rx][,E=(e)][,Q=(x,y)]
```

Figure 3-4 General Format of Utility Modifier Statement for Card-To-Disk Copy

In the general format illustrated above, it can be seen that the characters // must be in column 1 and column 2 of the card. The slashes are followed by a single blank and then the identifier UCD is punched on the card. This identifies the statement as the Utility Modifier Statement for the card-to-disk utility program. The identifier is followed by a single blank.

The function-type code for the card-to-disk copy program is TC, which indicates that the cards are to be directly copied onto the disk. The record format entry must specify FF, which indicates that fixed-length records are to be processed. The input record and block length must be equal to 80 bytes and, for the straight copy function, the output record and block length must be 80 bytes (in order to block card input files, the REBLOCK function must be utilized and is illustrated later in this chapter).

The remaining entries in the Utility Modifier statement are optional and, if they are not used, default values will be used by the program. The "I" entry is used to designate the format of the card input, that is, whether it is in EBCDIC format or binary format. In the example, the characters in the card are in EBCDIC format, which is the default value if this entry is omitted. The "O" entry indicates whether a write-disk check is to be performed for the records written on the disk output file. If this option is requested, a parity check is performed on all records written on the disk file to ensure that they were written correctly. The "Y" entry, such as is used in Figure 3-1, specifies that the check should be performed. An entry of "ON" states that a write-check should not be performed. The default value is "OY" because it is normally desirable to use a write-check.

The "Rx" entry is used to specify the first logical record to be processed by the program. The "x" entry may be any decimal value which is less than or equal to the number of input records. The default value is "R1", which indicates that the first record in the input file should be written on the output file. Note from Figure 3-1 that the value R1 is used.

The E=(e) entry is used to specify the type of device which is to be used for the output file. The "e" operand may be the value 2311 or the value 2314, indicating that a 2311 disk drive or a 2314 disk drive is being used for the output file. The default value is 2311. The "Q" operand is used to specify a sequence check of the input records. The "x" entry specifies the column in which the check should begin and the "y" entry specifies the number of bytes which are to be checked in the field. The maximum value of "y" is 10. If the operand is omitted, such as in Figure 3-1, no sequence checking is performed on the input records.

When the card-to-disk utility program is executed, the informational messages from the utility program are printed on SYSLST and the cards are read from the device assigned to SYS004 and stored on the disk file. In addition to printing the information which is derived from the Utility Modifier Statement, the messages contain information pertaining to the processing after the cards have been written on the disk. Specifically, it contains a block count relating the number of blocks which were written and also the cylinder, track, and record where the end of file marker was written. The listing printed as a result of processing the job stream illustrated in Figure 3-1 is illustrated below.

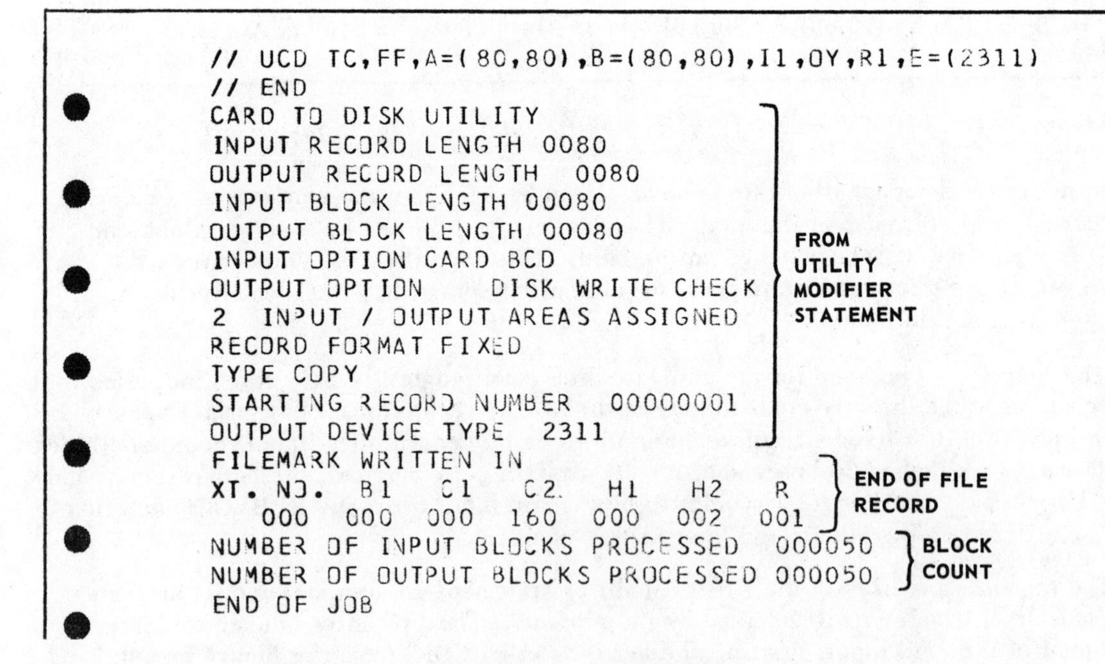
```
// UCD TC,FF,A=(80,80),B=(80,80),I1,OY,R1,E=(2311)
// END
CARD TO DISK UTILITY
INPUT RECORD LENGTH 0080
OUTPUT RECORD LENGTH    0080
INPUT BLOCK LENGTH 00080
OUTPUT BLOCK LENGTH 00080
INPUT OPTION CARD BCD
OUTPUT OPTION       DISK WRITE CHECK
2  INPUT / OUTPUT AREAS ASSIGNED
RECORD FORMAT FIXED
TYPE COPY
STARTING RECORD NUMBER    00000001
OUTPUT DEVICE TYPE   2311
FILEMARK WRITTEN IN
XT. NO.   B1    C1    C2.    H1    H2    R
    000   000   000   160   000   002   001
NUMBER OF INPUT BLOCKS PROCESSED   000050
NUMBER OF OUTPUT BLOCKS PROCESSED 000050
END OF JOB
```


Figure 3-5 Informational Messages from Card-To-Disk Utility

Note from the listing above that the information is presented in the same manner as that from the card-to-card and card-to-print programs. The options which were requested in the Utility Modifier Statement are listed together with the number of buffers (Input/Output areas) which will be used by the program.

When the utility program has completed writing the cards on the disk, three additional factors are included on the report. The first is the address on the disk where the end-of-file (Filemark) indicator was written. When a 2311 or 2314 disk unit is used, the cylinder number is printed under the "C2" heading, the track number is printed under the "H2" heading and the record number is printed under the "R" heading. Thus, in the listing illustrated in Figure 3-5, it can be seen that the filemark was written on Cylinder 160, Track 2, Record 1.

The second and third values included on the report are the number of input blocks processed and the number of output blocks processed. In the example above, it can be seen that 50 cards were read as input and 50 disk blocks were written as output. Since the disk records were written as unblocked records, the 50 input cards were written as 50 physical records or blocks on the output file. The "END OF JOB" message indicates that the program was normally terminated after completing its assigned task.

DISK-TO-PRINT Utility Program

After a disk file is created by means of the card-to-disk utility program, it may be processed by user programs or other utility programs. One of the frequently desired functions is to print the file on the printer. The Disk-To-Print Utility program can be used for this purpose. The following job stream could be used to print the file which was created in the previous card-to-disk example.

EXAMPLE

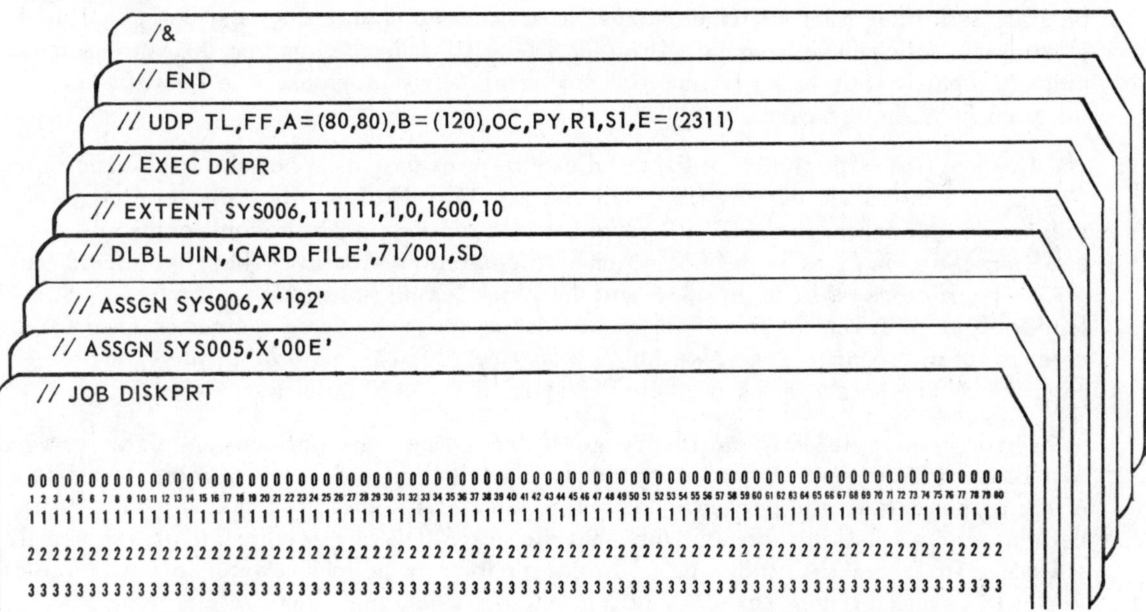

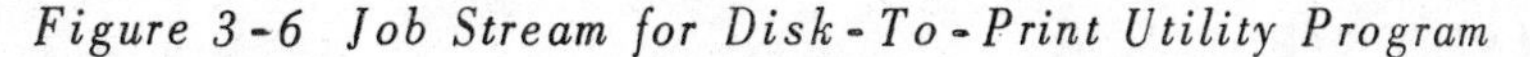

Figure 3-6 Job Stream for Disk-To-Print Utility Program

Note from the job stream above that the printer, X'00E', is assigned to symbolic unit SYS005. Whenever the printer is the output device for any of the file-to-file utility programs, it must be assigned to SYS005. The disk drive which contains the input file may be any programmer logical unit except SYS005, which must be assigned to the printer. In the example above, the disk input unit is assigned to SYS006. Note that the symbolic unit entry in the EXTENT card for the file is SYS006, which reflects the assignment of SYS006 to the disk drive.

The filename entry for a disk input file in the DLBL card MUST be UIN. This filename is used for all disk input files, regardless of the utility program which is being executed. The file-ID field contains the name 'CARD FILE', which is the same name as the file which was created in the previous card-to-disk program. The file-ID for the disk input file must correspond to the file-ID which was used when the file was created. The remainder of the DLBL and EXTENT cards must contain the same information as when the input file was created.

From the job stream in Figure 3-6 it can be seen that the sequence of cards in the job stream is identical to those used for other utility programs, that is, the Utility Modifier Statement follows the Execute card and it is followed by the END Statement. Note from the example that the phase name of the Disk-To-Print program is DKPR. This name must be included on the execute card in order for the program to be loaded into core storage and executed.

As with the other utility programs, the Utility Modifier Statement is used with the Disk-To-Print program to specify the options required by the user. The general format of the Utility Modifier Statement for the disk-to-print program is shown below.

```
// UDP TL,FF,A=(n,m),B=(p)[,Oy][,Px][,Rx][,Sx][,E=(e)]
```

Figure 3-7 General Format of Utility Modifier Statement

In the general format, it can be seen that the same format is used for the disk-to-print program as for the other utility programs. The first two characters must be // and these slashes are followed by a space. The identifier UDP must be used on the statement in order to identify it as being for the disk-to-print utility program. The identifier is followed by a single space.

The function-type indicator for the disk-to-print program when the unblocked input file is to be listed without any field selecting is TL. The records are fixed-length and this is indicated through the "FF" record format indicator. The record length and the block size are identical to the file which was created previously. Thus, the record length (n) is specified as 80 in Figure 3-6 and the block length (m) is also specified as 80. The length of the print line is specified by the "B" entry in the same manner as used for the card-to-print program. The value for "p" may be 120, 132, or 144. In the example in Figure 3-6, the length of the print line is specified as 120 columns.

The remaining fields in the Utility Modifier Statement are optional and default values will be used if the entries are not included. The "O" entry is used to indicate the format of the output data on the printed report. The value "OC" specifies that the data should be printed in a character format. Note that the entry "OC" is used in the utility modifier statement in Figure 3-6, indicating that the report is to be in character format. Character format is assumed if this entry is omitted, and must be used when the function-type indicator is TL.

Page numbering is controlled by the "Px" entry. If x is equal to Y, page numbering will be performed by the utility program and if x is equal to N, page numbering will not take place. If the page numbering entry is not included, page numbering will take place when the disk-to-print list program is executed.

The "Rx" entry indicates which logical record in the input file will be the first one processed. The x operand may be any value which is less than or equal to the number of logical records which are contained on the input file. The "Sx" entry is used to specify the spacing which will take place on the printed report. The x operand may contain the values 1, 2, or 3 to indicate single, double, or triple spacing. In addition, disk records may contain, as the first character in each record, a "carriage control character" which is used to control spacing. The use of records which contain carriage control characters is explained later in this chapter.

The "E" entry is used to specify the type of input device which is being used. The valid values for "e" are 2311 and 2314. The 2311 disk drive is assumed to be the input device if this operand is not used.

As a result of the job stream illustrated in Figure 3-6, the following report would be generated.

EXAMPLE

```
2001325HATFIELD, MARK I.   01511220020539100700184000022O    3
2007332HELD, ANNA J.       024400000295001009005115000926    3
1006409ICK, MICK W.        041012201950800502000590000590    1
1003487KING, MILDRED J.    185089601804291510038260005322    1
1008505LAMBERT, JERRY O.   015400100407451504002840000000    5
1005568LYNNE, GERALD H.    092448701333111007016530001428    1
1009574MELTZ, FRANK K.     075436600590301210015760000000    1
2002590NEIL,CLARENCE N.    075006500950231208014000002324    2
1006607ODELLE, NICHOLAS P. 058250701875070503006220006220    3
2007689OWNEY, REED M.      043778800530661009008635001132    1
2004721RASSMUSEN, JOHN J.  081000001000001208017680002346    1
1001730REEDE, OWEN W.      090055501051441008010520002310    1
2004739RIDEL, ROBERT R.    055750400825711207033303007224    3
2009740RIDGEFIELD, SUZY S. 180417801905061210033024002310    1
2002801SCHEIBER, HARRY T.  009523700325081203000520000520    3
2007802SHEA, MICHAEL H.    064203300820091008012920004680    1
2004806STOCKTON, NORMAN Q. 067250700725881209013380013380    1
1009820TELLER, STEPHEN U.  195704401770091210051920032222    3
1006825TILLMAN, DON M.     123044401995090505019230004665    1
2003834TRAWLEY, HARRIS T.  057732600550001510011000001329    2
1005909UDSON, DORIS M.     015449900303251004000780000000    1
1008921ULL, GEORGE A.      180200002000001509034090012080    3
2002956WANGLEY, THEO. A.   012000000150001207001720000275    2
1001960WINGLAND, KEITH E.  005000000350001002000000000000    1
 END OF DATA
PAGE  2
```

```
NUMBER OF INPUT BLOCKS PROCESSED  000050
NUMBER OF OUTPUT BLOCKS PROCESSED 000050
END OF JOB
```

Figure 3-8 Example of report from Disk-To-Print Utility Program

Note from the above listing that both double spacing and page numbering occurred as requested. In addition, note the end of job "statistics" which are printed on the page following the last page of the report. In the example it can be seen that fifty input blocks, or records, were read and 50 output blocks were written. The utility programs always give these processing statistics at the end of the job.

CARD-TO-DISK REBLOCKING

In many direct-access device applications, it is desirable to block the logical records so that more records may be stored and they may be processed more rapidly. The card-to-disk utility program can be used to read fixed-length unblocked card records and create a disk file of fixed-length blocked records. The job stream illustrated in Figure 3-9 could be used to read a card input file and create a blocked output file.

EXAMPLE

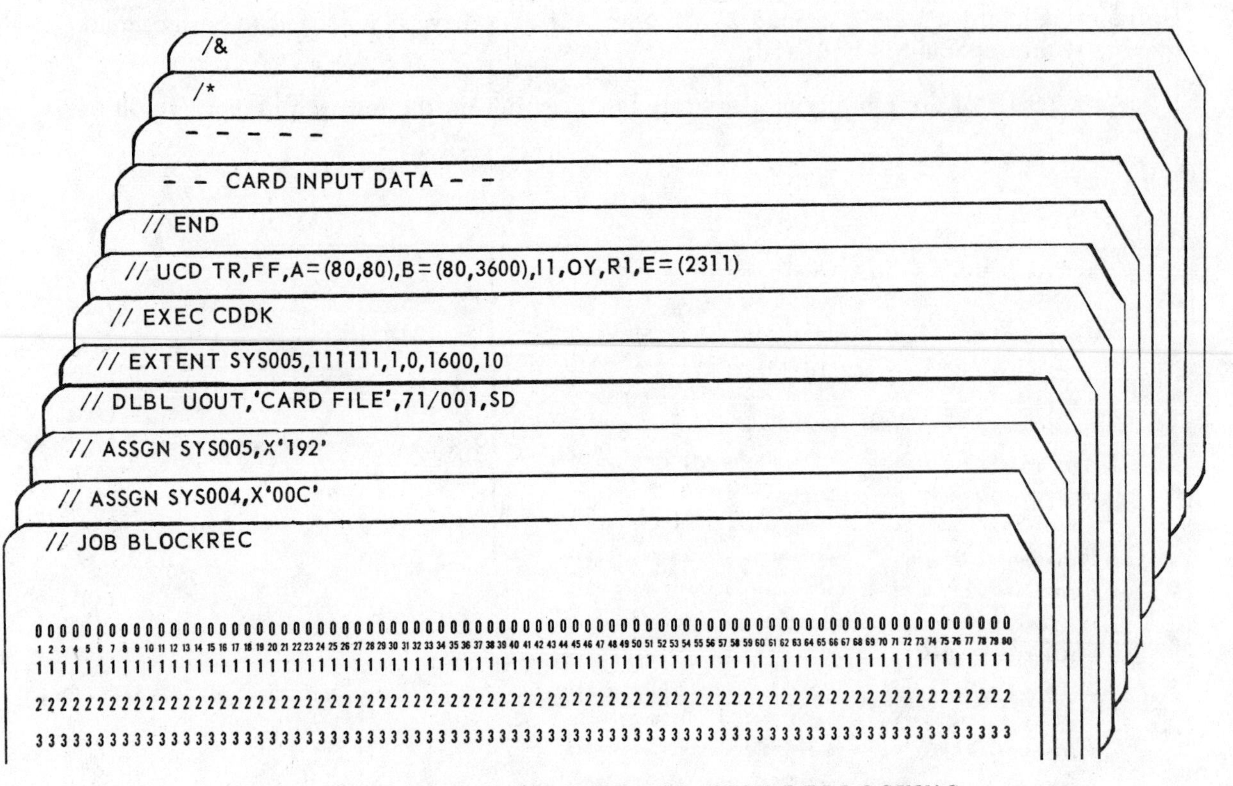

Figure 3-9 Job Stream for CARD-TO-DISK REBLOCKING

Note from the job stream illustrated in Figure 3-9 that the job control ASSGN, DLBL, and EXTENT cards are identical to those used when the disk file is not to be blocked. The program, CDDK, must be specified on the execute card in order to cause the card-to-disk program to be loaded into core storage and executed.

The Utility Modifier Statement contains two changes from the one used when the disk file is not blocked—the type indicator is changed to TR, which indicates reblocking, and the block size in the "B" parameter must be changed to the size of the block to be created. In the example above, the entry B=(80,3600) is used for the output file. Thus, 45 logical records will be stored in a physical record or block on the disk ($80 \times 45 = 3600$). The remainder of the utility modifier statement is the same as the one described for unblocked disk output files. The informational messages printed as a result of the job stream in Figure 3-9 is illustrated in Figure 3-10.

EXAMPLE

```
// UCD TR,FF,A=(80,80),B=(80,3600),I1,CY,R1,E=(2311)
// END
CARD TO DISK UTILITY
INPUT RECORD LENGTH 0080
OUTPUT RECORD LENGTH  0080
INPUT BLOCK LENGTH 00080      } NOTE:
OUTPUT BLOCK LENGTH 03600     } BLOCKED
                                RECORDS
INPUT OPTION CARD BCD
OUTPUT OPTION      DISK WRITE CHECK
2 INPUT,2 OUTPUT AREAS ASSIGNED
RECORD FORMAT FIXED
TYPE REBLOCK
STARTING RECORD NUMBER   00000001
OUTPUT DEVICE TYPE  2311
FILEMARK WRITTEN IN
XT. NO.  B1    C1    C2    H1    H2    R
   000   000   000   160   000   002   001
NUMBER OF INPUT BLOCKS PROCESSED   000050  — NUMBER OF CARDS
NUMBER OF OUTPUT BLOCKS PROCESSED 000002  — 2 OUTPUT BLOCKS REQUIRED
END OF JOB
```

Figure 3-10 Messages from Card-To-Disk Reblocking Utility Program

Note from the listing above that the disk file will be blocked and will contain 45 records in each block. The block size is 3600 bytes. Note also that 50 card "blocks" were read and only 2 disk blocks were written.

DISK-TO-PRINT REBLOCKING

The file which was created from the job stream illustrated in Figure 3-9 can be printed on the printer by merely specifying that a list-type function is to be performed. The following job stream could be used to print the file.

EXAMPLE

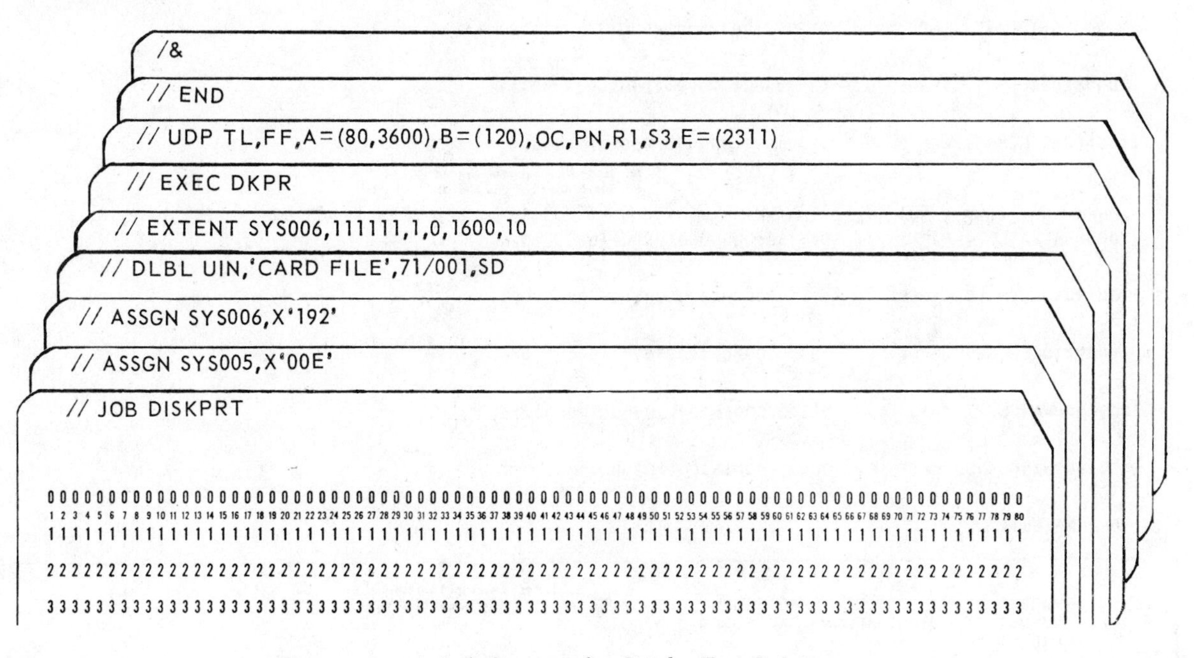

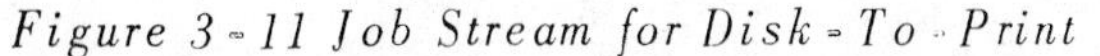

Figure 3-11 Job Stream for Disk-To-Print

Note from the job stream illustrated in Figure 3-11 that the printer, X'00E', is assigned to SYS005. This, as stated before, must always be true when the printer is used as the output device with utility programs. The disk input file will be stored on drive X'192', which is assigned to SYS006. Note the use of the filename UIN in the DLBL card for the disk input file. The filename in the DLBL card for a direct-access device which is input to the utility programs must always be UIN.

The Utility Modifier Statement specifies that a file listing is to be prepared with fixed-length records as input. The input record size is specified as 80 bytes and the input block size is specified as 3600 bytes (ie. A=(80,3600)). This is because the file which is to be input to the disk-to-print program was stored as a blocked file. Note that this is the only indication in the Utility Modifier Statement that the input file is blocked. Unlike the card-to-disk program, where reblocking must be specified, the disk-to-print program only requires that the size of the input record and input block be indicated. The remainder of the Utility Modifier Statement utilizes the same format as illustrated in Figure 3-7. Note that no page numbering is requested.

The listing produced from the job stream in Figure 3-11 is shown below.

```
2007689DWNEY, REED M.        04377880053066100900863500l132      1

2004721RASSMUSEN, JOHN J.    081000001000001208017680002346      1

1001730REEDE, OWEN W.        090055501051441008010520002310      1

2004739RIDEL, ROBERT R.      055750400825711207033303007224      3

2009740RIDGEFIELD, SUZY S.   180417801905061210033024002310      1

2002801SCHEIBER, HARRY T.    009523700325081203000520000520      3

2007802SHEA, MICHAEL H.      064203300820091008012920004680      1

2004806STOCKTON, NORMAN Q.   067250700725881209013380013380      1

1009820TELLER, STEPHEN U.    195704401770091210051920032222      3

1006825TILLMAN, DON M.       123044401995090505019230004665      1

BLOCK NO. 000002, INPUT AREA UNDERFLOW
2003834TRAWLEY, HARRIS T.    057732600550001510011000001329      2

1005909UDSON, DORIS M.       015449900303251004000780000000      1

1008921ULL, GEORGE A.        180200002000001509034090012080      3

2002956WANGLEY, THEO. A.     012000000150001207001720000275      2

1001960WINGLAND, KEITH E.    005000000350001002000000000000      1

END OF DATA
```

Note: Underflow because second block does not contain 45 records

```
NUMBER OF INPUT BLOCKS PROCESSED  000002
NUMBER OF OUTPUT BLOCKS PROCESSED 000050
END OF JOB
```

Note: Two input blocks and fifty output blocks

Figure 3-12 Listing from Card-To-Disk Utility Program

From the listing in Figure 3-12 it can be seen that the lines are triple-spaced as requested on the Utility Modifier Statement (S3). Note also that no page numbering occurred as requested (see Figure 3-8 for an example of page numbering). The message contained in the listing which states "Block No. 000002, Input Area Underflow" is a standard message which is written by the disk-to-print utility program whenever the last block on the input file does not contain the number of records which are contained within a "full" block, that is, when the last block has been truncated because the number of records in the file is not an even multiple of the number of records in each block. For example, fifty records were loaded onto the disk from the card input file (see Figure 3-10). Since 45 records will be placed in a single block, the second or last block of the file will contain only 5 records. This is a common occurrence with blocked files but the utility program will always print a message indicating the situation if it occurs.

Note that the end-of-job statistics, which are printed on the page following the report, indicate that 2 input blocks were read and fifty output "blocks", or printed lines, were processed. These end-of-job statistics are always included after a utility program completes the file-to-file processing so that the user can ensure that all records were processed in the proper manner.

CARD-TO-DISK FIELD SELECT

In Chapter 2 the process of field selection was illustrated for card input and printer output. This same option of field selection is available using the card-to-disk and disk-to-print utility programs. In addition, the data which is input on the card file in a zoned-decimal format may be stored on the disk file as packed-decimal data. The job stream to read the data cards, reblock them and to pack certain of the numeric fields on the disk output file is illustrated in Figure 3-13.

EXAMPLE

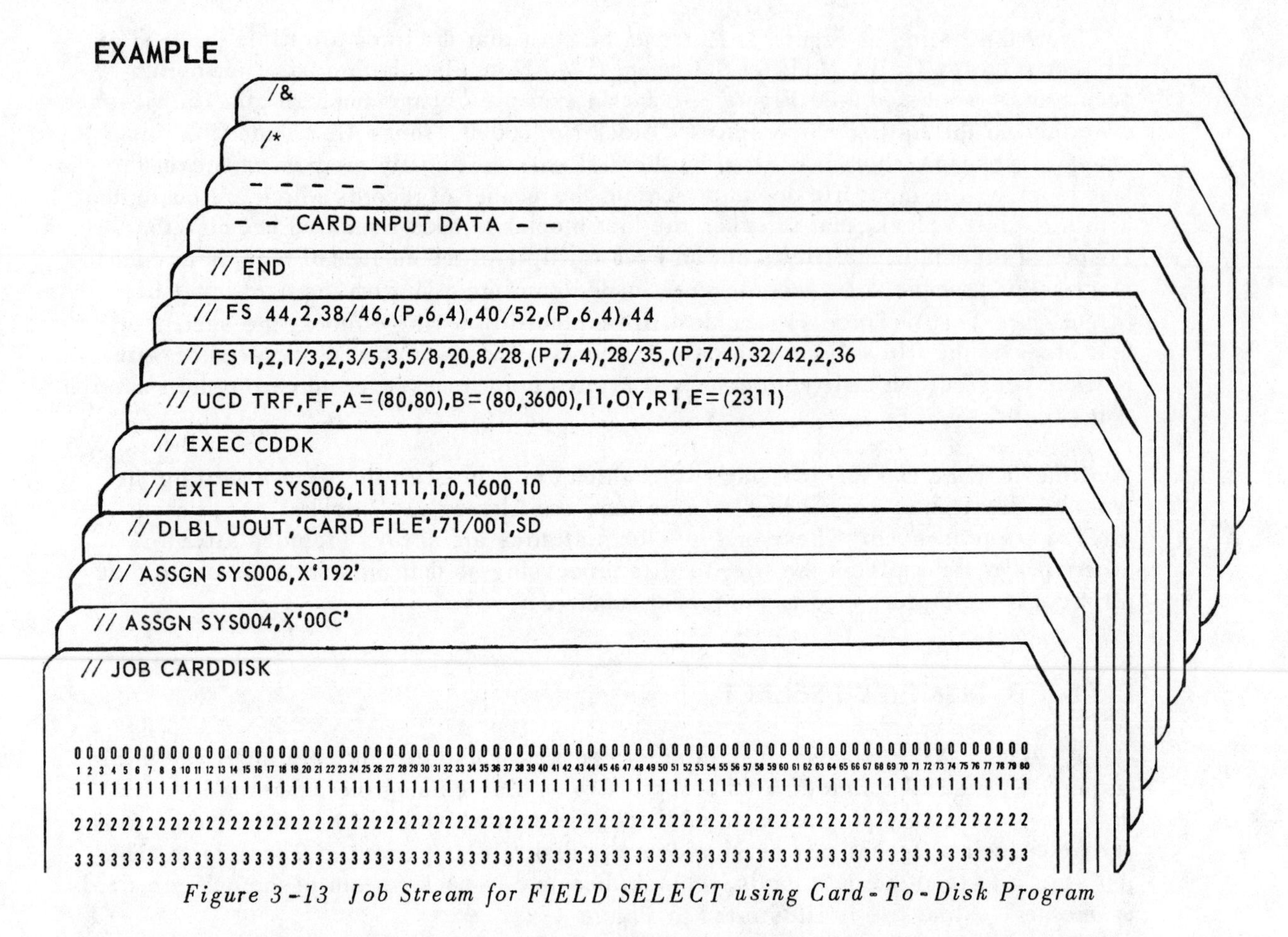

Figure 3-13 Job Stream for FIELD SELECT using Card-To-Disk Program

The card reader must be assigned to SYS004 when it is input to the card-to-disk program and the disk drive on which the output file is to be written may be assigned to any programmer logical unit except SYS004. The filename in the DLBL card for the output file must be UOUT. The remaining entries are at the programmer's discretion, although the output file must be a sequential file.

The Utility Modifier Statement which is used for reblocking and field select is in the same format as those used for a straight copy or for simply reblocking a file. The only difference is that the type of function indicator must be "TRF", which indicates that both reblocking and field select are to take place.

The field select cards illustrated in Figure 3-13 are used to indicate the fields which are to be copied from the input card data to the output file, and also the numeric fields which are to be stored in a packed-decimal format in the output record. Note the use of the two field select cards. As many field select cards as are necessary or desired may be used with the Utility programs. The only requirement is that the three entries for a single field, that is, the "r,s,t" entries, be included on the same card. The input data card which is to be read by the program is illustrated in Figure 3-14.

Input Data Card

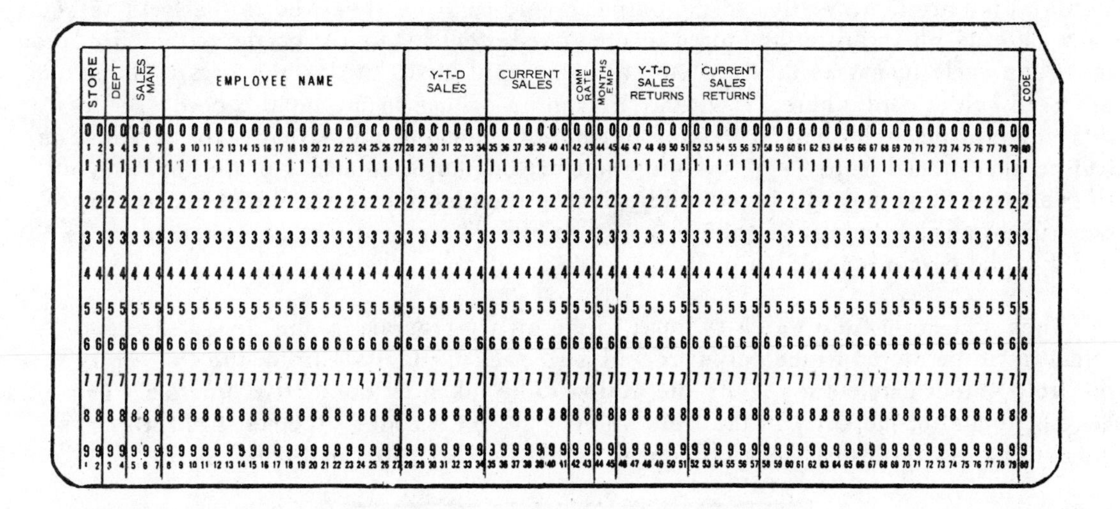

Figure 3-14 Format of Data Card

All of the fields in the card illustrated above except the code field are to be copied onto the disk file. The code field is to be omitted. The relationship of the field select cards with the input data is illustrated below.

First Field Select Card

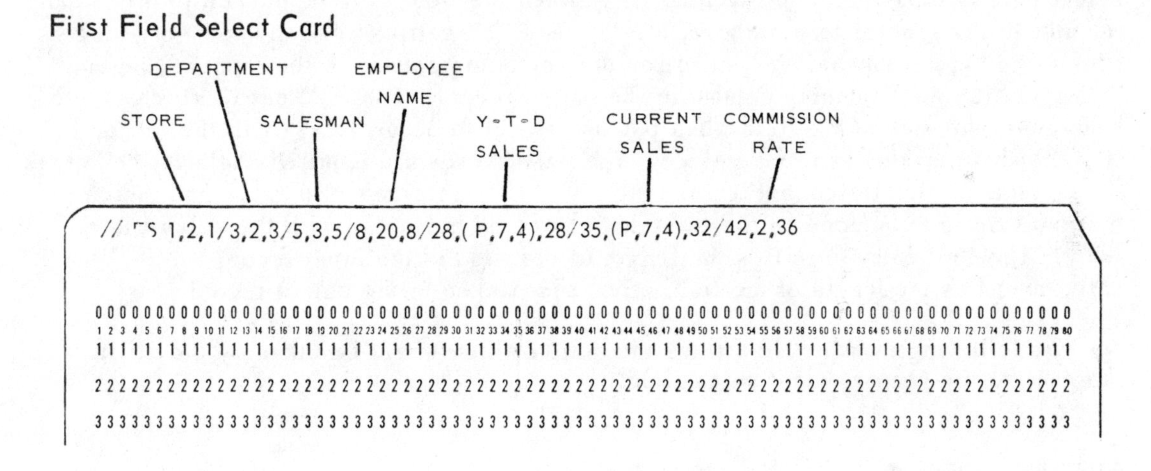

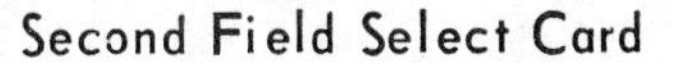

Second Field Select Card

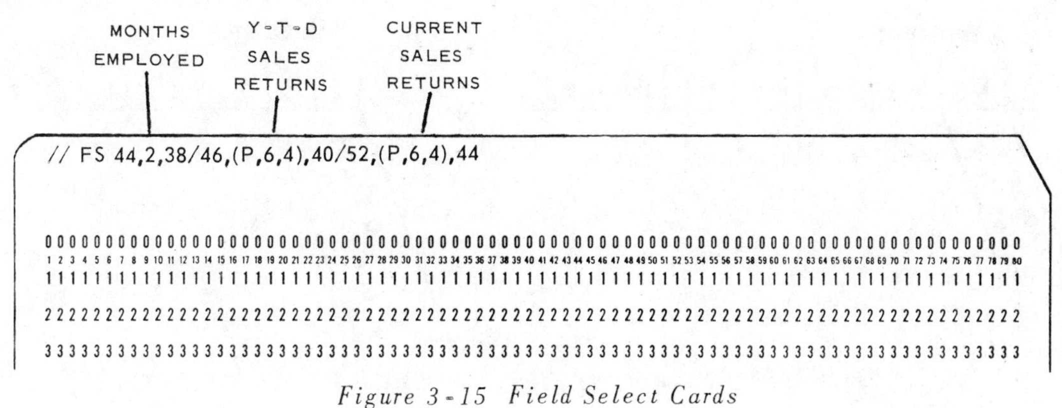

Figure 3-15 Field Select Cards

In the Field Select cards illustrated in Figure 3-15 it can be seen that each of the fields which are to be written in the output record must be specified in the field select card. Fields which are to be stored in the zoned-decimal format on the output file, that is, in the same format as they are stored on the card input, utilize the "r,s,t" format on the field select card where "r" is the beginning column in the input record, "s" is the length of the field, and "t" is the beginning column in the output record. Thus, as can be seen from Figure 3-15, the store number begins in column 1 of the input record, is two bytes long, and will be stored beginning in column 1 of the output record. The department number begins in column 3, is two bytes long, and will be stored in bytes 3 and 4 of the output record.

When a numeric field which is input to the utility program in the zoned-decimal format is to be stored in the output record as a packed-decimal field, the "s" entry in the field select card must specify the action to be taken by the utility program. The general format of the entry in the field select card for a zoned-decimal field which is input to be stored in the packed-decimal format is illustrated in Figure 3-16.

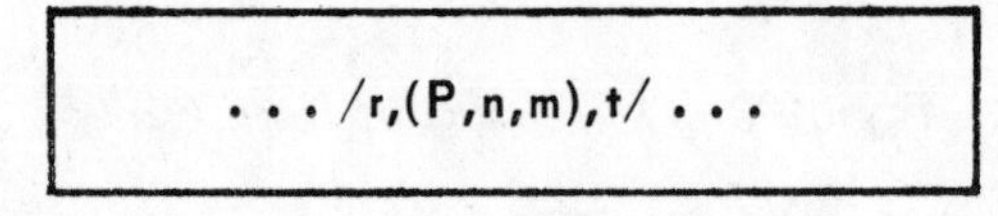

Figure 3-16 General format of entry in Field Select card to pack field

In the general format shown above, it can be seen that the "s" entry in the field select card is replaced by the (P,n,m) entry when the field is to be packed in the output record. In the general format above, the "r" and "t" entries have the same meaning as illustrated before, that is, "r" specifies the beginning column in the input record and "t" specifies the beginning column in the output record. The "s" entry, which is used to indicate the size of the field when packing is not to occur, must be in the format (P,n,m) when packing is to take place. The parentheses are required and must be between the commas as illustrated in Figure 3-16. The "P" is a required value and indicates that packing is to be done on the field designated as beginning with the column in the "r" entry. The "n" entry specifies the length of the field in the input record and the "m" entry specifies the length of the field after it is packed in the output record.

The following example illustrates the action which would be taken by the utility program for the year-to-date sales field.

Field Select Entry . . . /28,(P,7,4),28/ . . .

Card Input

F2	F3	F5	F9	F1	F6	F7

Y-T-D SALES COLUMN 28

Disk Output

2 3	5 9	1 6	7 F

Y-T-D SALES COLUMN 28

Figure 3-17 Example of FIELD-SELECT PACKED FIELD

Note from the example in Figure 3-17 that the card input field (Year-To-Date Sales), which begins in column 28, is to be packed and stored in the disk output record beginning in column 28. The input field is 7 bytes in length, as indicated by the "n" entry, and the output field after the value in the year-to-date sales field has been packed is to be 4 bytes, as indicated by the "m" entry. Thus it can be seen that the seven zoned-decimal numeric digits which are stored in the input field are packed and stored in four bytes in the output record.

The year-to-date sales returns field is also to be stored as a packed-decimal field in the output record. The field select entry for the year-to-date sales return field is illustrated below.

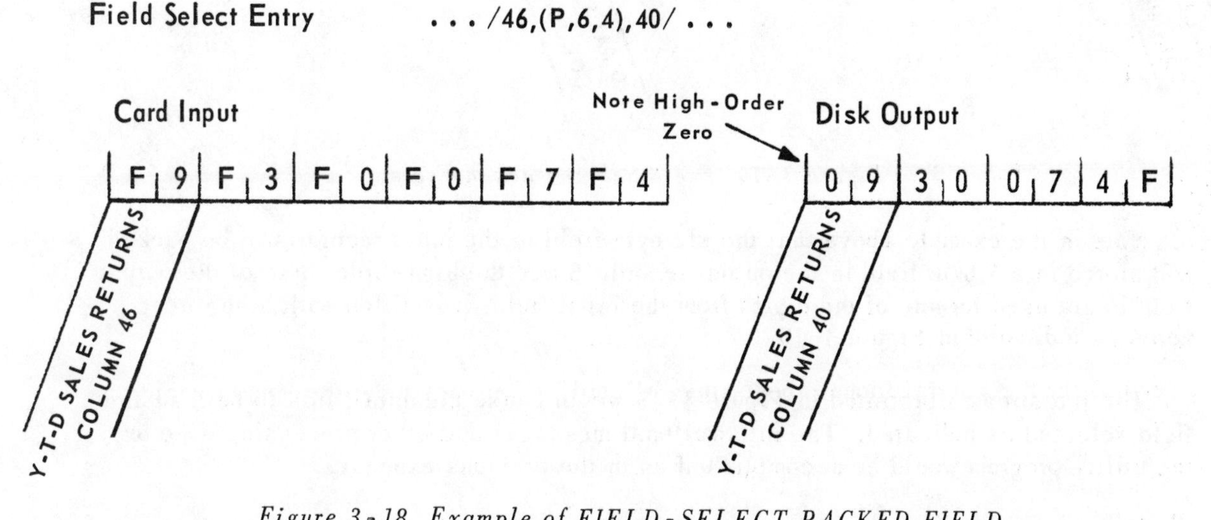

Figure 3-18 Example of FIELD-SELECT PACKED FIELD

In the above example, a six-digit zoned-decimal field is to be packed in the output record. The field select entry specifies that the length of the output field is to be 4 bytes. This is because the output field must be long enough to contain all of the digits from the input field. A packed-decimal field always contains an odd number of digits; thus, a high-order zero is inserted in the packed-decimal field in the output record since that digit is not used by the input record. If the output field had been specified as 3 bytes in length, the high-order 9 in the input field would have been truncated, that is, it would not have been included in the output field. Thus, the output field must be large enough to contain all of the digits in the input field.

In some applications, it may be desirable to have the output field contain more digits than are contained in the input field. This may be the case if the field in the disk record is going to be incremented from values in another program. The field must be large enough to contain the largest possible number. The field select entry to cause the year-to-date sales returns output field to be large enough to contain 9 digits is illustrated in Figure 3-19.

Field Select Entry . . . /46,(P,6,5),40/ . . .

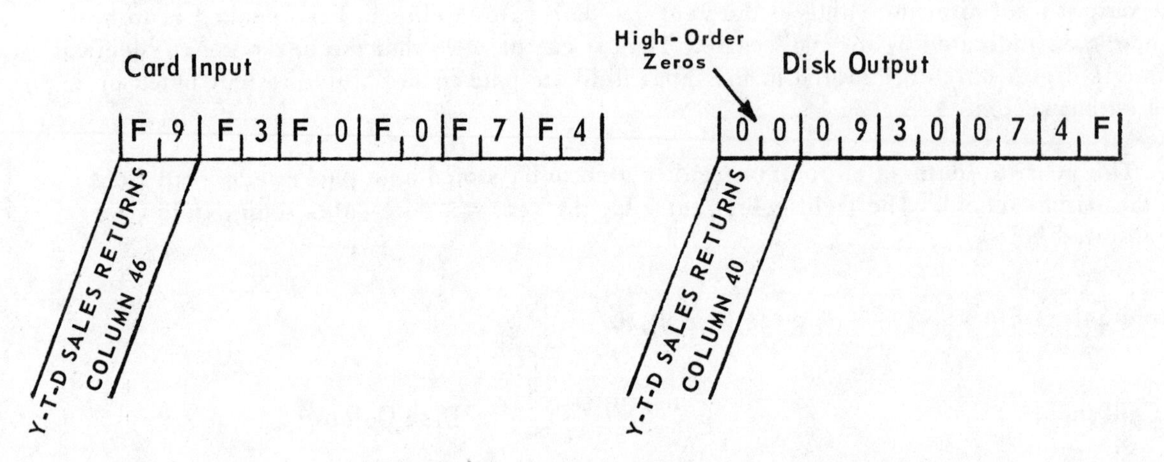

Figure 3-19 Field Select Packed Field

Note in the example above that the six byte field in the input record is to be packed and stored in a 5 byte field in the output record. Since the high-order byte of the output field is not used for any of the digits from the input field, it is filled with high-order zeros as indicated in Figure 3-19.

The job stream illustrated in Figure 3-13 would cause the data cards to be read and field selected as indicated. The informational messages and other processing done by the utility program would be accomplished as in the previous examples.

DISK-TO-PRINT FIELD SELECT

When a disk output file is created with packed-decimal numeric data, it is more efficient in terms of disk storage space. In addition, if the numeric fields are to be used for arithmetic opera.ions in application programs, the data need not be packed before being processed. When a packed-decimal field is to be printed on the printer in a disk-to-print operation, however, it must be converted to a printable form. This is because numeric digits which are in a packed decimal format many times do not form printable characters. For example, in Figure 3-19, the hexadecimal value '07' is not a printable character, that is, there is no alphabetic, numeric, or special character which has the hexadecimal value '07'. Thus, data which is stored in a packed-decimal format must either be unpacked to be printed or it must be displayed in a hexadecimal format. The field select statement which is used with the disk-to-print program can specify the format in which packed-decimal fields are to be printed.

The following job stream could be used to print the file created by the job stream in Figure 3-13.

EXAMPLE

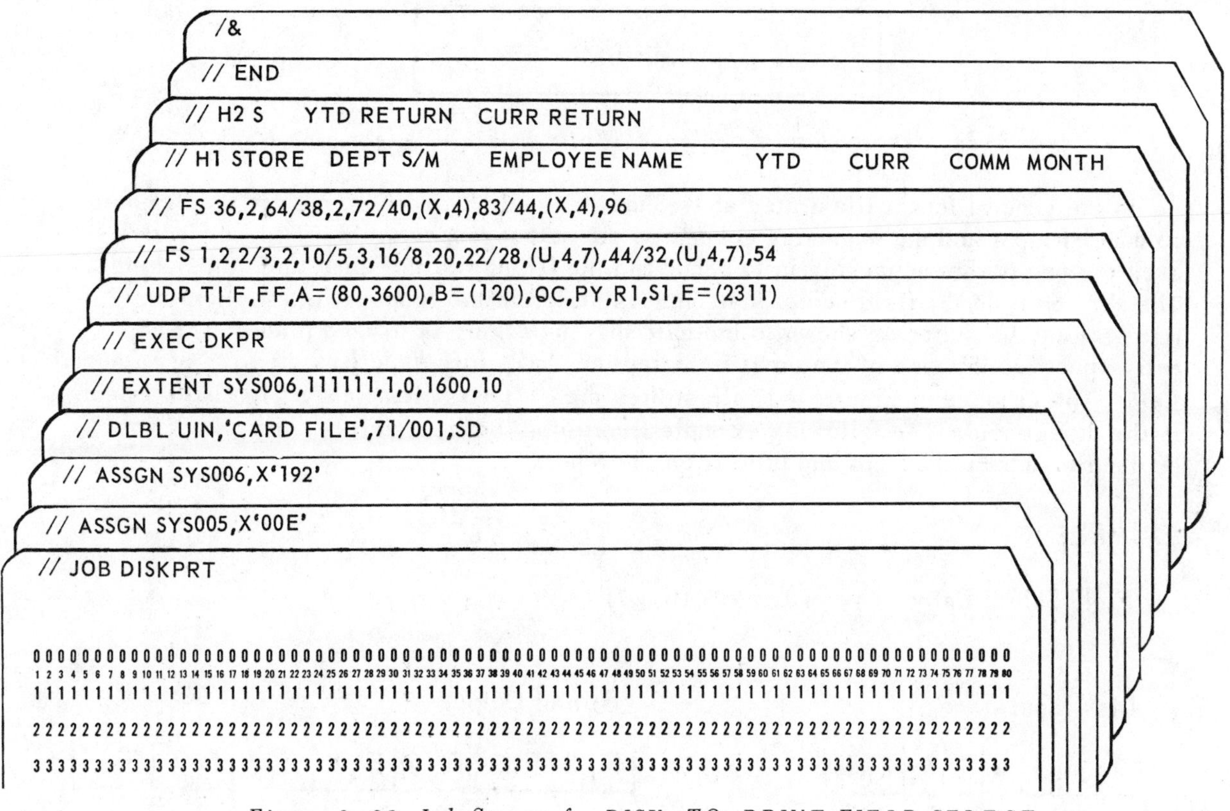

Figure 3-20 Job Stream for DISK-TO-PRINT FIELD SELECT

Note in the above job stream that the printer is assigned to SYS005 as is required for all printer output files used with the utility programs and the filename used in the DLBL card for the input file is UIN, which is also required for direct-access input files. The utility modifier statement is identical to that used to print the blocked disk file except that the function type indicator must contain the value "TLF", which indicates that the field select option is to be used.

As can be seen from Figure 3-20, two field select cards are used to select the fields from the disk file and place them on the printed report. As with the card-to-disk utility program, there is no specified number of field select cards which are to be used with the program. The only requirement is that the "r,s,t" entires for a single field must be on one card.

As noted, a field which is stored in a disk record in the packed-decimal format must either be unpacked before printing or it must be printed in a hexadecimal format. The general format of the field select entry to unpack a packed decimal field for printing on the report is illustrated below.

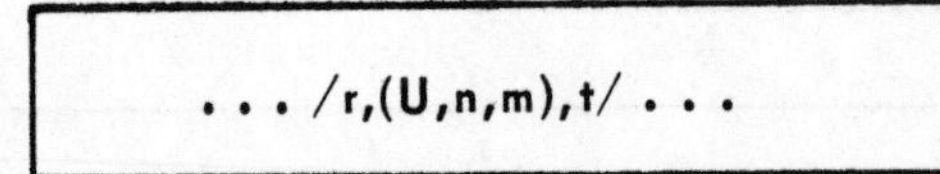

Figure 3-21 General format of Unpack Entry in Field Select Statement

In the general format illustrated above, the "r" and "t" entries specify the beginning column on input and the beginning column on the output respectively. The "s" entry is used to specify that unpacking is to occur and the lengths of the input and output fields. The "s" entry in the field select statement is stated in the format (U,n,m) where the letter U must be stated as shown to indicate that unpacking is to take place. The "n" entry specifies the size of the input field and the "m" entry specifies the size of output field. The example in Figure 3-17 illustrates the Y-T-D Sales field as it would appear in the disk record. The following example illustrates the field select entry which would be used to unpack the field and print it on the report.

EXAMPLE

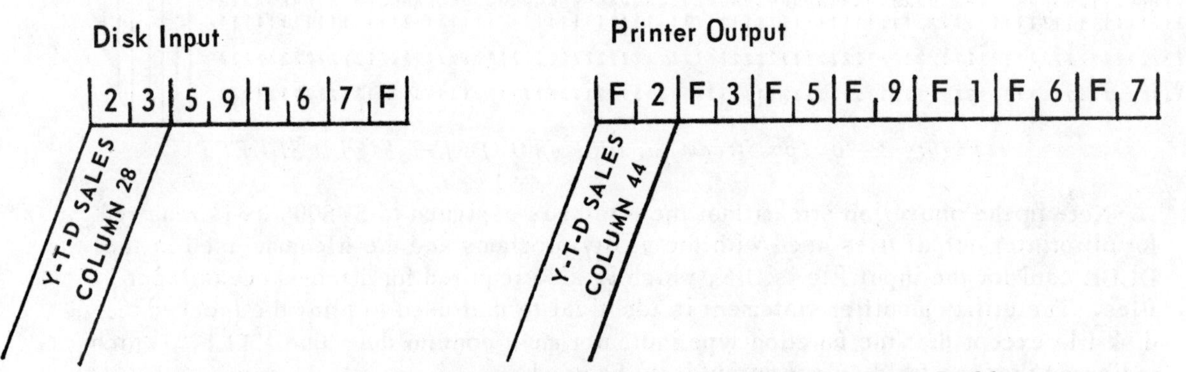

Figure 3-22 Example of Field Select Entry to Unpack Data Field

Note in the above example that the data to be unpacked begins in column 28 of the input record, which is the Y-T-D Sales field. The "r" entry in the field select entry specifies 28 as the beginning column in the input record. The Y-T-D Sales field is to be printed on the report beginning in column 44 as specified by the "t" entry. The "s" entry in the field select entry indicates that the input data is to be unpacked by specifying the "U" as the first character in the entry. The "4" states that the input field is four bytes in length and, since the field is to be unpacked, it will contain packed numeric data. Thus, this entry indicates that columns 28, 29, 30, and 31 contain packed decimal data which is to be unpacked on the printed report.

The "7" in the entry states that 7 columns on the printed report are to be used for the Y-T-D Sales field. This length is used because the input field contains a 7-digit numeric value. Thus, the 7 digits which are contained in the input field will be printed as seven zoned-decimal digits on the report.

The Y-T-D Sales Returns field was also packed in the disk record. It contains six numeric digits. The field select which could be used to print the Y-T-D Sales Returns field in a zoned-decimal, or unpacked, format is illustrated below.

EXAMPLE

Field Select Entry . . . /40,(U,4,6),83/ . . .

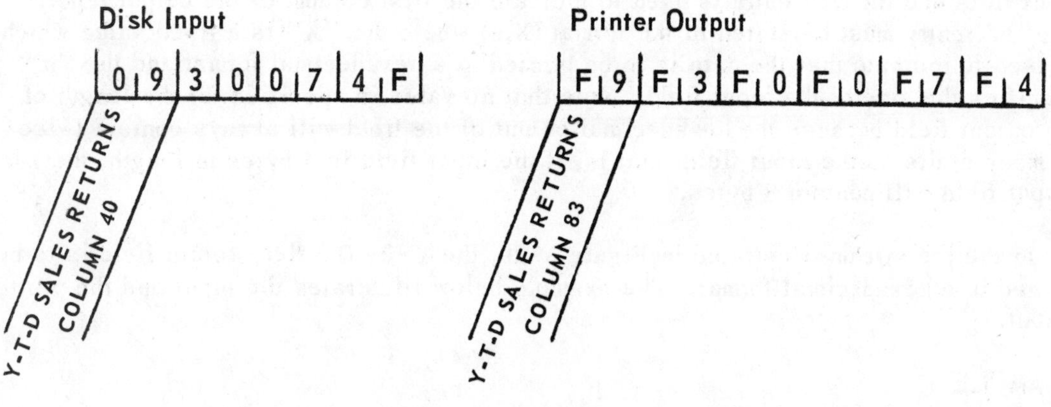

Disk Input

0	9	3	0	0	7	4	F

Y-T-D SALES RETURNS
COLUMN 40

Printer Output

F	9	F	3	F	0	F	0	F	7	F	4

Y-T-D SALES RETURNS
COLUMN 83

Figure 3-23 Example of Field Select to Unpack Data Field

Note in the example above that the printer output field contains six digits, not seven as illustrated in Figure 3-22. Thus, when the data is unpacked, the high-order digit in the packed field will be truncated, that is, it will not be included in the output field. For the sales return field this is permissable because only six digits were placed in the field when it was created in the card-to-disk program. Note that if this was done for the Y-T-D Sales field, the high-order digit would be truncated and this would be an error because seven digits are used for the field.

It was noted previously that data which is stored in a packed-decimal format will not always contain printable characters and must, therefore, be changed into zoned-decimal or hexadecimal characters for printing. In the previous examples, the conversion of packed data to the zoned-decimal format was illustrated. It should be noted, however, that data which is unpacked by the utility program is not edited. Therefore, if the packed-decimal field contained the value -324, which is stored in a packed-decimal format as X'324D', it would be printed as '32M' on the printed report. This is because when the X'4D' is unpacked the value obtained is X'D4', which is the hexadecimal representation of the letter of the alphabet "M". Since the utility program does no editing of the data, negative values would be printed as letters of the alphabet. An alternative to this is presented by the capability of printing a packed-decimal field in a hexadecimal format on the report. The use of the field select statement to print data in a hexadecimal format is discussed in the following section.

Hexadecimal Output

The general format of the field select entry which is used to indicate that data is to be printed in a hexadecimal format is illustrated below.

. . . /r,(X,n),t/ . . .

Figure 3-24 General Format of Hexadecimal Field Select Entry

In the general format of the field select entry illustrated above it can be seen that the "s" entry is used to indicate that the input data is to be printed on the report in a hexadecimal format. The "r" entry is again used to specify the first column of the input field and the "t" entry is used to indicate the first column of the output report. The "s" entry must be stated in the format (X,n) where the "X" is a fixed value which is used to indicate that the data is to be printed in a hexadecimal format and the "n" specifies the size of the input field. Note that no value is specified for the length of the output field because the hexadecimal output of the field will always contain twice as many digits as the input field, that is, if the input field is 4 bytes in length, the output field will contain 8 bytes.

In the job stream illustrated in Figure 3-20, the Y-T-D Sales Return field is to be printed in a hexadecimal format. The example below illustrates the input and the printed output.

EXAMPLE

Field Select Entry . . . /40,(X,4),83/ . . .

Disk Input **Printer Output**

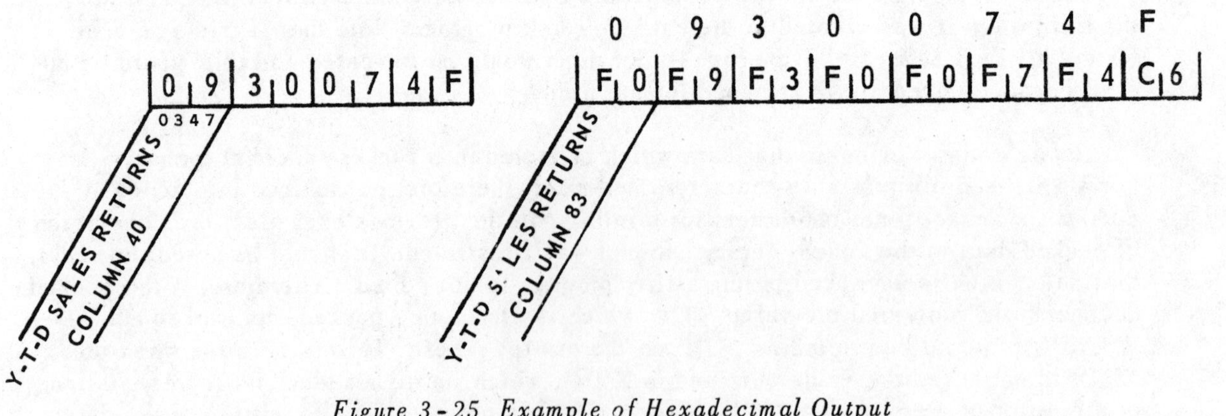

Figure 3-25 Example of Hexadecimal Output

Note in the example in Figure 3-25 that bits 0-3 of the first byte in the input record is printed as the first byte of the output record and bits 4-7 of the first byte in the input record is printed as the second byte in the output record. This sequence is continued for the complete input field, thus causing the output field to be twice the length of the input field. Note that the last byte of the printed output contains the sign (F) of the packed decimal field. If the input field had been a negative number, the report would have the value "0930074D" in the sales returns field. Thus, the negative sign would be explicitly printed instead of being implied by the letter "M" being printed in the low-order position as is the case when the field is merely unpacked.

The determination of whether to print the field in the hexadecimal format or in the zoned-decimal format must be made by the programmer. As noted, the zoned-decimal format, which is obtained when the packed-decimal data is unpacked, is more "readable" because it is presented as actual numeric values. The hexadecimal format, on the other hand, represents the data as it is stored in the input field. If the data in the input field is signed with either a positive sign (C) or negative sign (D), the sign will be printed as the low-order byte in the output field whereas with the zoned-decimal output, the signed value is represented as a letter of the alphabet.

In the job stream used in Figure 3-20, two heading statements are used for the report. These heading statements are again illustrated below.

EXAMPLE

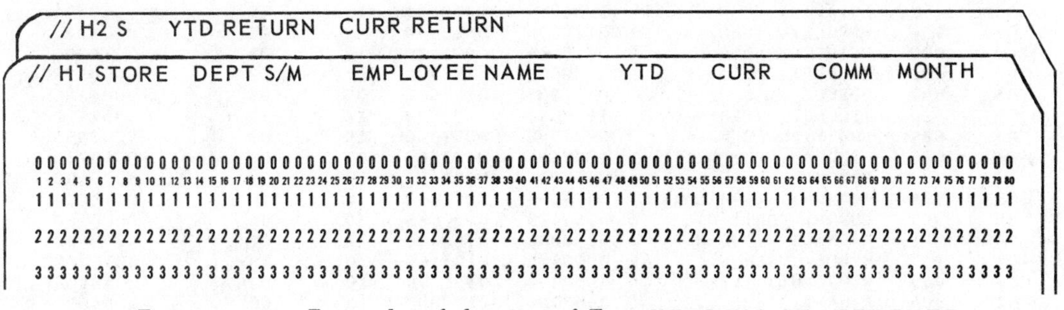

Figure 3-26 Example of the use of Two HEADING STATEMENTS

Note in the example above that two heading statements are used and they are identified by the values // H1 and // H2 in the first five columns of the card. An additional blank must follow the H1 and H2 identifiers and then the constant values to be used as headings on the reports are punched in the card. Note that the maximum number of characters which may be punched in the H1 card is 74 (all 80 columns of the card are used). If the heading on a report is to contain more than 74 columns, the H2 heading card must be used. The first constant value which is punched on the H2 card, that is, the value which is punched in column 7 of the card, will be contained in column 75 of the report heading line. Thus, as can be seen from the example above, the word MONTH is contained in columns 76-80 of the first heading statement and the letter "S" is contained in column 7 of the second heading statement. On the report, the word "MONTHS" will be printed in the heading line in columns 70-75. The report which is generated as a result of the job stream in Figure 3-20, which uses the heading statements in Figure 3-26, is illustrated in Figure 3-27.

LISTING

STORE	DEPT	S/M	EMPLOYEE NAME	YTD	CURR	COMM	MONTHS	YTD RETURN	CURR RETURN
20	03	311	GROLER, GRACE B.	2300643	0205420	15	10	0260480F	0048369F
20	07	802	SHEA, MICHAEL H.	0642033	0082009	10	08	0012920F	0004680F
20	04	806	STOCKTON, NORMAN Q.	0672507	0072588	12	09	0013380F	0013380F
10	01	004	ACHER, WILLIAM C.	0190726	0057524	10	03	0000850F	0000000F
10	04	027	ALHOUER, ELAINE E.	0112573	0022066	12	05	0001880F	0000400F
10	03	030	ALLOREN, RUTH W.	2132020	0000000	15	10	0114000F	0020933F
20	06	100	BATES, TONY F.	0819066	0207645	05	04	0088340F	0005170F
20	03	318	HANEY, CAROL S.	0975000	0145000	15	05	0012020F	0006019F
10	08	322	HARLETON, JEAN H.	0780899	0120089	15	06	0010180F	0000000F
20	01	325	HATFIELD, MARK I.	0151122	0020539	10	07	0001840F	0000220F
20	07	332	HELD, ANNA J.	0244000	0029500	10	09	0005115F	0000926F
10	02	111	CARTOLER, VIOLET B.	0280998	0075006	12	04	0003020F	0001140F
20	05	122	CENNA, DICK L.	0355777	0044001	10	08	0003270F	0000000F
20	08	102	BELLSLEY, ARTHUR A.	0883000	0099000	15	09	0019033F	0002762F
20	09	315	HALE, ALAN A.	1274000	0168000	12	08	0025954F	0003829F
20	04	317	HANBEE, ALETA O.	0385000	0039500	12	10	0007530F	0001180F
10	04	171	COSTA, NAN S.	0580356	0056002	12	10	0013020F	0000903F
20	06	179	DAMSON, ERIC C.	0352502	0180888	05	02	0000223F	0000223F
10	03	487	KING, MILDRED J.	1850896	0180429	15	10	0038260F	0005322F
10	08	505	LAMBERT, JERRY O.	0154001	0040745	15	04	0002840F	0000000F
20	05	207	EBERHARDT, RON G.	0564007	0094009	10	06	0006300F	0000980F
10	06	409	ICK, MICK W.	0410122	0195080	05	02	0000590F	0000590F
10	05	568	LYNNE, GERALD H.	0924487	0133311	10	07	0016530F	0001428F
10	09	574	MELTZ, FRANK K.	0754366	0059030	12	10	0015760F	0000000F
10	03	181	DELBERT,EDWARD D.	1250659	0130554	15	09	0037540F	0000000F
10	01	185	DONNEMAN, THOMAS M.	0650423	0090019	10	07	0006700F	0001020F
10	07	214	EDMONSON, RICK T.	0079067	0033057	10	03	0000290F	0000000F
10	09	215	EDSON, WILBUR S.	0607050	0082005	12	07	0011885F	0001575F
10	09	105	BOYLE, RALPH P.	0878044	0143055	12	06	0016070F	0007787F
20	08	282	ESTABAN, JUAN L.	1984055	0000000	15	10	0040505F	0005050F
20	02	956	WANGLEY, THEO. A.	0120000	0015000	12	07	0001720F	0000275F
20	02	590	NEIL,CLARENCE N.	0750065	0095023	12	08	0014000F	0002324F
10	06	292	EVERLEY, DONNA M.	0332000	0177500	05	02	0000322F	0000322F
10	06	607	ODELLE, NICHOLAS P.	0582507	0187507	05	03	0006220F	0006220F
10	09	820	TELLER, STEPHEN U.	1957044	0177009	12	10	0051920F	0032222F
20	07	689	OWNEY, REED M.	0437788	0053066	10	09	0008635F	0001132F
10	06	825	TILLMAN, DON M.	1230444	0199509	05	05	0019230F	0004665F
20	01	300	FELDMAN, MIKE R.	0250000	0030000	10	09	0000600F	0000000F
20	03	834	TRAWLEY, HARRIS T.	0577326	0055000	15	10	0011000F	0001329F
10	05	909	UDSON, DORIS M.	0154499	0030325	10	04	0000780F	0000000F
20	04	721	RASSMUSEN, JOHN J.	0810000	0100000	12	08	0017680F	0002346F
10	01	730	REEDE, OWEN W.	0900555	0105144	10	08	0010520F	0002310F
10	01	960	WINGLAND, KEITH E.	0050000	0035000	10	02	0000000F	0000000F
10	02	304	FROMM, STEVE V.	0650000	0120000	12	05	0006090F	0001823F
10	08	921	ULL, GEORGE A.	1802000	0200000	15	09	0034090F	0012080F

BLOCK NO. 000002, INPUT AREA UNDERFLOW

STORE	DEPT	S/M	EMPLOYEE NAME	YTD	CURR	COMM	MONTHS	YTD RETURN	CURR RETURN
20	04	739	RIDEL, ROBERT R.	0557504	0082571	12	07	0033303F	0007224F
20	09	740	RIDGEFIELD, SUZY S.	1804178	0190506	12	10	0033024F	0002310F
20	02	801	SCHEIBER, HARRY T.	0095237	0032508	12	03	0000520F	0000520F
10	05	308	GLEASON, JAMES E.	0145000	0039000	10	04	0000925F	0000000F

PAGE 1

Figure 3-27 Example of Disk-To-Print Listing

Note in the example above that the headings on the report extend across the page so that each field on the report is identified. Since the heading is longer than 74 bytes, the H2 heading statement was used. Note also that the Y-T-D Sales and the CURRent Sales are displayed in a zoned-decimal format after being unpacked from the packed input field and the YTD Returns and the CURRent Returns are displayed in a hexadecimal format as a result of field select statements read by the Disk-To-Print Utility program. Thus, using the field select feature of the disk-to-print utility program, a report may be formatted in any manner desired by the user.

DISK - TO - CARD UTILITY PROGRAM

In some applications it may be necessary to copy a disk file to punched cards in order to create a backup file or for some other reason. The disk - to - card program provides a means by which a disk file may not only be copied but the data on the file may also be field selected so that only certain fields are copied. In addition, packed - decimal fields may be punched as zoned - decimal fields on the cards. The following job stream could be used to copy the file created by the job stream in Figure 3 - 13.

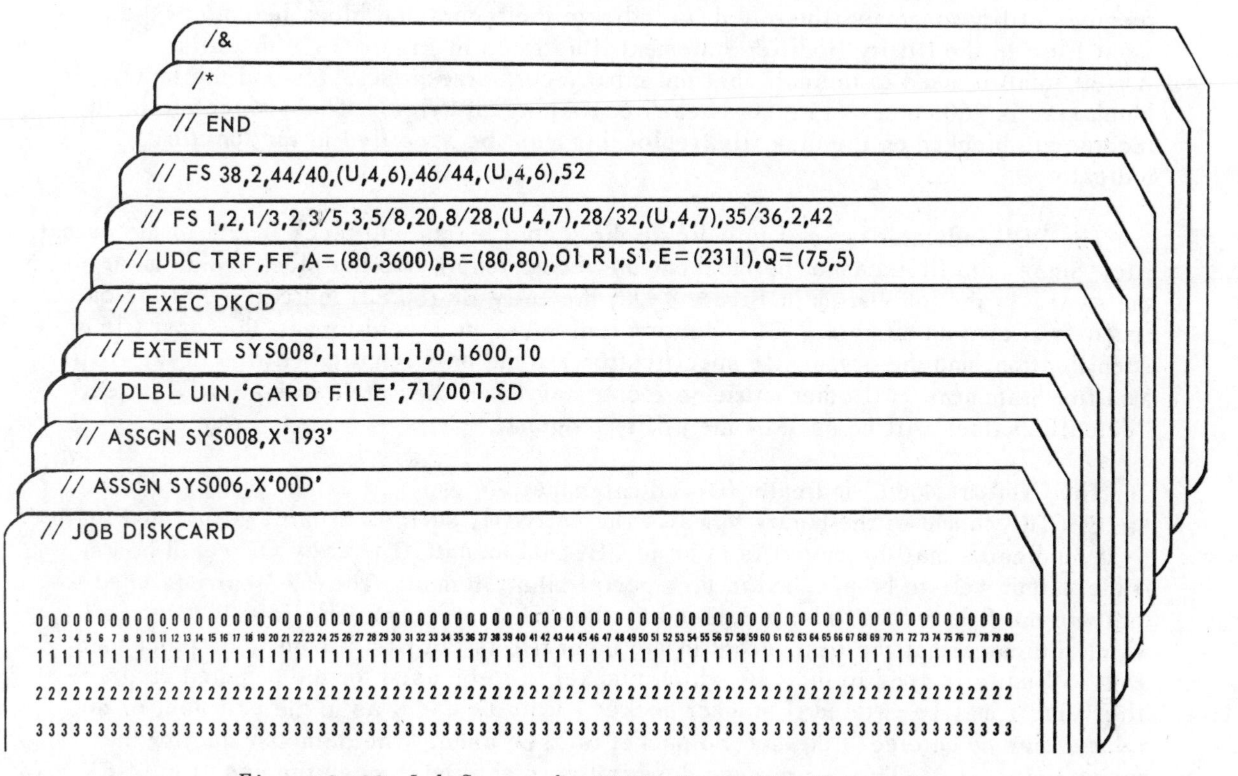

Figure 3 - 28 Job Stream for DISK - TO - CARD Utility Program

In the example above it can be seen that the card output device, X'00D', is assigned to SYS006. As with the card - to - card utility program, the card output device must be assigned to SYS006. The disk input device, which may be assigned to any programmer logical unit except SYS006, is assigned to SYS008. As noted for previous utility programs, the filename in the DLBL card for the disk input file must be UIN, as is illustrated above. The phasename of the disk - to - card utility program is DKCD, and this phasename must be stated in the execute card which precedes the utility control cards.

The Utility Modifier Statement must contain the identifier "UDC", which indicates that the card for the disk - to - card program. The function type code in the example above is "TRF", which indicates that reblocking and field select is to take place when the utility program is executed. If a straight copy of the disk record to the punched card is to occur, the value must be "TC". If the TC code is used, the disk records must be 80 characters in length and cannot be blocked. The copy program will reproduce the entire 80 character disk record on the card with no changes.

If a disk file contains blocked records with a fixed-length of 80 bytes, the 80 byte records may be "unblocked" and punched on cards by using the Reblock option of the utility program which is indicated through the use of the TR function indicator. Note again that the records must be 80 bytes in length.

If the disk records do not contain 80 bytes, or if field selection is to take place in order to unpack packed-decimal fields, the TRF indicator must be used as illustrated in Figure 3-28. All records which are input to the disk-to-card program must be fixed-length; thus, the format indicator FF is used. The "A" entry is used, as with previous utility programs illustrated, to indicate the record and block lengths of the input file. In the Utility Modifier Statement illustrated in Figure 3-28 the value A=(80,3600) is used to indicate that the input record size is 80 bytes and the input block size is 3600 bytes (45 records each containing 80 bytes). Thus, since the input records are blocked on the disk file, reblocking must be specified in the function indicator.

The "B" indicator is used to indicate the record length and block length of the output file. Since card files cannot be blocked, the record length and the block length must be the same. In the job stream in Figure 3-28, the entry B=(80,80) indicates that the output record will contain 80 bytes. The function indicator, the record format, the input file specification, and the output file specification are the only required entries in the Utility Modifier Statement. All other entries are optional and if they are not explicitly stated, "default" values will be used by the utility program.

The "Output Mode" indicator (O) indicates that the punched output will be either in the EBCDIC format or the binary format. The entry O1, such as in the example in Figure 3-28, indicates that the output is to be in EBCDIC format. The entry O2 would be used if the output were to be punched in the special binary format. The "R" entry is used to indicate the first record to be processed by the utility program. The default value is R1, which indicates that the first record in the input file is the first record to be processed. The "S" entry is used to indicate which stacker is to be used for the punched cards. The value 1 may be entered if stacker pocket 1 is to be used, as in the example, or the value 2 may be entered if the second pocket is to be used. The default value for the pocket select is 2. The device type description is entered through the use of the "E" parameter. This parameter is used to indicate the type of direct-access device which is to be used for the input file. The entry E=(2311) specifies that a 2311 device is to be used. The value 2314 could be used to specify a 2314 disk drive. The 2311 disk drive is the assumed device if this field is not included on the Utility Modifier Statement.

Sequence numbering in the output card is controlled by the "Q" entry. The disk-to-card utility program will automatically punch an ascending sequence number in the field requested if the sequence-numbering parameter is specified. In the example, the value Q=(75,5) is included on the Utility Modifier Statement. This indicates that the sequence number is to begin in column 75 and is to be 5 columns in length. Thus, columns 75-79 will contain the sequence number. The number generated by the program begins with the value 1 and is incremented by 1 for each card that is punched. An example of the punched output that would be produced with the sequence number in columns 75-79 is illustrated in Figure 3-29.

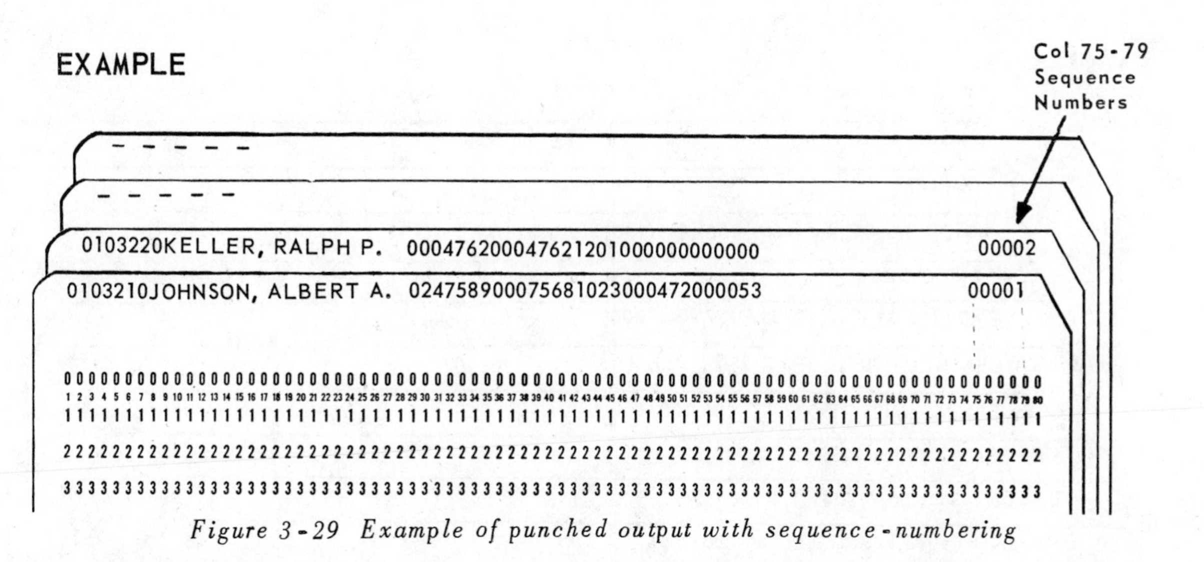

Figure 3-29 Example of punched output with sequence-numbering

In the above example it can be seen that the punched output of the disk-to-card utility program contains the unpacked numeric fields in a zoned-decimal format and also an ascending sequence number in columns 75-79. This sequence number is automatically incremented by 1 for each card that is punched and placed in the columns specified by the "Q" entry in the Utility Modifier Statement. The field select cards as illustrated in Figure 3-28 control the unpacking of the data from the disk input record to the card output record in the same manner as the field select cards used with the disk-to-print utility program.

DISK-TO-DISK UTILITY PROGRAM

Files which are stored on direct-access devices may be copied from one disk pack to another or to the same disk pack through the use of the Disk-To-Disk Utility Program. The job stream in Figure 3-30 illustrates an application where a file which is contained on multiple extents on two different disk packs is to be copied to a disk pack as a single extent file.

EXAMPLE

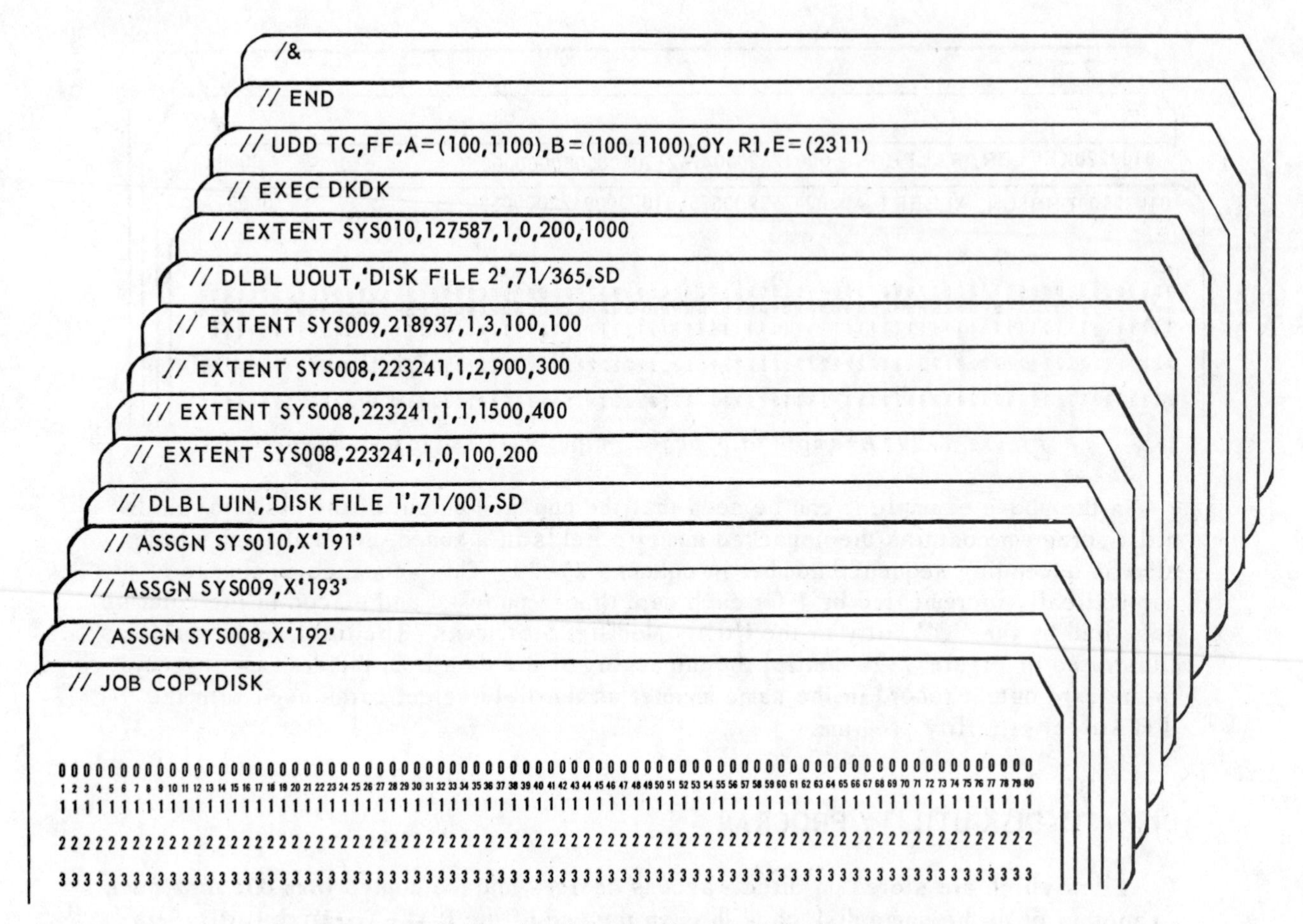

Figure 3-30 Example of Disk-To-Disk Utility Program

In the above example it can be seen that the file 'DISK FILE 1' is stored on three different extents on pack 223241 and on one extent on pack 218937. This multiextent file is to be copied to disk pack 127587 as a single extent file. Since no reblocking or field selecting is to take place, the function indicator contains the value TC, which indicates a straight copy function. If a reblocking and copy function were to be performed, the function code TR would be used. Field select with no reblocking requires the function code TF and reblocking and field selecting require the entry TRF.

Since no reblocking is to occur, the input record and block length and the output record and block length are the same and are specified in the "A" and "B" entries in the Utility Modifer Statement. The disk write-check option is specified as "OY" in the example above indicating that a check is to be performed on each block that is written on the output file to ensure that it is written properly. The "R1" entry requests that copying begin with the first record in the input file and the output device is a 2311 disk drive.

It should be noted that the function codes and the required entries on the Utility Modifier Statement, such as the input and output record and block lengths, are specified in a standard manner. This is true for all of the file-to-file utility programs.

INDEXED SEQUENTIAL DISK-TO-PRINT

For the most part, the DOS file-to-file utility programs process sequential data files. The one exception is that an indexed sequential file may be processed by the disk-to-print utility program. The job stream to read an indexed sequential file and print it on the printer using the disk-to-print Utility program is illustrated below.

EXAMPLE

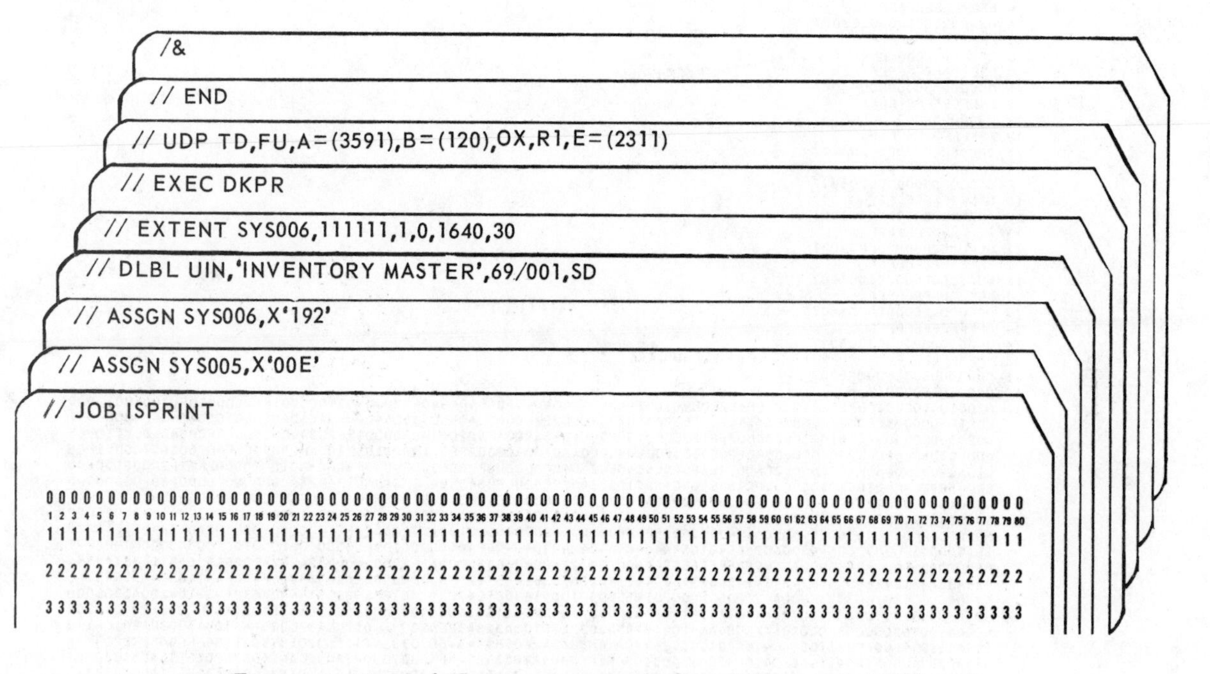

Figure 3-31 Job Stream to print Indexed Sequential File

In the job stream illustrated above, it can be seen that the indexed sequential file 'INVENTORY MASTER' is defined as a sequential file (SD entry in DLBL card). The extents defined are the prime data area extents only. Thus, the prime data area of the indexed sequential file is treated as a sequential input file by the disk-to-print utility program. Note that the track indexes are stored within the prime data extents. Thus, the track indexes will be printed together with the data stored in the file.

The name of the utility program, DKPR, is the same as is used for sequential file input. The Utility Modifier Statement is used to indicate the proper processing for the indexed sequential file. The function indicator is TD when an indexed sequential file is to be printed. The TD stands for a Display function. When data is "displayed" in the disk-to-print utility program, it may be printed in a character or hexadecimal format. In the example above, a hexadecimal format is requested by the entry "OX". Character format could be requested by the entry "OC". In addition, file characteristics such as block length, etc., are included on the report. The printout in Figure 3-32 illustrates a display report.

EXAMPLE

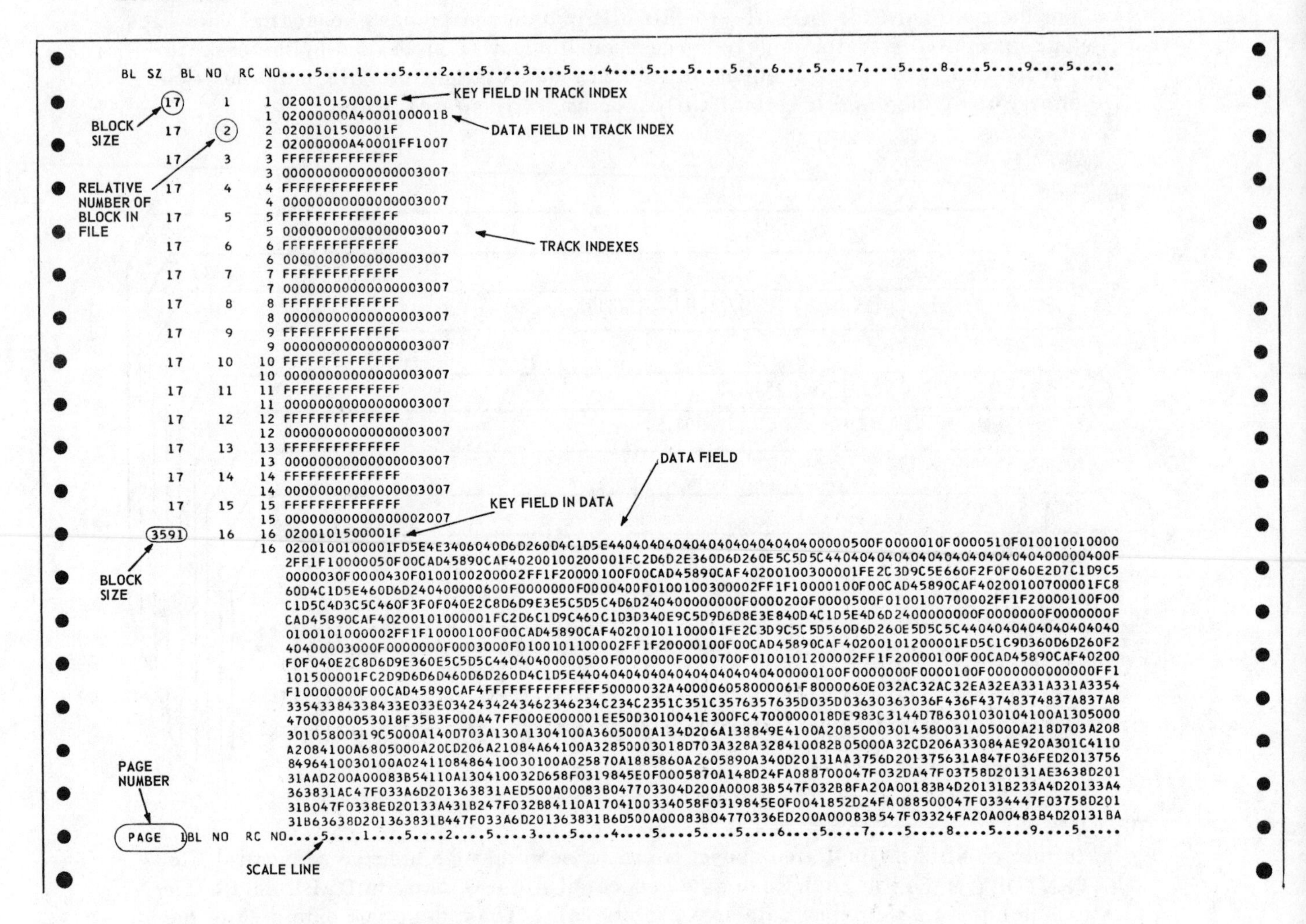

Figure 3-32 Example of "Display" listing of Indexed Sequential File

Note in the above listing obtained from the "display" function of the disk-to-print program that different size records are processed by the program. This is indicated by the "FU" record format entry in the Utility Modifier Statement which specifies that undefined-length records are to be input to the program. Whenever undefined-length records are used for the disk-to-print program, the "A" entry which is used to define the length of the input records must contain a single value which is the length of the largest record which will be read. When reading indexed sequential files, the length stated must be equal to the length of the data block which is stored in the prime data area plus the length of the key. In the example, the block of data contains 3584 bytes (64 bytes/rec × 56 recs/block) and the key is 7 bytes in length. Thus, the length stated for the input record is 3591 bytes. It should be noted again that when using undefined length records, the length of the largest record to be processed is the length stated in the utility modifier statement. For an indexed sequential file, this will always be the length of the block of data in the prime data area plus the length of the key in the record.

From the report in Figure 3-32 it should be noted also that the block size and the relative number of the block within the file are included in the first twenty columns of the report. Thus, for the track index, it can be seen that each block contains 17 bytes, 7 of which are in the key field in the index and 10 of which are in the data field in the track index. Block number 16 contains 3591 bytes, 7 of which are in the key field and the remainder are in the data block. The report, as specified by the "OX" entry in the utility modifier statement, contains a hexadecimal representation of the data in the file. The scale line at the top and bottom of the page may be used to find the relative position of a byte within the block of data. The page number, which is always included when the "display" function of the disk-to-print program is used, is contained at the bottom of the page.

Unlike the list function, which prints only a single line on the printer for each input record, regardless of the length of the input record, the display function is used to print the entire data block as well as the key areas on the disk. Thus, the display function may be used for any indexed sequential or sequential file in order to print all of the data contained in the file, regardless of the size of the input block or record. Field select cannot be used with the display function. Thus, data is printed in a character or hexadecimal format as specified by the "O" parameter in the Utility Modifier Statement.

CARRIAGE CONTROL CHARACTERS

In the previous examples of the list function of the disk-to-print program, carriage spacing has been controlled through the use of the "S" parameter in the utility modifier statement and a decimal number 1, 2, or 3 which indicated single, double, or triple spacing. When processing data from disk input files, it is possible to use the first character in the record as a "carriage control character", that is, a unique character which will cause spacing to take place on the printer. Although there are four different sets of control characters which may be used with the utility programs, the most common are the ASA control characters. The ASA carriage control characters and their meanings are illustrated below.

CODE	ACTION
blank	Space one line before printing
0	Space two lines before printing
–	Space three lines before printing
+	Suppress space before printing
1	Skip to channel 1 before printing
2	Skip to channel 2 before printing
3	Skip to channel 3 before printing
4	Skip to channel 4 before printing
5	Skip to channel 5 before printing
6	Skip to channel 6 before printing
7	Skip to channel 7 before printing
8	Skip to channel 8 before printing
9	Skip to channel 9 before printing
A	Skip to channel 10 before printing
B	Skip to channel 11 before printing
C	Skip to channel 12 before printing

Figure 3-33 ASA Carriage Control Characters

Note from Figure 3-33 that each character or code indicates a skipping or spacing action which is to be taken by the printer when that character is found in the first byte of the input record. In order to utilize these carriage control characters, the proper entry must be made in the Utility Modifier Statement. The following job stream illustrates the use of the disk-to-print utility program utilizing the ASA carriage control characters.

EXAMPLE

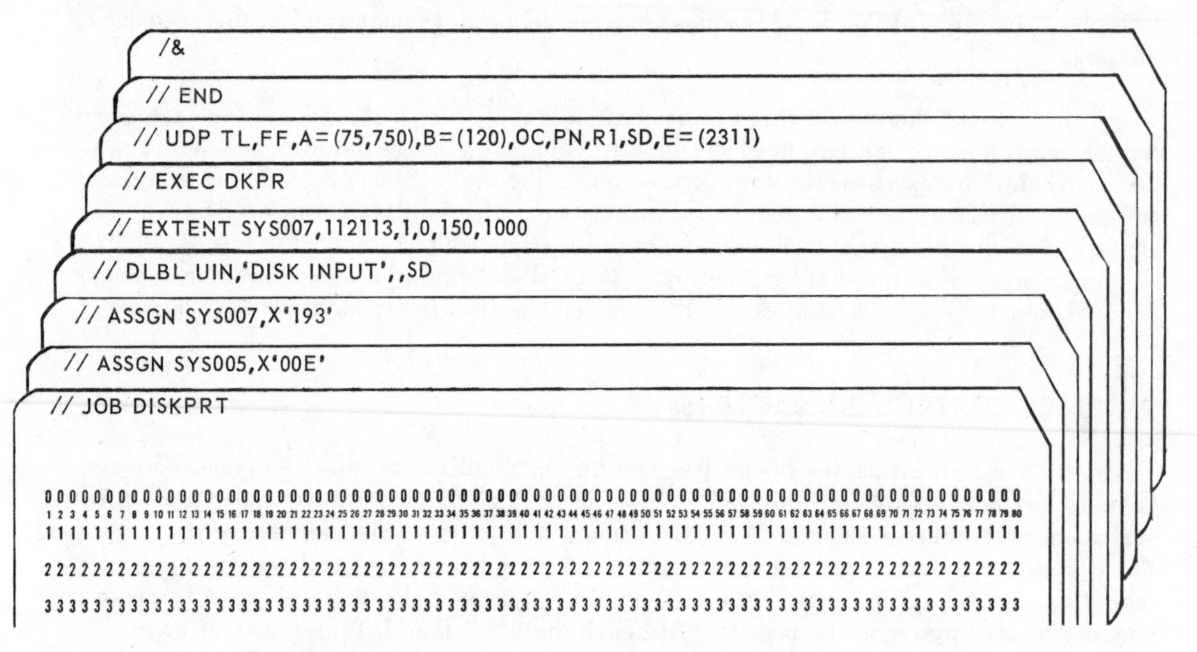

Figure 3-34 Example of ASA Carriage Control Characters

Note in the example above that the utility modifier statement indicates that the list function is to be performed. Carriage control characters cannot be used with the display function. The indication of which set of carriage control characters are to be used is made through the use of the "S" parameter. In the example, the value "SD" is entered in the utility modifier statement, indicating that the "D" set of carriage control characters are to be used for the spacing and skipping of the printer. The "D" set of control characters is the ASA set of characters.

When carriage control characters are used for the disk-to-print program, each of the characters in the first position of a logical record is used to control the spacing. If the first character in the record is not a valid control character for the type specified in the utility modifier statement, then a single line will be printed after being spaced once and no error indication is given. It should be noted that when a carriage control character is the first byte in the record, it is not printed on the report; it is used to control the spacing only.

In the example above note that page numbering is not requested. This parameter is forced if carriage control characters are to be used for the report. Thus, it is not possible for the utility program to number pages when the first byte of the input record is to be used as a carriage control character. In most cases, when carriage control records are used, the heading lines and page counting has taken place when the record was formatted and written on the disk prior to being printed using the disk-to-print utility program. Thus, page numbering, if desired, should be in the data contained on the disk file.

STUDENT ASSIGNMENTS

1. Write the job stream to load a file of punched cards onto the disk. The disk file is to be stored on a 2311 disk drive and the disk output should be checked. The card records should be sequence-checked in columns 2-7. The disk output device is X'193' and the card input device is X'00C'. The extents used for the file are to be determined by the programmer. The card file contains approximately 2000 cards.

1	2	3	4	5	6	7	8	9	10	11	12	13	14	15	16	17	18	19	20	21	22	23	24	25	26	27	28	29	30	31	32	33	34	35	36	37	38	39	40	41	42	43	44	45	46	47	48	49	50	51	52	53	54	55

2. Write the job stream to list the data stored in the disk file in problem #1 on the printer. The report should be double-spaced and be printed on device X'00E'. Page numbering should be included on the report.

1	2	3	4	5	6	7	8	9	10	11	12	13	14	15	16	17	18	19	20	21	22	23	24	25	26	27	28	29	30	31	32	33	34	35	36	37	38	39	40	41	42	43	44	45	46	47	48	49	50	51	52	53	54	55

3. Write the job stream to copy the file created in Problem #1 to another disk file. The input file will be stored on drive X'193' and the output file will be stored on drive X'195'. The disk output should be checked.

4. Write the job stream to load a file of punched cards onto a disk file. The input device is X'00C' and the output device is X'193'. The disk file should be checked when it is written. The file should be blocked 10 records per block. The following fields from the card are to be copied.

Col 1-3: Salesman Number – Packed-Decimal
Col 5-9: Account Number – Packed-Decimal
Col 10-25: Account Name – Zoned-Decimal
Col 30-37: Sales Amount – Packed-Decimal
Col 40-44: Net Amount – Packed-Decimal

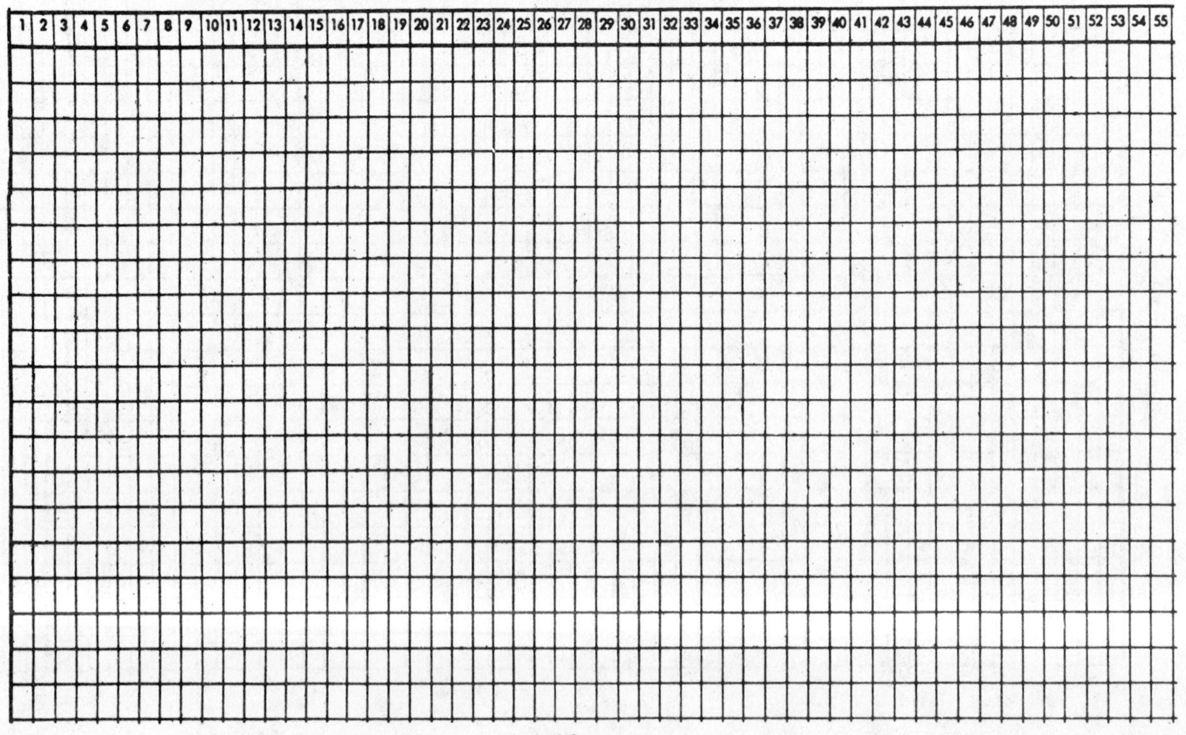

5. Write the job stream to print the file created in Problem #4. All numeric fields should be printed in a zoned-decimal format and the pages should be numbered. The output device is X'00E'. Headings should be included on the report.

1	2	3	4	5	6	7	8	9	10	11	12	13	14	15	16	17	18	19	20	21	22	23	24	25	26	27	28	29	30	31	32	33	34	35	36	37	38	39	40	41	42	43	44	45	46	47	48	49	50	51	52	53	54	55

6. Write the job stream to print the file created in Problem #4 with all of the numeric fields displayed in a hexadecimal format. Headings should be included on the report as well as page numbers. The printer is device X'00E'.

1	2	3	4	5	6	7	8	9	10	11	12	13	14	15	16	17	18	19	20	21	22	23	24	25	26	27	28	29	30	31	32	33	34	35	36	37	38	39	40	41	42	43	44	45	46	47	48	49	50	51	52	53	54	55

7. Write the job stream to load the cards contained in Appendix E onto a blocked disk file. The blocking factor is 10. All of the numeric fields should be stored in a packed-decimal format on the disk file. Also write the job stream to print the disk file after it is created. All of the numeric fields should be printed in the zoned-decimal format. Keypunch the job stream and process the jobs on a System/360 operating under DOS.

1	2	3	4	5	6	7	8	9	10	11	12	13	14	15	16	17	18	19	20	21	22	23	24	25	26	27	28	29	30	31	32	33	34	35	36	37	38	39	40	41	42	43	44	45	46	47	48	49	50	51	52	53	54	55

CHAPTER 4

MAGNETIC TAPE UTILITIES

INTRODUCTION

As with direct-access devices, magnetic tape devices play an important role in the use of the System/360. Magnetic tapes are normally used to store master files which contain considerable data or for back-up files to direct-access files. In addition, tape can be used for many applications in which files must be created for interim use and then scratched.

The DOS file-to-file utility programs provide for processing tape files together with direct-access, card, printer and other tape files. Again, all of the utility programs are used to copy files from one storage medium to another. The programs which process tape files allow standard labels, non-standard or user labels, unlabelled tapes, or standard and user labels to be stored on the tape files. User-written routines may be included in the utility programs to process user-labels. In addition, multi-file volumes or multi-volume files may be processed by the tape utility programs.

The following sections illustrate the use of the magnetic tape utility programs.

CARD-TO-TAPE UTILITY PROGRAM

The card-to-tape utility program is used to copy a card file from the card reader to a tape file. The job stream illustrated below could be used to copy a group of cards to magnetic tape.

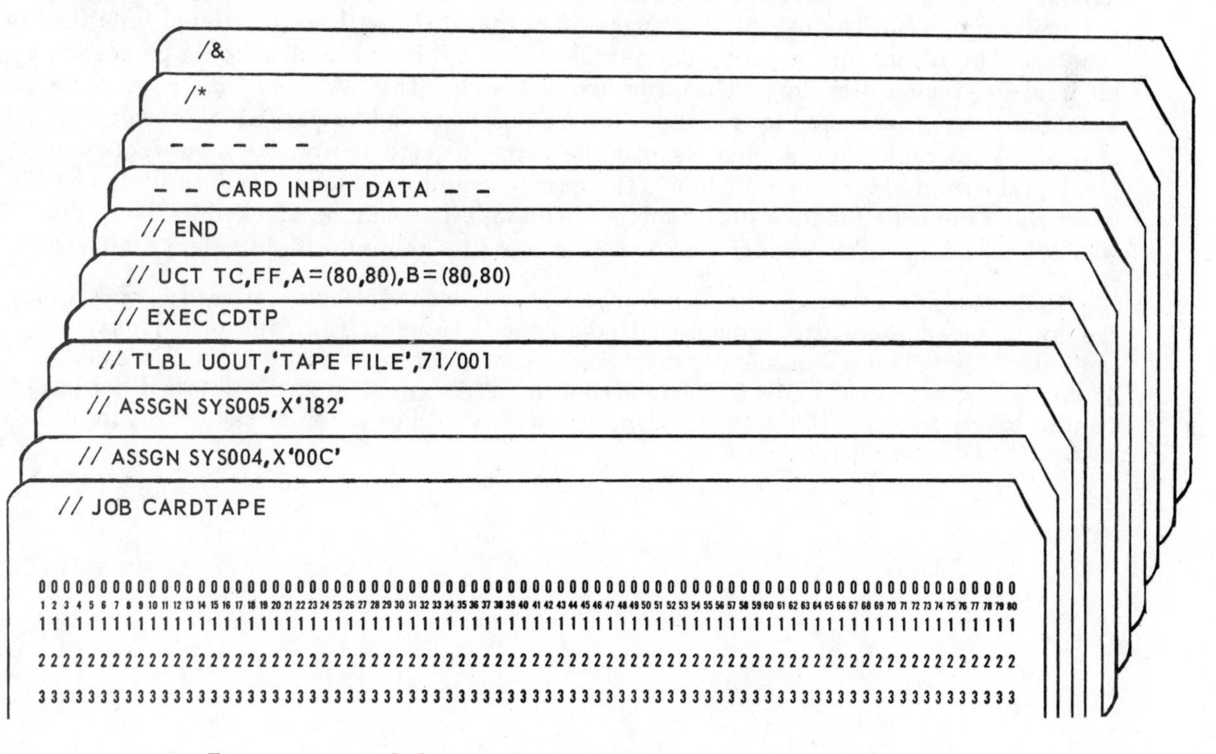

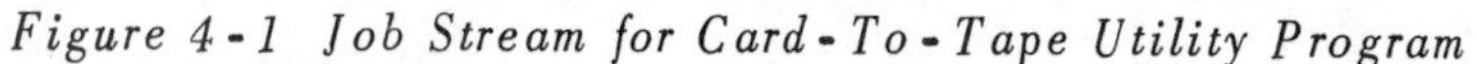

Figure 4-1 Job Stream for Card-To-Tape Utility Program

In the job stream illustrated in Figure 4-1 it can be seen that the card input device, X'00C', is assigned to SYS004 and that the tape output device, X'182', is assigned to SYS005. As with all of the utility programs, the card input device must be assigned to SYS004. Tape output devices must always be assigned to SYS005. The tape file contains standard labels and the TLBL card is included in the job stream to provide label information for the tape file. Note that the filename in the TLBL card is UOUT, which must be used for all tape output files. This entry is followed by the file-ID and the expiration date of the file.

The name of the card-to-tape utility program is CDTP. Thus, the execute card containing // EXEC CDTP causes the card-to-tape program to be loaded into core storage for execution. The Utility Modifier Statement which is used with the card-to-tape program utilizes the same format as the modifier statements used with the other utility programs. The general format of the Utility Modifier Statement used with the card-to-tape program is illustrated in Figure 4-2.

```
// UCT Tt,FF,A=(n,m)[,Ix][,Ox][,Q=(x,y)][,Rx]
```

Figure 4-2 General format of Utility Modifier Statement

In the above general format it can be seen that the first two columns of the card must contain the values //, which are used to indicate to the utility program that a control card is being read. The identifier UCT uniquely identifies the statement as the Utility Modifier Statement for the card-to-tape program. The function-type code is illustrated in the format "Tt", where "t" may have the values C, F, R, or RF. The "C" indicates a straight copy, such as in the example in Figure 4-1. The "F" indicates that the field select option is to be performed, the "R" value indicates that reblocking is to be accomplished, and the "RF" specifies that both field select and reblocking should be done by the utility program. The use of the entries other than "C" will be illustrated later in this chapter. The record format entry contains the value FF because all records processed by the card-to-tape utility program must be fixed-length. The "A" and "B" entries are used to indicate the record and block lengths for the input (A) and output (B) files which are processed. When card input is used, as with the card-to-tape utility, the input record and block sizes must always be 80 bytes. The output record and block sizes must also be 80 bytes when the copy function (TC) is used. The tape file may be blocked by using the Reblock function and the record size may be altered by using the field select statement.

The remaining entries in the utility modifier statement are optional entries and if they are not included, the utility program will use default values. The card input format indicator, "Ix", is used to specify the format of the card input data. If "x" is equal to 1, the data will be in the EBCDIC format and if "x" is equal to 2, the data will be in the column binary format. The default value, which is used in the job stream in Figure 4-1, is the EBCDIC format.

The "O" entry specifies the rewind output option, that is, the action which will be taken with the output tape before and after it is processed. The value "OR" specifies that the output tape should be rewound both before and after processing has occurred. Thus, the file which is written on the tape will be written immediately after the load point on the tape and the tape will be rewound to the load point after the file has been written. The value "ON" in this entry specifies that the tape should not be rewound either before or after the file has been processed. This value is normally used with a multi-file volume when the file which is being created will be the second or subsequent file on the tape. The use of a multi-file volume with the card-to-tape program is illustrated later in this chapter. The entry "OU" specifies that the tape should be rewound to the load point prior to the file being processed and should be rewound and unloaded after the file has been loaded. This entry is normally used when a tape file is to be created but is not to be used in a job or job step which follows in the job stream. The "OR" option is normally used when a file created by the card-to-tape program is to be used immediately because the tape will be rewound to the load point and will still be loaded in a ready status rather than being unloaded as is the case with the "OU" entry. The default value used by the utility program if this optional entry is not included on the Utility Modifier Statement is "OU".

The sequence numbering entry Q=(x,y) is used to specify the location and length of the field which is to be sequence-checked in the card input data. If this parameter is used, the "x" entry specifies the column in the card where the field to be checked begins, and the "y" entry specifies the length of the field. A maximum length of 10 bytes can be specified for the sequence check. No sequence check is performed if this parameter is omitted such as in the example in Figure 4-1.

The first record indicator (R) is used to specify the first logical record which is to be processed by the program. The "x" entry is a number less than or equal to the number of logical records contained in the input file. If this parameter is omitted from the Utility Modifier Statement, such as in Figure 4-1, it is assumed by the program that the first record in the input file is to be the first record processed.

As with the utility programs discussed previously, the card-to-tape program produces an informational messages listing. The listing produced from the specifications in the example in Figure 4-1 is illustrated in Figure 4-3.

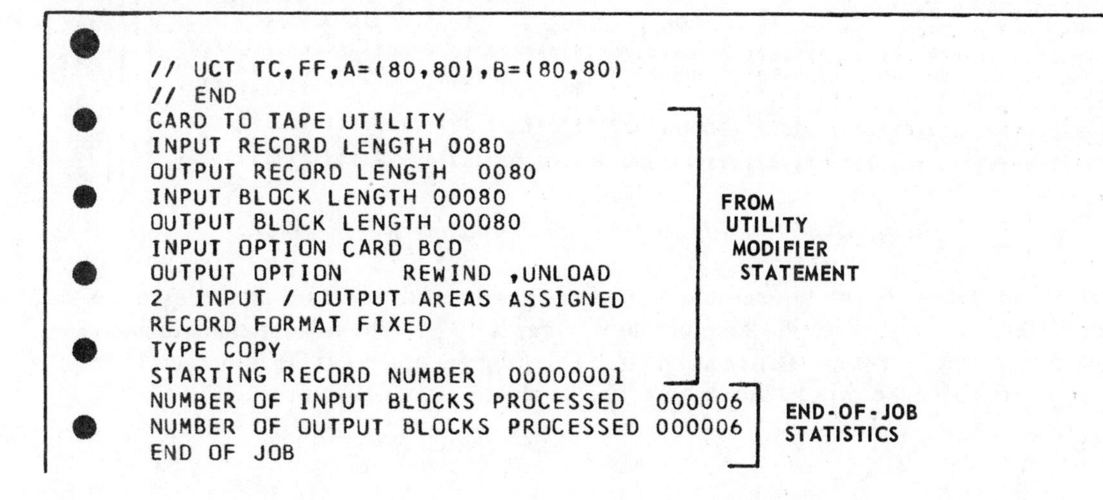

Figure 4-3 Card-To-Tape Utility Informational Messages

In the example in Figure 4-3 it can be seen that the listing contains the options which were either specified in the Utility Modifier Statement, such as the input record and block length, or which were assumed by the utility program since no parameter was included in the modifier statement, such as the output option to rewind and unload the tape output file. The end-of-job statistics are included on the listing after the program has completed processing to indicate the number of blocks processed and also to indicate that the program was successfully terminated. Note in the example that six cards were read and six tape records were written.

CARD-TO-TAPE REBLOCKING

As with the card-to-disk utility program, the card-to-tape program can reblock the input card in order to form blocked records on the tape. This is normally desirable because more records can be stored on the tape when they are blocked and they can be processed more rapidly. The job stream illustrated below could be used to reblock a card input file.

EXAMPLE

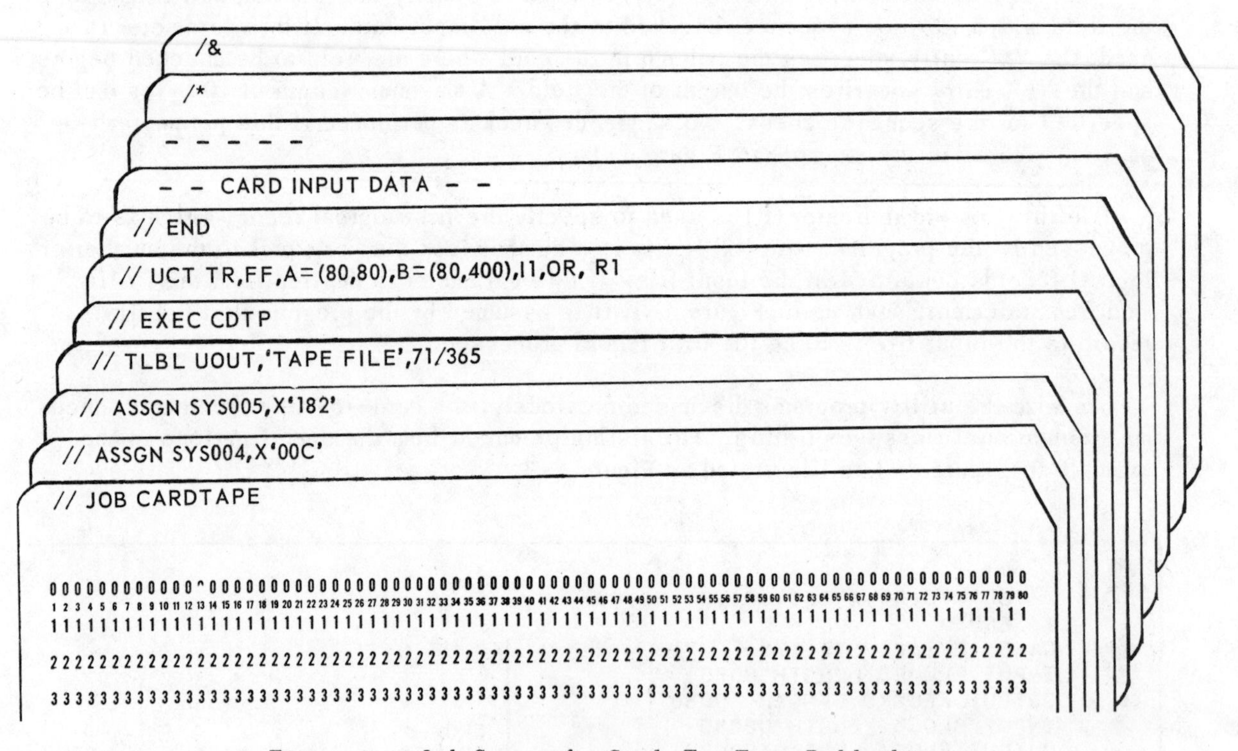

Figure 4-4 Job Stream for Card-To-Tape Reblocking

Note in the above job stream that SYS004 is assigned to the card input device, X'00C', and SYS005 is assigned to the tape output device, X'182'. These assignments must be made for the card-to-tape utility program. The program name CDTP is placed in the execute card because this is the name of the card-to-tape utility program.

The Utility Modifier Statement is used to indicate both that reblocking is to occur and the size of the output records and blocks, in addition to the other standard information supplied by it. The function-type indicator "TR" is used to indicate that reblocking is to occur and the "B" parameter is used for the output record and block sizes. In the example in Figure 4-4 it can be seen that 5 logical records are to be included in a block which means that the block length is 400 bytes. Note also that the output tape will be rewound at the conclusion of processing but that it will not be unloaded. Thus, it is ready to be processed by another program after the card-to-tape program is complete.

CARD-TO-TAPE FIELD SELECT

Fields in the card input files may be selected for omission or inclusion in the tape output record by the field select feature of the card-to-tape utility program. The job stream below illustrates the statements which would be used to field select the store number, department number, salesman number, and employee name from the cards used in Chapter 2 (see Figure 2-1).

EXAMPLE

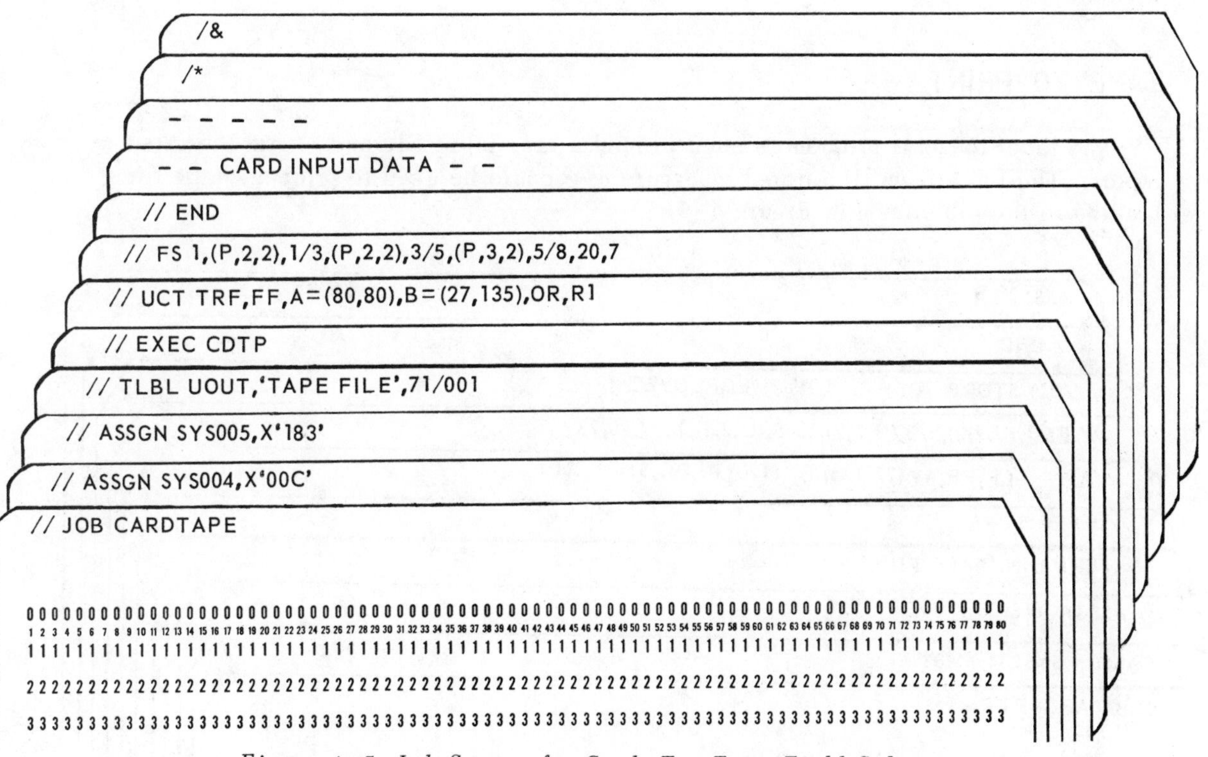

Figure 4-5 Job Stream for Card-To-Tape Field Select

In the above example the function-type is indicated as "TRF", which specifies that both reblocking and field select is to take place. Together with reblocking, the size of the record is to be altered as indicated by the "A" and "B" parameters. Note that the input record length is 80 bytes and the output record length is 27 bytes, thus indicating that 53 bytes of the input card record are not to be included in the output record. The tape output file is to be blocked 5 records per block ($5 \times 27 = 135$).

The field select options include packing the numeric data from the input record before it is written in the output tape record. Thus, the store number, the department number, and the salesman number will be included in the tape record in a packed-decimal format. The entries in the field select card for selecting data in the output record are used in the same manner as with the card-to-disk utility program, that is, the entries for each of the fields is in the "r,s,t" format where "r" indicates the beginning column in the input record, "t" indicates the beginning column in the output record, and "s" indicates the length of the field. Thus, from the example, it can be seen that the employee name, which begins in column 8 of the card input record, will be placed in the tape output record beginning in column 7 and the field is 20 bytes in length.

When the data is to be packed, the "s" operand assumes the format "(P,n,m)" where "n" is the length of the input field and "m" is the length of the output field. The letter "P" is included to indicate that the pack operation is to take place. Thus, for example, the store number, which is stored in columns 1 and 2 of the input record, will be packed and stored in column 1 and column 2 in the tape output record. As noted previously, the two advantages to packed decimal numeric data in disk or tape records is that, in many instances, it requires fewer bytes in the record to store the same data and also the data need not be packed in order to be processed by arithmetic statements in a program using the data.

TAPE-TO-PRINT

The tape-to-print program is used to read a tape input file and create a printed report. The job stream illustrated in Figure 4-6 could be used to print the tape file created in the job stream in Figure 4-5.

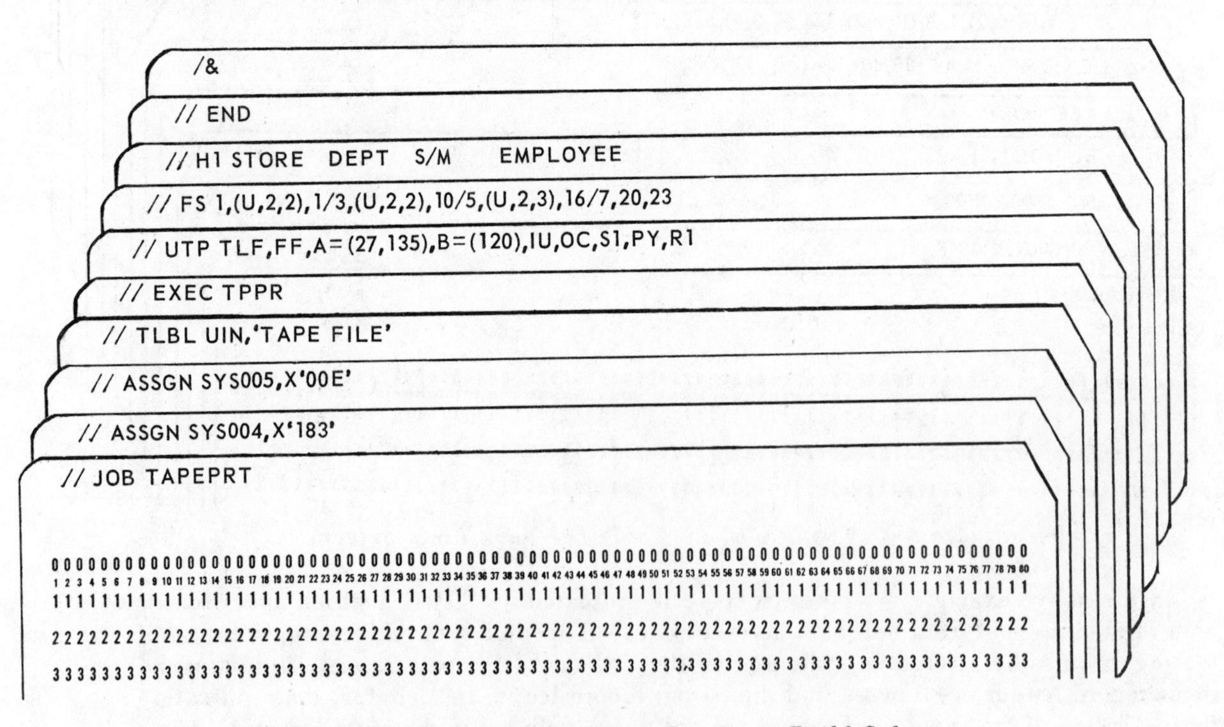

Figure 4-6 Job Stream for Tape-To-Print Field Select

Note from the job stream illustrated in Figure 4-6 that the tape input device, X'183', is assigned to SYS004 and the printer output device, X'00E', is assigned to SYS005. These assignments are always used for the tape-to-print utility program. The filename used in the TLBL card must be UIN, which is always the filename used for tape and disk input files when used with the utility programs. The program name, TPPR, must be included in the execute card.

The general format of the Utility Modifier Statement which is used with the tape-to-print utility program is illustrated below.

```
// UTP Tt,FF,A=(n,m),B=(m)[,Ix][,Ox][,Sx][,Rx][,Px]
```

Figure 4-7 General Format of Utility Modifier Statement for Tape-To-Print

In the general format above it can be seen that the identifier for the card contains the value UTP which is always required on the utility modifier statement for the tape-to-print utility program. The function-type indicator may have the value TD, TL, or TLF. The "TD" entry indicates that a "display" of the tape data in a hexadecimal format is to take place (for an example of the output of the display function, see Figure 3-32). The TL value indicates that a straight list is to occur and the TLF specifies that a list function with field select is requested. The record format indicator of "FF" specifies that fixed-length records are to be processed by the program.

The "A" and "B" fields are used for the same purpose as for all other utility programs, that is, to specify the input record and block length and the output record length. The "n" entry specifies the input record length and the "m" indicates the input block length. The "m" entry in the "B" parameter may have the value 120, 132, or 144, depending upon the length of the printer line desired.

The remaining entries in the utility modifier statement are optional and if they are not included, the program uses assumed values. The Rewind Input option (I) is used to specify the disposition of the input tape after processing has been completed. The value "IU", such as in the job stream in Figure 4-6, requests that the tape be rewound before processing and be rewound and unloaded when the program is complete. If an entry of "IR" is used, the tape is rewound before and after the processing and the value "IN" indicates that the tape should not rewind either before or after the tape file has been printed. If the Rewind Input option is not included on the utility modifier statement, the tape will be handled in the same manner as if the IU entry had been specified.

The "Ox" entry in the general format of the utility modifier statement shown above is used to specify the type of printing that is to appear on the report, that is, either character or hexadecimal format. When the list function is requested, the format must be character unless this is overridden through the use of the field select statement. Thus, the value "OC" is entered on the utility modifier statement to indicate a character format on the report.

Spacing on the printed report is controlled through the use of the "S" parameter. As with the disk-to-print program, the number of lines to be spaced or skipped may be specified using the values 1, 2, or 3 for single, double, or triple spacing or carriage-control characters may be used. In the example in Figure 4-6, single spacing is requested through the use of the "S1" entry in the utility modifier statement. The "R" operand is used to indicate the record number of the first record to be processed by the program. The "x" value may be any decimal number which is less than or equal to the number of logical records contained in the file. The default value is "R1", which indicates that processing should begin with the first record in the file.

The pages on the report will be numbered by the utility program if the entry "PY" is placed in the utility modifier statement and page numbering is suppressed by the entry "PN". If this parameter is not included, the pages will be numbered unless carriage control characters are to be used, in which case page numbering will not be performed by the utility program. As a result of the job stream illustrated in Figure 4-6, the following report would be printed.

EXAMPLE

```
STORE   DEPT   S/M     EMPLOYEE

01      01     025    JONES,ROBERT
01      01     056    ARNETT, RALPH M.
01      01     068    YANCEY, JAMES   .
02      03     560    KILLIET, VERNON M.
02      09     897    GRANGER, THOMAS H.
END OF DATA

PAGE  1


          NUMBER OF INPUT BLOCKS PROCESSED  000001
          NUMBER OF OUTPUT BLOCKS PROCESSED 000005
          END OF JOB
```

Figure 4-8 Example of Tape-To-Print Report

In the example above it can be seen that the store number, the department number, and the salesman number, which were packed when they were written on the tape, are unpacked for the report. Each of the fields is identified by a heading as specified by the // H1 card in Figure 4-6.

TAPE-TO-DISK UTILITY PROGRAM

As noted previously, the function of the general utilities is to copy files. Copying files from tape-to-disk and disk-to-tape is a common requirement in a data processing environment and the utility programs provide a convenient means for doing this copying. In addition, the options of reblocking and field select are available to enable the programmer to reformat the files if this is necessary.

The following job stream could be used to copy the tape file created in Figure 4-6 to a disk file.

EXAMPLE

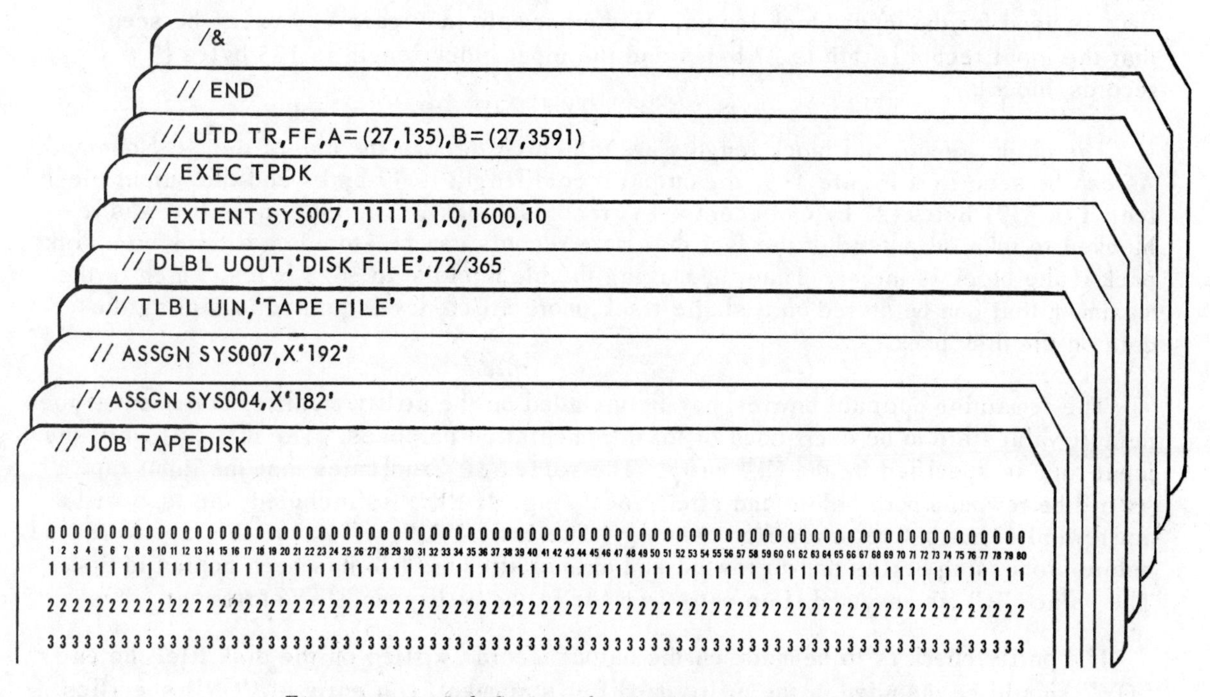

Figure 4-9 Job Stream for TAPE-TO-DISK Utility

Note from the above example that the tape input file will be mounted on drive X'182', which is assigned to SYS004. The tape input drive must always be assigned to SYS004 in any of the tape utility programs. The disk drive, which may be assigned to any programmer logical unit except SYS004, is assigned to SYS007. The filename entry in the TLBL card for the input file is UIN, which is always the filename entry in the label card for tape or disk input. The filename UOUT is used for the disk output file, and again, this filename is used for the entry in the label card for tape or disk output. The phasename for the tape-to-disk utility program is TPDK, and this name must be specified in the Execute card.

The utility modifier card for the tape-to-disk utility program is similar in format to the utility modifier statements used for the other file-to-file utility programs. The general format is illustrated below.

```
// UTD Tt,Ff,A=(n,m),B=(n,m)[,Ix][,Ox][,Rx][,E=(e)]
```

Figure 4-10 General Format of Utility Modifier Statement

The identification for the tape-to-disk utility modifier statement is UTD. The function-type indicator may contain the values "TC" for a straight copy, "TR" for a reblock and copy function, "TF" for a field-select and copy function, or "TRF" for field select, reblocking and copying.

The record format entry, Ff, is used to indicate the format of the records, that is, whether they are fixed-length, variable-length, or undefined length. The entry "FF", such as in the example in Figure 4-9, indicates that the records to be processed are fixed-length. The entry "FV" would indicate variable length, and "FU" would indicate undefined length records. The input record and block lengths are specified in the "A" entry. The value "n" is used to indicate the length of the input record and the value "m" is used for the input block length. In the example in Figure 4-9 it can be seen that the input record length is 27 bytes and the input block length is 135 bytes (5 records/block).

The output record and block lengths are indicated through the use of the "B" operand. As can be seen from Figure 4-9, the output record length is 27 bytes and the output block length is 3591 bytes (27 bytes/record × 133 records/block). The tape input file was reblocked to take advantage of the fact that more records can be stored on a track on a disk pack if the block is larger. Thus, by making the block close to 3625 bytes, which is the maximum that can be stored on a single track, more efficiency is gained in storing the data on the disk pack.

The remaining optional entries may be included on the utility modifier statement if the default values are to be overridden or for documentation purposes. The disposition of the input tape is specified by the "I" entry. The value "IR" indicates that the input tape should be rewound both before and after processing. If "IN" is included, the tape will not rewind either before or after processing. The entry "IU" will cause the tape to rewind before processing begins and to rewind and unload after the file has been copied to disk. The value "IU" is assumed if no entry is made in the utility modifier statement.

If a parity check is to be made on the output records written on the disk file, the entry "OY" should be included in the utility modifier statement. An entry of "ON" specifies that no disk write-check should be performed. The default value is "OY". The first record entry, "R" is used to specify the first record to be processed and the device type entry specifies whether a 2311 or 2314 disk drive is to be used for the disk output file.

In the example in Figure 4-9, no field selecting is to take place. If field select were to be performed, however, the function-type indicator would be specified as "TF" or "TRF" and the field select control card would be included in the utility control cards as illustrated in previous examples.

The informational messages which are printed as a result of the job stream in Figure 4-9 are illustrated in Figure 4-11.

```
// UTD TR,FF,A=(27,135),B=(27,3591)
// END
TAPE TO DISK UTILITY
INPUT RECORD LENGTH 0027
OUTPUT RECORD LENGTH  0027
INPUT BLOCK LENGTH 00135
OUTPUT BLOCK LENGTH 03591
INPUT OPTION REWIND ,UNLOAD
OUTPUT OPTION    DASD WRITE CHECK
2 INPUT,2 OUTPUT AREAS ASSIGNED
RECORD FORMAT FIXED
TYPE REBLOCK
STARTING RECORD NUMBER  00000001
OUTPUT DEVICE TYPE  2311
FILEMARK WRITTEN IN
XT. NO.  B1   C1   C2   H1   H2   R
    000  000  000  160  000  001  001
NUMBER OF INPUT BLOCKS PROCESSED  000001
NUMBER OF OUTPUT BLOCKS PROCESSED 000001
END OF JOB
```

Figure 4-11 Utility Informational Messages

Note from the listing illustrated in Figure 4-11 that the assumed values for the tape-to-disk program are used when they are not explicity stated in the utility modifier statement, that is, the tape is to be rewound and unloaded after processing, the write check is to take place, the first record to be processed is record 1 and the device to be used is a 2311 disk drive. Note also that the end-of-file marker is written on Cylinder 160, Track 1, Record 1.

DISK-TO-TAPE UTILITY PROGRAM

The Disk-To-Tape utility program is used to copy a file from a disk to a magnetic tape. This program is used many times to store "backup" files of a disk file, that is, to copy a disk file to a tape so that if the disk file is destroyed for some reason, the tape file can be used to recreate the disk file. The following job stream could be used to copy the file created from the job stream in Figure 4-9.

EXAMPLE

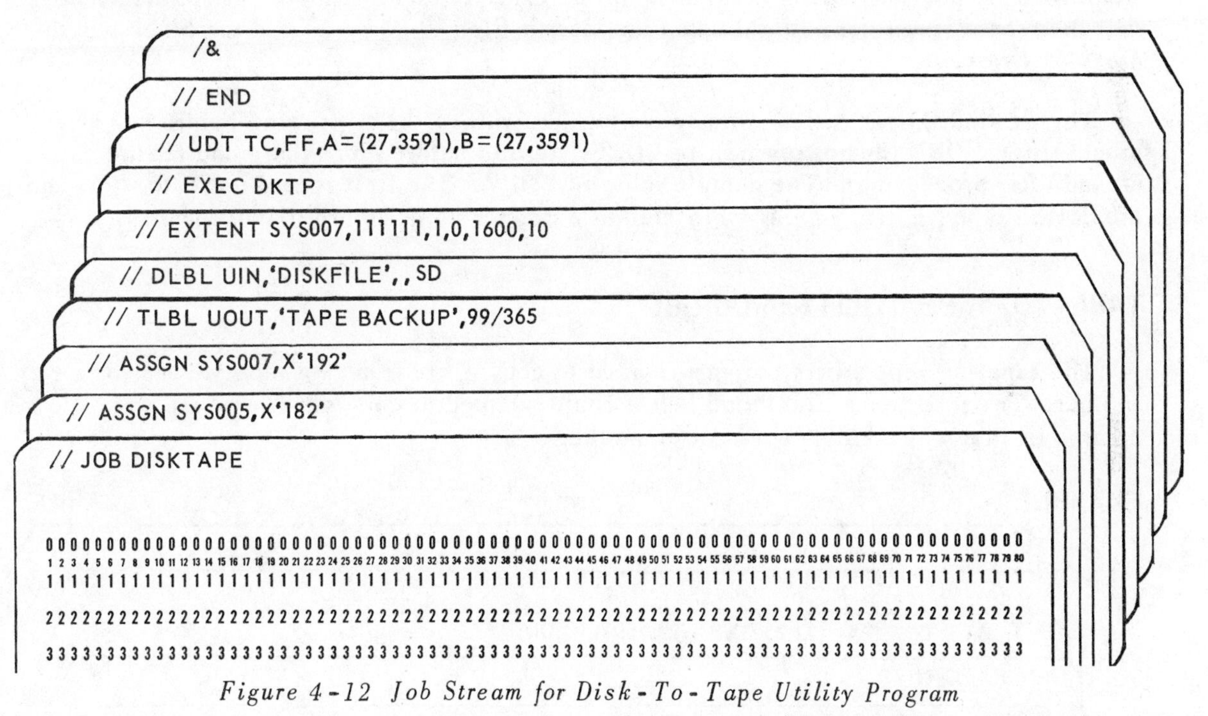

Figure 4-12 Job Stream for Disk-To-Tape Utility Program

In the example above it can be seen that the tape output file will be stored on drive X'182' which is assigned to SYS005. The tape output file in any of the file-to-file utility programs must be assigned to SYS005. The disk input file will be stored on drive X'192' which is assigned to SYS007. The filename in the TLBL card for the tape output file is UOUT and the disk input filename in the DLBL card is UIN, and these names must always be used for labelled input and output files. The phasename of the disk to tape program, DKTP, must be included on the Execute card to cause the program to be loaded into core storage and executed.

The general format of the utility modifier statement for the disk-to-tape program is illustrated below.

```
// UDT Tt,Ff,A=(n,m),B=(n,m)[,Ox][,Rx][,E=(e)]
```

Figure 4-13 General Format of Disk-To-Tape Utility Modifier Statement

The statement identification for the disk-to-tape program must be UDT. The function-type indicator can have the values "TC" for a straight copy, "TF" for copy and field select, "TR" for copy and reblock, or "TRF" for copy, reblock, and field select. If field select is to take place, the field select control card must be included following the utility modifier statement.

The input record and block sizes are specified by the "A" parameter and the output record and block sizes are specified by the "B" parameter in the same formats as illustrated for previous utility programs. In the example in Figure 4-12 it can be seen that the record length for both the input and output files is 27 bytes and the block lengths are 3591 bytes.

The Rewind Output option (O) may contain the values "OR" (rewind before and after processing), "ON" (do not rewind), or "OU" (rewind before processing and rewind and unload after processing). The default value is "OU". The first record indicator (Rx) and the device type (E=(e)) are used in the same manner as in other utility programs.

TAPE-TO-TAPE UTILITY PROGRAM

The tape-to-tape utility program is used to copy a file from one tape volume to another. The job stream illustrated below could be used to copy the file which was created in Figure 4-12 from one reel to another.

EXAMPLE

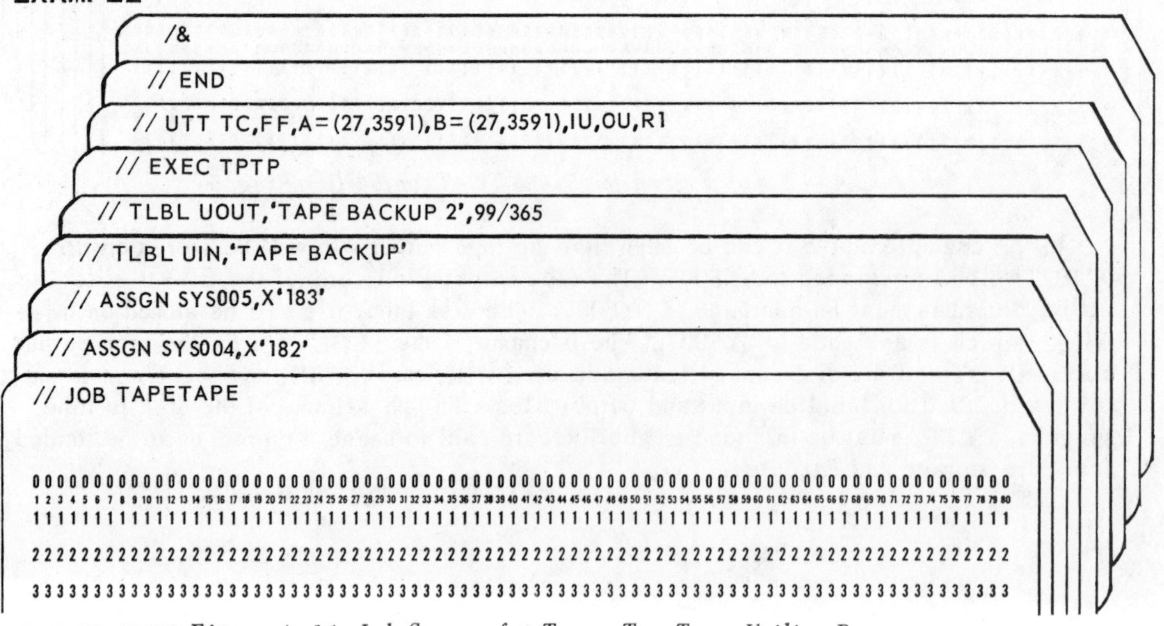

Figure 4-14 Job Stream for Tape-To-Tape Utility Program

Note in the job stream in Figure 4-14 that the tape input file will be mounted on drive X'182', which is assigned to SYS004 as all tape input files must be. The tape output device is X'183' and is assigned to SYS005, again as all output tapes must be. The filename in the TLBL card for the input file is UIN and for the output file the filename is UOUT. Again, these standard names must be used for all labelled input and output files. The program name which is contained in the execute card is TPTP.

The Utility Modifier Statement for the tape-to-tape program has the same general format as the ones used with the other utility programs. The function-type may indicate a straight copy (TC), a copy with field select (TF), a copy and reblock operation (TR), or a copy with reblocking and field select (TRF). The record format may be fixed-length (FF), variable-length (FV), or undefined (FU). The input and output record and block lengths are indicated through the use of the "A" and "B" parameters.

Since tape is used for both input and output, the disposition of both the input and the output tapes must be specified. The "I" operand is used for the input tape and the "O" operand is used for the output tape. The values "IR" and "OR" specify that the input and output tapes will rewind both before and after processing. "IN" and "ON" indicate that neither the input nor the output should rewind either before or after processing, and the values "IU" and "OU" indicate that the tapes should be rewound before processing and should be rewound and unloaded after processing. These parameters may be used in any combinations desired for the input and output tapes.

LABEL CHECKING

In the previous tape examples, it was assumed that all tapes, both input and output, contained standard DOS labels. If non-standard or unlabelled tapes are to be processed, this must be indicated through the use of the UPSI job control card. The following combinations of values in the UPSI card may be used to indicate the label processing to be done by the utility program.

1. // UPSI 00000 – This indicates that standard input and output labels are to be processed with no user labels. If this processing is desired, such as in the previous examples, no UPSI card need be submitted because all bits are set to zero when a job is begun.

Input

2. // UPSI 10000 – This value indicates that the input tape contains no labels. Therefore, the utility program will not check labels for this tape and either a tapemark or the first data record will be the first record read by the program.

3. // UPSI 01000 – This value indicates that standard labels and user labels are contained on the tape. When user labels are contained on an input or output tape, user label processing routines must be included in the utility program. For the technique of processing user labels with the utility programs, consult IBM SRL GC24-3465 DISK AND TAPE OPERATING SYSTEMS UTILITY PROGRAM SPECIFICATIONS.

4. // UPSI 11000 – This UPSI card specifies that only non-standard labels are contained on the input tape. A user-written label processing routine must be included in the program to check these labels.

Output

5. // UPSI 00100 – This value indicates that no labels are to be written on the output tape from the utility program. In addition, it requests that the first record written on the unlabelled tape be a tape mark, which is desirable in some applications.

6. // UPSI 00101 – This UPSI card indicates to the utility program that no labels are to be written on the output tape and that no tapemark is to be written. Thus, the first record on the tape will be the first data record in the file.

7. // UPSI 00010 – This combination indicates that standard labels and user-written labels are to be placed on the output tape. When this option is requested, a routine to process the labels must be incorporated into the utility program.

8. // UPSI 00110 – This UPSI card specifies that only non-standard labels are to be written on the output tape. When this option is used, the only labels which are written on the tape are the labels which are written by a user-written routine which is included in the utility program.

As can be seen from the above examples, the input tapes utilize the first two bits of the UPSI byte and the output tapes utilize the next 3 bits of the byte. The options for input and output are independent options and may be used in any allowable combinations. For example, the card // UPSI 01100 indicates that the input tape contains standard labels and user labels and the output tape will contain no labels. The decision on tape labels normally rests with the installation standards which indicate the type of tape labels which are to be used.

MULTI-FILE VOLUMES

When processing magnetic tape files which contain standard labels, it is possible to store more than one logical file on a tape volume, or reel of tape. When this occurs, each file contains a unique header label which identifies the particular file. The concept of a multi-file volume is illustrated in Figure 4-15.

Multi-File Volume

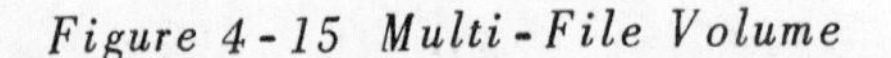

Figure 4-15 Multi-File Volume

In the drawing in Figure 4-15 it can be seen that three files are stored on a single tape volume. The volume label precedes the files and indicates that the volume serial number is 111111. The first tape file has a file identification TAPE FILE 1, the second TAPE FILE 2 and the third TAPE FILE 3. Note that each file on the tape contains its own header label which indicates the file-identification and also contains a file-sequence number which indicates the relative position of the file on the volume, that is, the first file has a file sequence number of 0001, the second file has a file sequence number of 0002, etc. Note that tape marks are placed on the tape following the header label for the first file (1), after the file itself (2), and following the trailer label for the first file (3). The header label for the second file is then written on the tape. Following the trailer label for the last file on the tape are two tape marks (not illustrated in Figure 4-15).

In order for the second or subsequent files to be written on the tape using a file-to-file utility program, the tape volume must be positioned in the proper position for the tape label to be written. The job stream illustrated in Figure 4-16 illustrates a job used to create the first file on a tape.

EXAMPLE

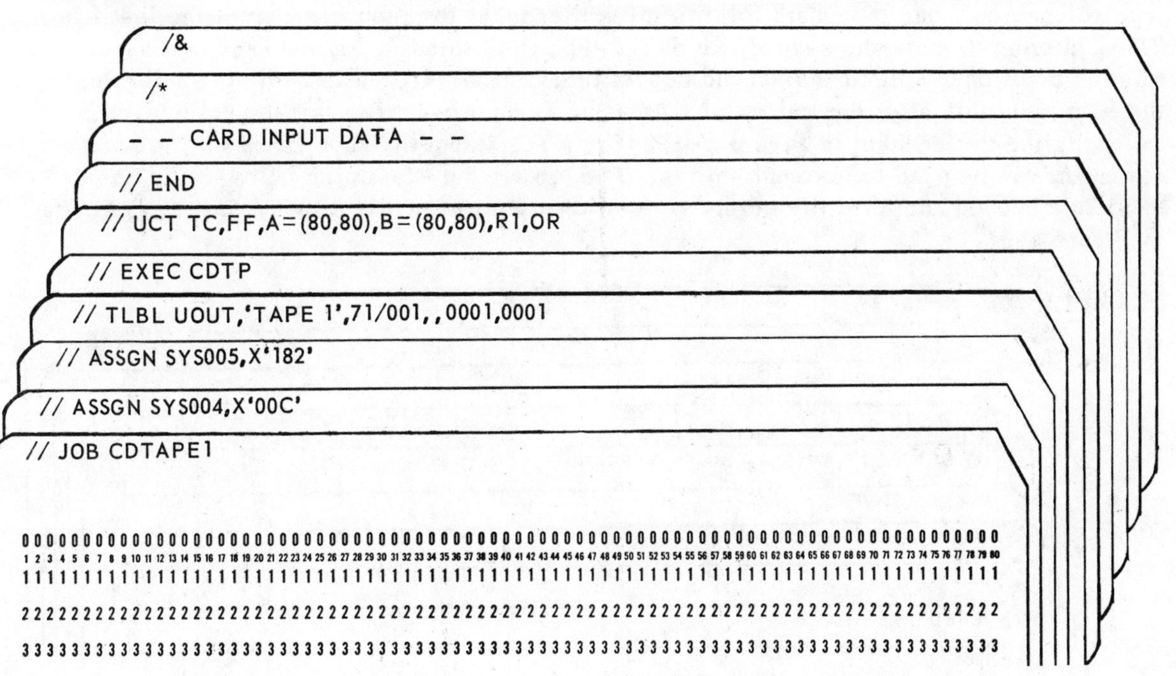

Figure 4-16 Job Stream to Load First File on Tape

As can be seen from the job stream in the above example, the card file is to be copied to the tape. The tape file is to be the first file on the tape volume because of the file-sequence-number entry in the TLBL card. Note that the tape will be rewound at the end of the job step.

The file created in the job stream in Figure 4-16 is the first file on the tape. After the job stream in Figure 4-16 has been executed, the tape would appear as illustrated below.

EXAMPLE

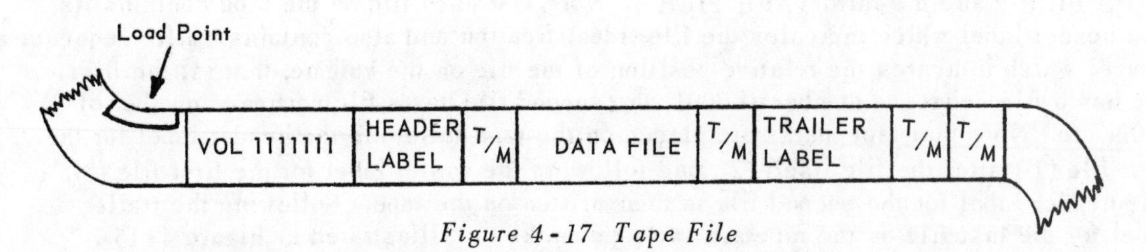

Figure 4-17 Tape File

Note in the above diagram that a tape mark follows the header label, a tape mark follows the data file, and two tape marks follow the trailer label. Whenever a single file is stored on magnetic tape, utilizing standard labels, the tape is organized as shown above.

Note from Figure 4-15, however, that when more than one file is stored on a single tape volume, only one tape mark follows the trailer label for each file except the last file. Thus, in order to write the second file on the tape illustrated in Figure 4-17, the tape must be positioned so that the second header label can be written immediately following the first tape mark after the trailer label, that is, it must overwrite the second tape mark following the trailer label in Figure 4-17. The MTC (Magnetic Tape Command) job control statement can be used to accomplish this. The job stream illustrated below could be used to write a second and third file on the tape following the first file created in the job stream in Figure 4-16.

EXAMPLE

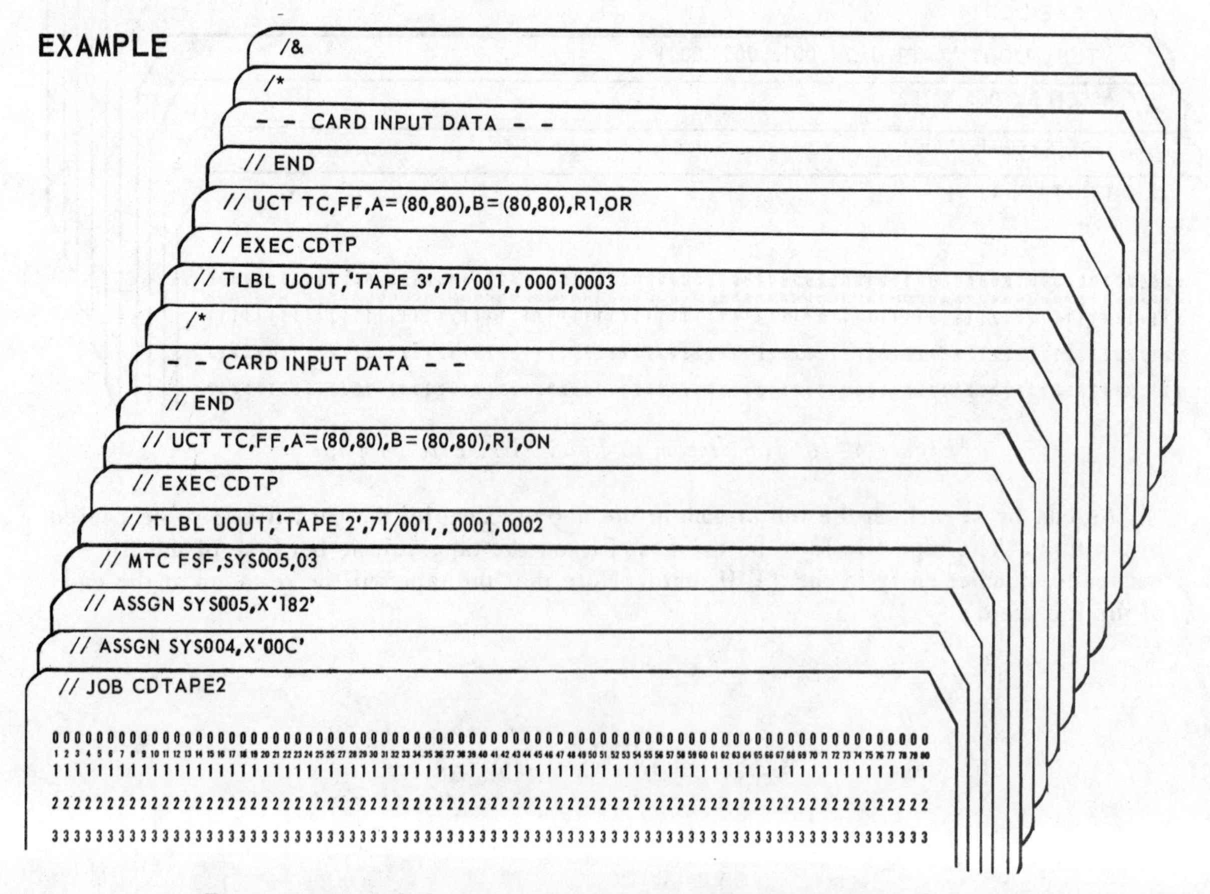

Figure 4-18 Job Stream to Load Second and Third Files

Note in the job stream in Figure 4-18 that two card-to-tape program are to be executed. The first loads the second file and the second job step loads the third file on the tape volume. The tape is positioned to write the second file on the tape through the use of the MTC Statement. The general format of the MTC statement is illustrated below.

Figure 4-19 General Format of MTC Statement

In the general format illustrated above, note that the identifier for the statement is MTC. The opcode entry is used to specify the action which is to be taken for the tape mounted on the device assigned to SYSxxx, where xxx is any valid programmer logical unit number. In the example in Figure 4-18 note that the opcode entry is FSF, which means Forward Space File. Other opcodes which may be used with the MTC Statement are explained in IBM SRL C24-5036. The optional "nn" entry is used to indicate the number of times the action is to take place. Thus, in Figure 4-18, three files will be skipped, which means that three tape marks will be bypassed. The file is positioned following the last tape mark indicated. Therefore, as a result of the MTC command in the job stream, the first three tape marks on the tape (see Figure 4-17) will be bypassed and the tape will be positioned to write or read the fourth tape mark. This is the desired position of the tape so that another file may be written on the tape. When the second tape file is opened by the card-to-tape utility program, the header label will be written over the second tape mark following trailer label for the first file.

Note in the utility modifier statement for the second file that no rewind is requested after the second file is created. Thus, the tape will be in the correct position for the third file to be written without the use of the MTC statement. After the third file has been written, the tape will be rewound as requested by the "OR" entry in the utility modifier statement.

In order to read the tape files created on the tape, it is only necessary to make the proper entries in the TLBL card for the file to be read. The following job stream illustrates the entries which could be used to read the third file on the tape.

EXAMPLE

```
/&
// END
// UTP TL,FF,A=(80,80),B=(80,80),IU,PN
// EXEC TPPR
// TLBL UIN,'TAPE 3',,,0001,0003
// ASSGN SYS005,X'00E'
// ASSGN SYS004,X'182'
// JOB READ3
```

Figure 4-20 Job Stream to Read Third File on Tape

Note in the job stream in Figure 4-20 that the file-sequence-number specified in the TLBL card contains the value 0003. This indicates to Logical IOCS that the third file on the tape is to be read and when the input file is opened in the tape-to-print program, the first two files will automatically be skipped and the header label for the third file will be read and checked against the information supplied in the TLBL card. Thus, when a multi-file volume is to be input to a utility program, it is only necessary to indicate which file is to be read by the program through the use of the file-sequence-number. It is not neccesary to position the tape by using the MTC statement prior to it being opened.

SUMMARY

As has been illustrated, the DOS Utility Programs may be used whenever file-to-file copying must be performed. The ability to reblock and field select allows some file manipulation to be performed. The utility programs are, however, limited to only these activities. No data editing may be performed and there are no user-exits, other than for tape labels, which allow any specialized processing to be performed on a file. Thus, for file copying, reblocking, or field selecting data from a record for a list report or storing particular data, the utility programs provide easy-to-use methods. For processing data which must be edited, checked, or otherwise processed in a unique manner, the file-to-file utility programs cannot be used. The user should normally write a program to accomplish the desired task.

STUDENT ASSIGNMENTS

1. Write the job stream to copy a card file to a tape file. The tape device to be used for the output file is X'184'. The tape should be rewound after it is processed.

1	2	3	4	5	6	7	8	9	10	11	12	13	14	15	16	17	18	19	20	21	22	23	24	25	26	27	28	29	30	31	32	33	34	35	36	37	38	39	40	41	42	43	44	45	46	47	48	49	50	51	52	53	54	55

2. Write the job stream to print the file created in Problem #1 using the tape-to-print utility program. The report should be single spaced and contain page numbers. Rewind and unload the input tape when processing is complete.

1	2	3	4	5	6	7	8	9	10	11	12	13	14	15	16	17	18	19	20	21	22	23	24	25	26	27	28	29	30	31	32	33	34	35	36	37	38	39	40	41	42	43	44	45	46	47	48	49	50	51	52	53	54	55

3. Write the job stream to copy the file created in Problem #1 to another tape for back-up. The output file should be blocked 20 records per block. Rewind and unload both tapes when processing is complete. The output file will be written on device X'180'.

1	2	3	4	5	6	7	8	9	10	11	12	13	14	15	16	17	18	19	20	21	22	23	24	25	26	27	28	29	30	31	32	33	34	35	36	37	38	39	40	41	42	43	44	45	46	47	48	49	50	51	52	53	54	55

4. Write the job stream to copy the file created in Problem #3 to a disk file. The disk file is to be blocked 10 records per block and written on drive X'194'. Only the following fields should be contained in the disk record.

Col 4-9: Part Number – packed-decimal
Col 10-20: Part Description
Col 21-25: Quantity – packed-decimal
Col 30-32: Re-order quantity – packed-decimal

The record format of the output file is to be designed by the programmer.

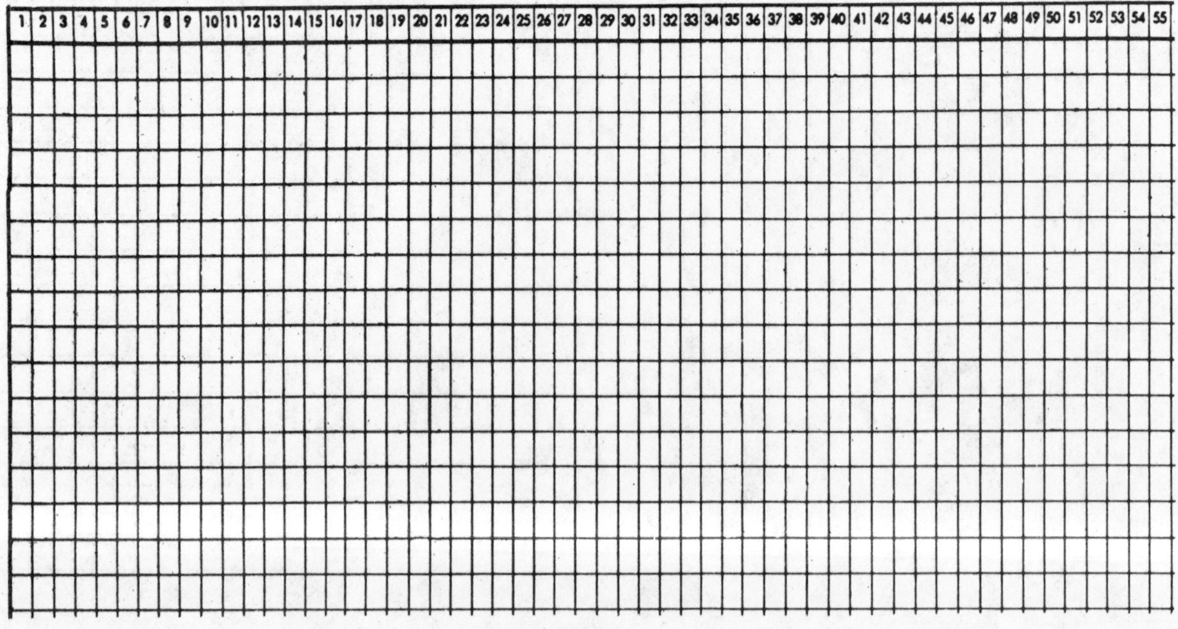

1	2	3	4	5	6	7	8	9	10	11	12	13	14	15	16	17	18	19	20	21	22	23	24	25	26	27	28	29	30	31	32	33	34	35	36	37	38	39	40	41	42	43	44	45	46	47	48	49	50	51	52	53	54	55

5. Using the tape-to-print program print the same information which was copied to the disk file in Problem #4 on the printer from the tape created in Problem #3. The report should contain headings and be double-spaced with page numbers. The output printer is X'00E'.

6. Write the job stream to take a back-up tape of the disk file created in Problem #4. The input disk device is X'194' and the output tape device is X'185'. The output tape should contain 40 records per block and all data should be stored in the format used on the disk.

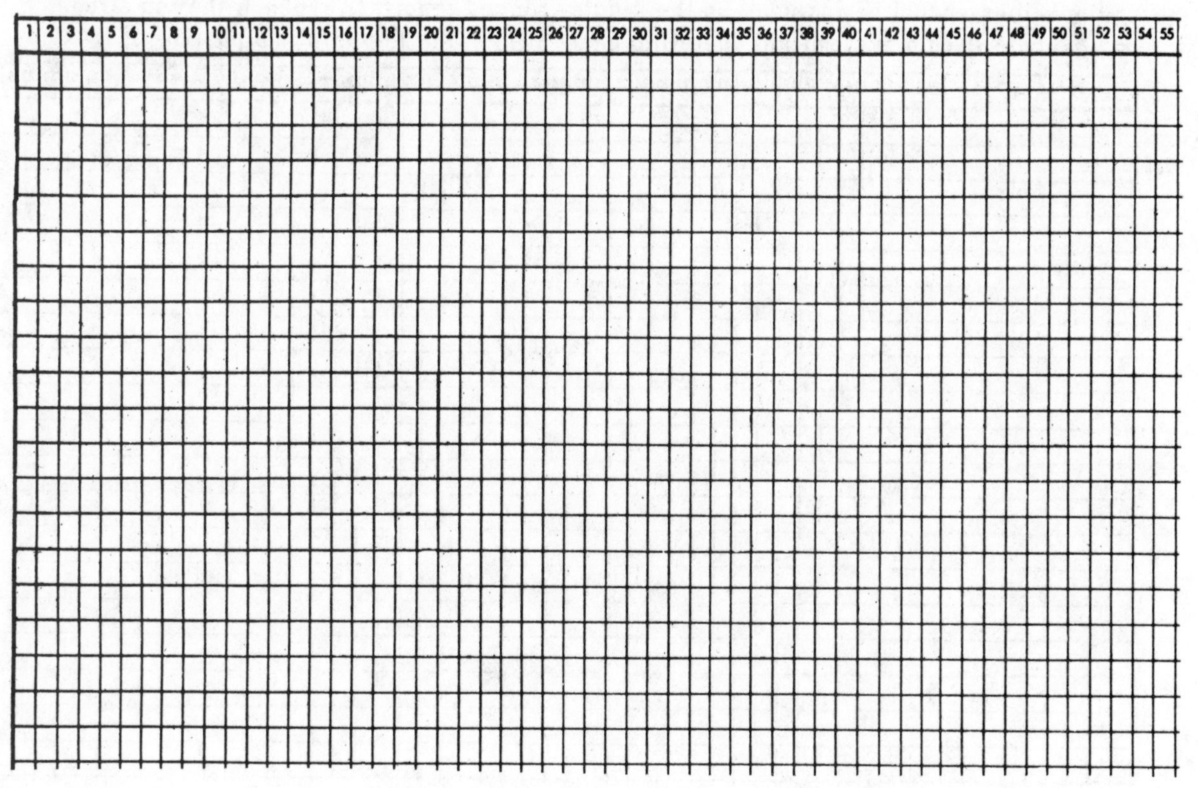

7. Write the job stream to print the records stored on the tape created in Problem #6. The data which is stored in a packed-decimal format on the tape should be printed in a hexadecimal format on the report. The report should be triple spaced on device X'00E'. The input tape should be rewound and unloaded at the conclusion of processing.

8. Write the job stream to load the cards contained in Appendix E onto a tape file. The following fields should be included on the tape: Invoice Number (packed-decimal), Customer Number (packed-decimal), and Description (zoned-decimal). Write, as part of the same job, the statements necessary to use the tape-to-print program to print the file just created. The packed-decimal fields should be printed in a zoned-decimal format and headings should be included on the single-spaced report. Keypunch the job stream and execute it on a System/360 operating under DOS using the data in Appendix E.

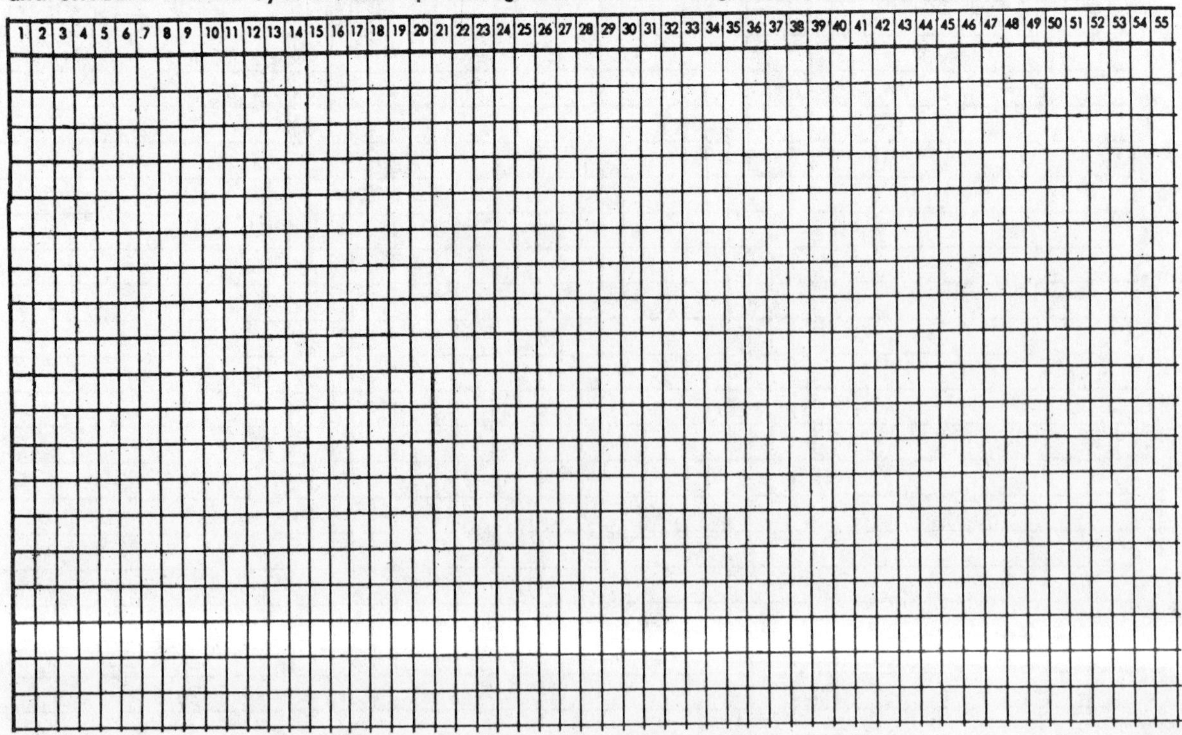

CHAPTER 5

TAPE AND DISK SORT/MERGE PROGRAM

INTRODUCTION

One of the most common operations in data processing is the sorting of data. Sorting refers to the process of placing records in a prescribed sequence based upon the value contained in a "control field". The following example illustrates records before and after they are sorted on the "salesman number" control field.

EXAMPLE

Before Sorting

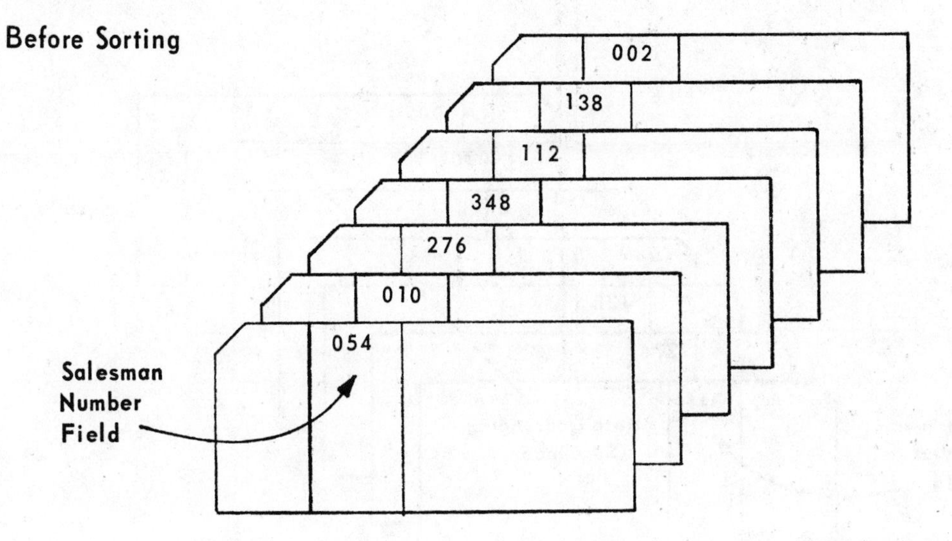

After Sorting

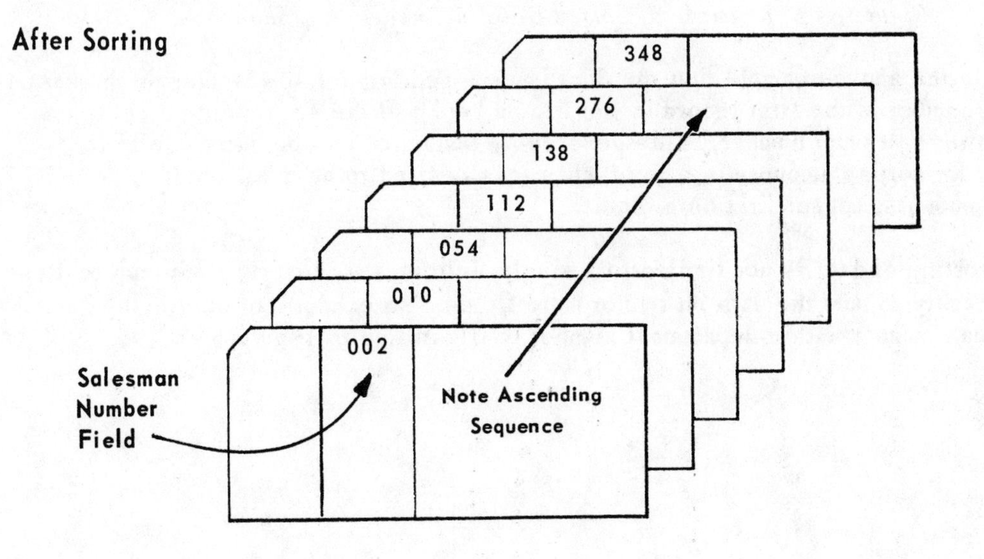

Figure 5-1 Example of SORTED DATA in Ascending Sequence

Note from the example in Figure 5-1 that prior to being sorted the salesman numbers on each of the cards were in no particular order. After they were sorted, the salesman numbers are in an ascending sequence, that is, each of the salesman numbers is higher in value than the one preceding it. This process of placing data is a prescribed sequence, or sorting it, is done so that printed reports will be more easily readable, so that update programs can process master records by performing high-low-equal comparisons, and for a myriad of other reasons.

As can be seen from the example in Figure 5-1, the records are sorted in an ascending sequence. Many times it is necessary to sort records in a descending sequence, that is, so that the first record in the file contains the greatest number in the control field. If the cards illustrated in Figure 5-1 were sorted in a descending sequence, they would appear as shown below.

EXAMPLE

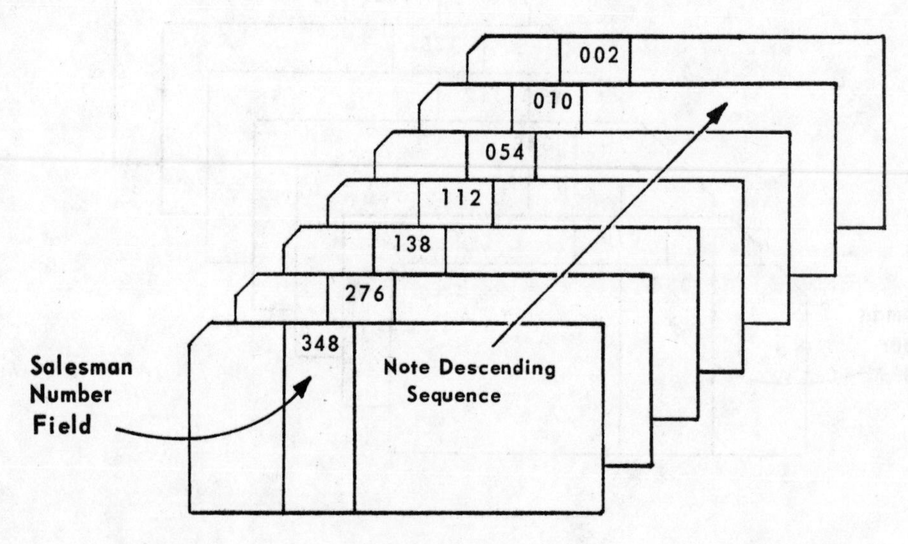

Figure 5-2 Example of Sorted Data in Descending Sequence

Note in the above example that the data is arranged in such a way that the highest salesman number is the first record in the file and each of the succeeding records contains a lower salesman number. This descending sequence can be quite useful for reports or for sorting amounts in a field when it is desired to have all credits, or negative amounts, appear first on a report.

The sorting of data is not limited to a single field in a record. In many applications it is necessary to sort the data on two or more fields. An example of data being sorted on salesman number within department number is illustrated in Figure 5-3.

EXAMPLE

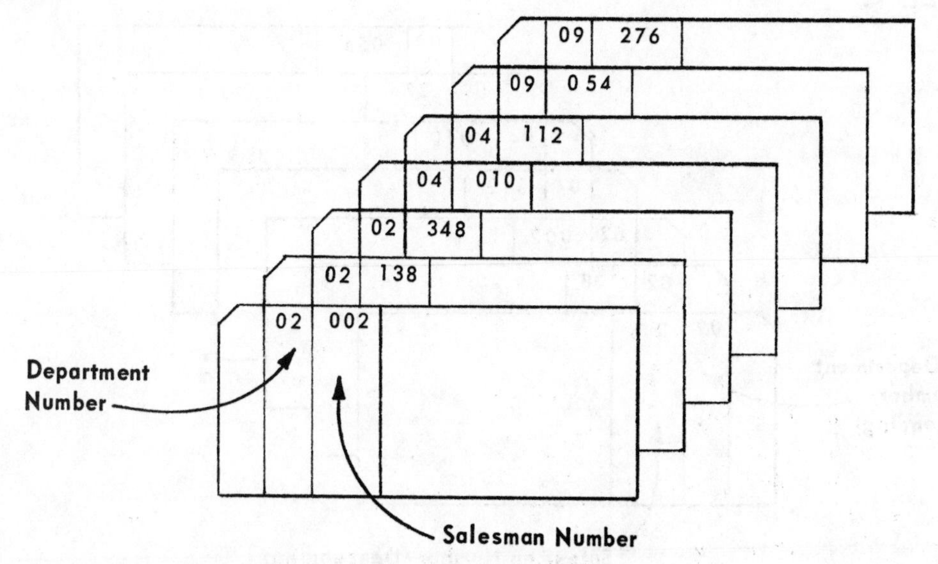

Figure 5-3 Example of Major, Minor Sort

In the example above, note that the cards are sorted in an ascending sequence by department number, that is, the department numbers are in the order 02, 04, and 09. The salesman numbers are in sequence within each department number, that is, for department 02, the salesman numbers are in the order 002, 138, and 348. For department 04, the salesman numbers run 010 and 112. It is important to note that for the complete file the salesman numbers are not in order, that is, they run 002, 138, 348, 010, 112, 054, and 276, but that they are in order within each department. The department number in the above example is said to be the MAJOR field in the sort and the salesman number is said to be a MINOR field in the sort which means that the entire file is in order based upon the value in the department number field and the salesman numbers are in order within each department.

Sorts which utilize minor fields do not necessarily need to be all in an ascending sequence. The example in Figure 5-4 illustrates a file which is in ascending sequence by department number and descending sequence by salesman number within department number.

EXAMPLE

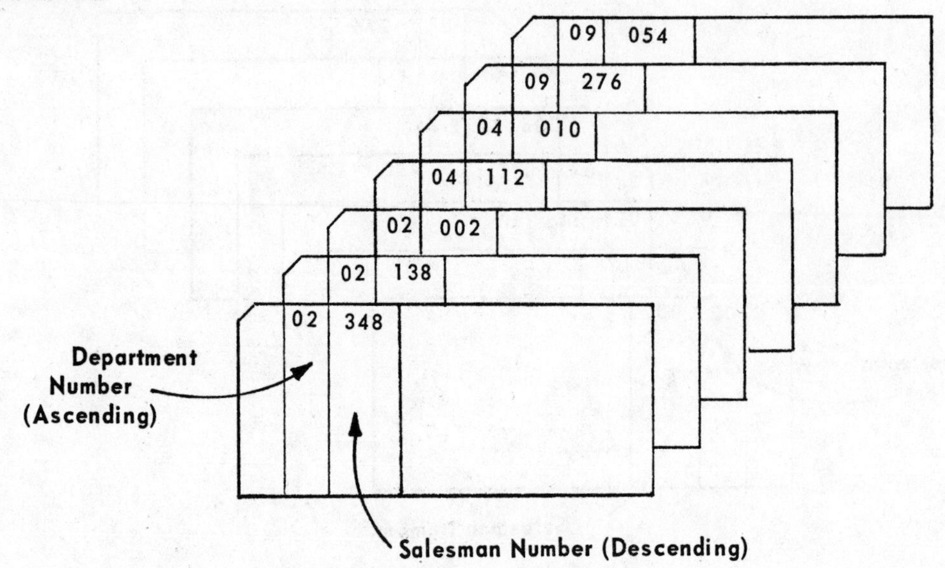

Figure 5-4 Descending Salesman Number within Ascending Department Number

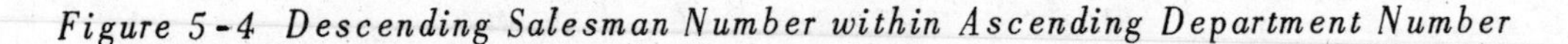

Note in the example above that the department numbers are in an ascending sequence and the salesman numbers are in a descending sequence within department number. The sequences which are utilized within a sort are totally dependent upon the application and needs of the programmer or the user. It should be noted also that the number of fields which are to be sorted in a record may vary also. The previous examples illustrated sorting on one or two control fields. As many as 12 fields may be used as control fields in order to place the records in the proper sequence.

DOS TAPE AND SORT PROGRAM

The sorting of data has been accomplished through many means during the past decade. Machines, such as the IBM 83 Sorter, were developed which did nothing but sort cards. Programs were written which would sort unique files or which were general enough to sort many different files. The Tape and Disk Sort Program (Program Number 360N-SM-483) is a program which is a part of the Disk Operating System intended for use in sorting all files which must be in a prescribed sequence for processing.

As noted previously, sorting is performed on fields within a record in order to arrange the records in a specific sequence. The fields which are used within the record are called CONTROL FIELDS. The sort program places the records in the prescribed sequence based upon the values in the control fields. The sort program can process one to twelve control fields within a record. The total number of bytes in all of the control fields combined must not exceed 256 bytes. Each individual control field may be any length desired within the 256 byte limit. In the examples presented previously in this chapter, it can be seen that the salesman number was a 3 byte control field and the department number was a 2 byte control field. A CONTROL WORD is the sum of all of the control fields.

Each of the control fields within a record may be sorted in an ascending or descending sequence, based upon the needs of the user. The sort program requires that each of the control fields always be located in the same relative position in each record, that is, if a control field is to begin in column 7 and be 4 bytes in length, all records will be sorted on that control field.

The input file for the sort program may be stored on either magnetic tape or on a direct-access device, such as a 2311 disk unit. It must be a sequential file which cannot be organized as a split-cylinder file. Up to nine separate input files may be read by the sort program and be treated as one "logical" file for sorting purposes. The records in the input files may be blocked or unblocked, and fixed- or variable-length. The output file which is created by the sort program contains all of the records which were read into it from the input files in a prescribed sequence. The output file may be stored on tape or a direct-access device and the records may be blocked or unblocked. The output records are in the same format, that is, fixed- or variable-length, as the input records

In addition to the input and output files, the sort utilizes a "work file" for processing the records during the actual sorting. This work file may be stored on magnetic tape or on a disk unit.

The DOS Sort program is a generalized program which is designed to sort various types of data dependent upon the needs of the user. Each execution of the sort program is "tailored" to the needs of the user through the use of control cards which specify the particular circumstances which are to be found in the problem. The example in Figure 5-5 illustrates a job stream which could be used to sort the cards which were used in Chapter 3 and Chapter 4 into an ascending sequence by salesman number (card columns 5, 6, and 7).

EXAMPLE

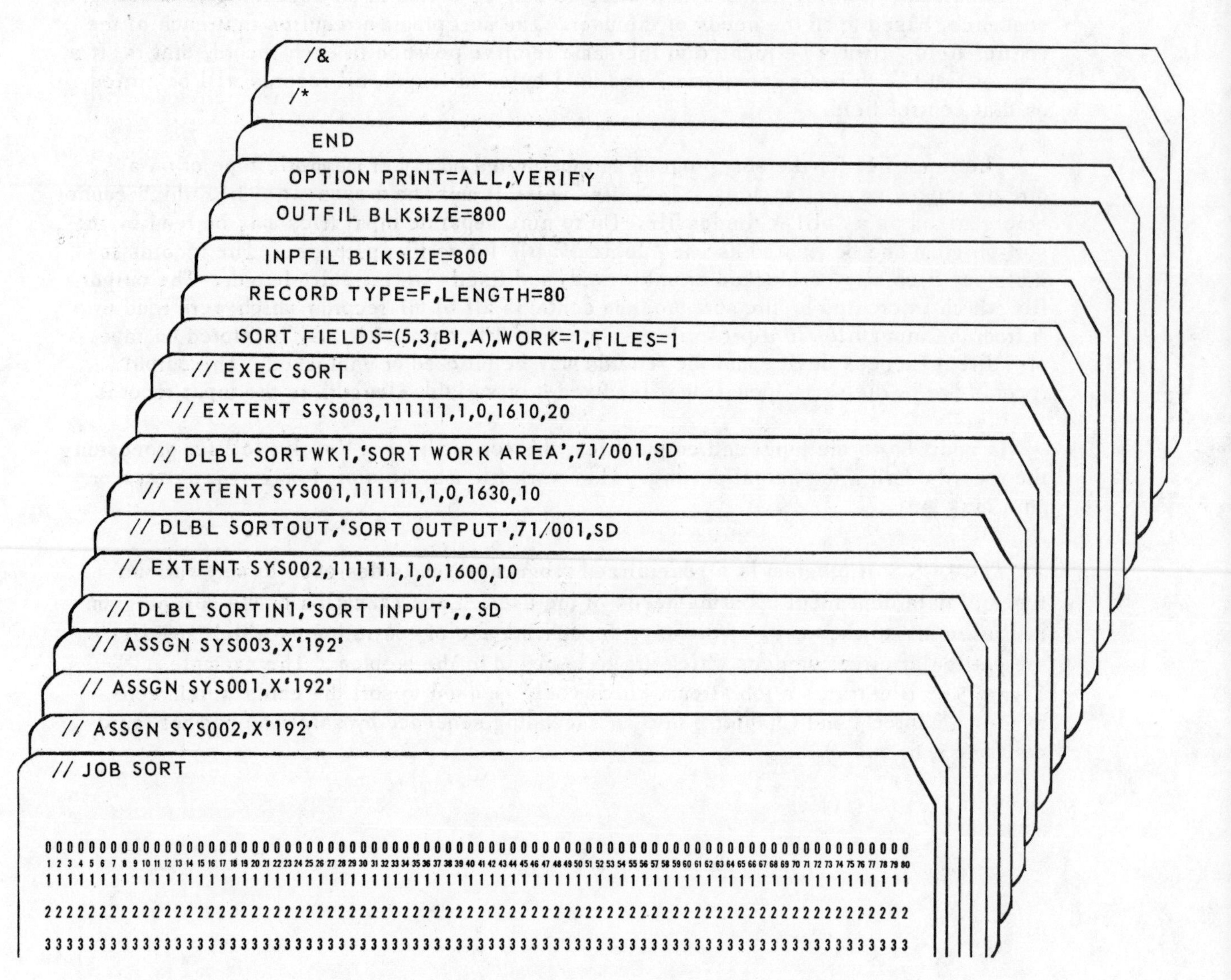

Figure 5-5 Job Stream for DOS Sort Program

The job stream above will sort the input file which is stored on a disk pack into an ascending sequence by salesman number. As can be seen from the example above, unique entries must be made in the job control statements which are utilized with the sort program and sort control cards must be included which are read by the sort program and which indicate the type of processing which is to occur. The job control statements and sort control statements in the above job stream are explained on the following pages.

Sort Input Files

The input file to the sort program in Figure 5-5 is stored on a 2311 disk file. Thus, as with other programs which are executed under DOS control, the device must be associated with a symbolic-device address and label information must be included in the job stream to identify the file on the disk pack. When the sort program is utilized, standard assignments must always be made and the filename entry in the DLBL card must be a standard name. The ASSGN, DLBL, and EXTENT cards used for the input file are illustrated below.

EXAMPLE

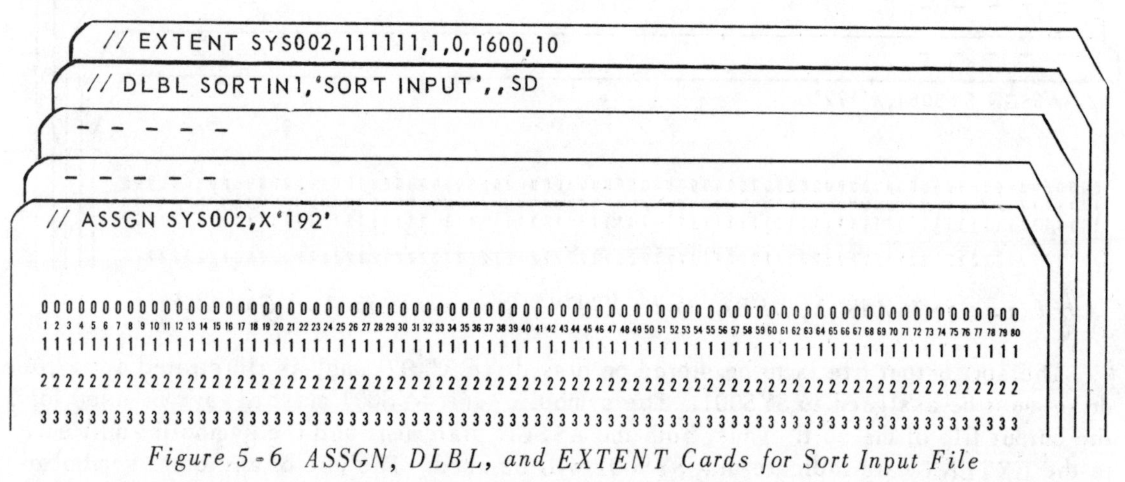

Figure 5-6 ASSGN, DLBL, and EXTENT Cards for Sort Input File

Note in the example above that SYS002 is assigned to disk drive X'192'. The first or only input file must ALWAYS be stored on the device assigned to SYS002. Note that this requirement differs from that of the Utility programs where a disk file may be assigned to any programmer logical unit available to the programmer. The disk input device for the sort program must be SYS002.

Note also that the filename entry in the DLBL card for the input file is SORTIN1. The filename for the first or only input file to the sort program must be SORTIN1. As with other programs executed under DOS, it is through this name that the file in the program is related to the label for the file on the disk, and therefore, this name must always be used. The file-ID, which in the above example is 'SORT INPUT', can be any name of the programmer's choosing and is not related to the sort program. The input file is a sequential file and all files which are input to the sort program are sequential files. Direct-access files, indexed sequential files, or split-cylinder sequential files may not be used as input or output with the sort program.

The EXTENT card is used with the sort program to define the extents of the disk file. In the example above, the symbolic-unit name used is SYS002, which must always be used for the first or only sort input file. The remaining entries in the EXTENT card are the same entries which are normally used for any files defined with the EXTENT statement. In the example in Figure 5-6, the disk pack used contains the serial number 111111 and the file is 10 tracks in length beginning on Cylinder 160, Track 0 (relative track 1600).

Sort Output Files

The sort output file in the job stream illustrated in Figure 5-5 will, like the input, be stored on a 2311 disk drive. The ASSGN, DLBL, and EXTENT statements for the output file are illustrated in Figure 5-7.

EXAMPLE

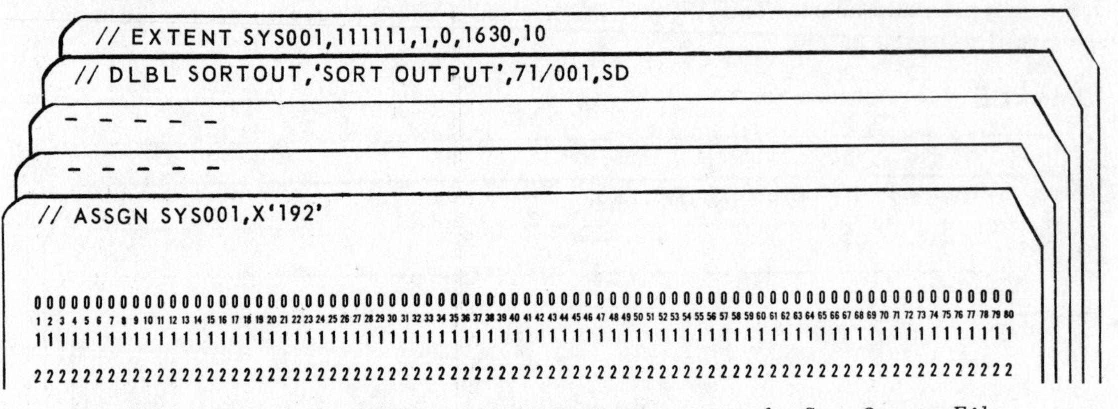

Figure 5-7 ASSGN, DLBL, and EXTENT Statements for Sort Output File

The sort output file is to be stored on disk drive X'192' and, as illustrated above, the drive must be assigned to SYS001. The symbolic-unit SYS001 must always be used for the output file of the sort. Thus, both the ASSGN statement and the symbolic-unit entry in the EXTENT card indicate that SYS001 will be used. The use of any other symbolic unit will cause an error when the output file is opened.

The filename entry in the DLBL card for the output file MUST contain the entry SORTOUT. This filename is used in the sort program for the output file and must be included in the DLBL card for the output file so that when the output file is opened, a relationship between the file and the label information stored on the standard label cylinder may be established. The file-ID, 'SORT OUTPUT', is not related to the sort program and may be any 44 byte or less name desired by the programmer. The dating information is determined by the programmer and the file must be a sequential file. The standard input and output files for the sort program must be sequential files which are not stored in a split-cylinder format.

The remaining information in the EXTENT card for the output file contains whatever information is desired by the programmer. Note, however, that the type field in the Extent card must contain the value "1" indicating that a standard sequential file is being defined.

Sort Work Files

When the sort program processes the input data and places it in the prescribed sequence, a work file is necessary for the intermediate steps which are done by the sort program. This work area, as with the input and output files, may be either on disk or tape. The ASSGN, DLBL, and EXTENT cards illustrated in Figure 5-8 are used in Figure 5-5 to define the work area, which is stored on disk, to be used for the sort.

EXAMPLE

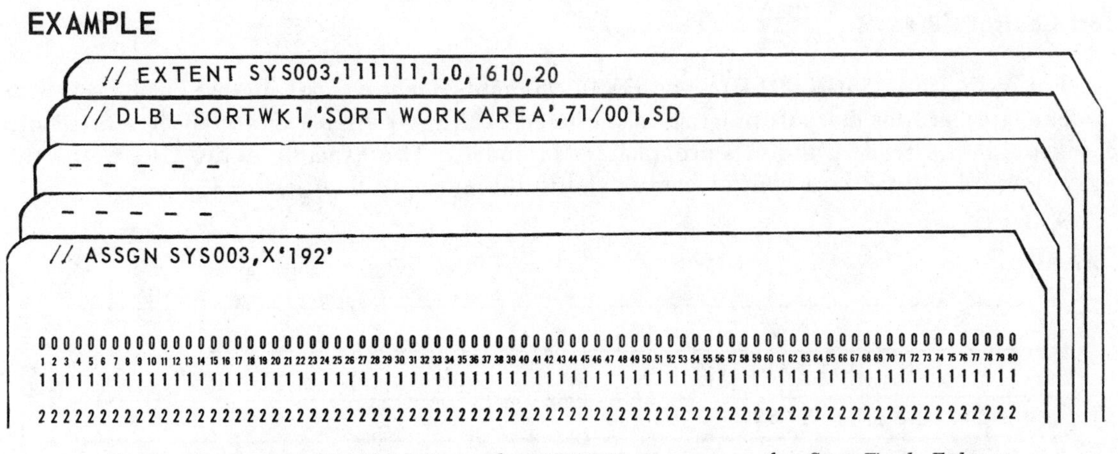

Figure 5-8 ASSGN, DLBL, and EXTENT Statement for Sort Work File

Note in the above example that the work file will be stored on drive X'192' which is assigned to symbolic-unit SYS003. The "SYS" number which is used for the work file(s) must be one greater than the last number used for the input file(s). Thus, as can be seen from Figure 5-6, the "SYS" number used for the input file is SYS002. Therefore, SYS003 must be used for the first or only work file to be used for the sort.

The DLBL card which is used to define the work file must have the filename SORTWK1 for the first or, in this example, the only work file. Again, this is the filename used in the sort program and must be specified on the DLBL card in order to identify the labels to be used for the work files. The file-ID, as with the other files used in the sort program, is of the programmer's choosing. Note that the expiration data on the work file is 71/001. It is normally not necessary that the sort work area be saved after the sort processing has been completed. Therefore, with very few exceptions, a date greater than the current date will not be specified. The sort work file(s) are sequential.

The EXTENT statement must contain the same symbolic-unit entry as is assigned to the device containing the disk pack. In the example above, SYS003 is used because it is the SYS number which is to be used for the first work file. The number of tracks to be allocated for the work area is very important for the sort program. There must be enough room in the work files to store twice the number of records which are stored in the input file. Thus, in the example in Figure 5-5, since the number of input tracks specified in the EXTENT card is 10, the work area must be at least 20 tracks in length, as illustrated in Figure 5-8. Note again that the area available in the work file area must be able to hold twice the number of records which can be stored in the input area. Thus, the tracks requested for the work area must be twice the number of tracks requested for the input area.

Sort Control Cards

Following the ASSGN, DLBL, and EXTENT cards for the input, output, and work files, the execute card for the sort program is included in the job stream and then the sort control cards which are read by the sort program are included. The example below illustrates the Execute card and the sort control cards used in the example in Figure 5-5.

EXAMPLE

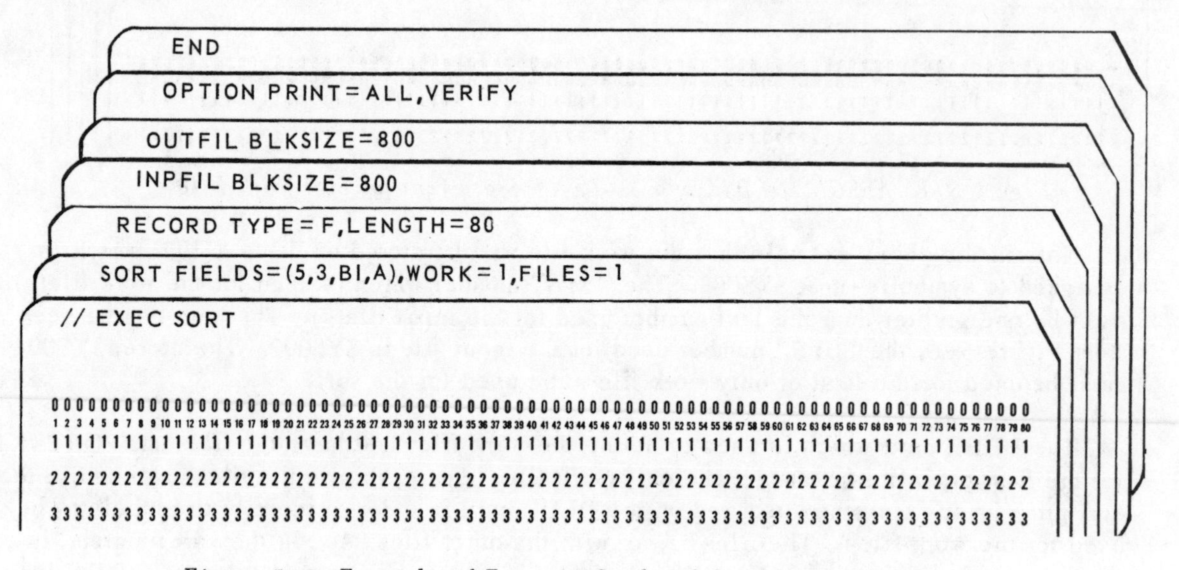

Figure 5-9 Example of Execute Card and Sort Control Cards

Note in the above example that the name of the sort program which is stored in the core image library is SORT and that this name must be indicated on the execute card in order that the sort program be loaded into core storage for execution. After the program is loaded and receives control, it reads the sort control cards which immediately follow the execute card in the job stream. These control cards are used to submit the unique specifications for the particular application so that the generalized sort program can be "tailored" to the needs of the user.

The SORT control card is the first control card which is to be read by the sort program. The general format of the sort control card is illustrated in Figure 5-10.

SORT FIELDS=(p_1,m_1,f_1,s_1,p_2,m_2,f_2,s_2, . . . ,p_{12},m_{12},f_{12},s_{12}),WORK=n[,SIZE=n][,CHKPT][,FILES=n]

Figure 5-10 General format of SORT Control Card

The SORT Control card is used to specify the fields which are to be used in the input records as control fields and also the number of work files which are to be used by the sort program. The word SORT must be written on the card as shown and must begin to the right of column 1, that is, it cannot begin in column 1 of the card. One or more blanks must follow the word SORT.

The "fields" entry is used to specify which fields in the input record are to be the sort control fields. The values "FIELDS=(" must be included exactly as shown in Figure 5-10. The entries within the parentheses are used to indicate the fields to be used. The "p" entry specifies the beginning location of the field to be sorted, that is, the high-order byte, relative to 1, of the field to be sorted. In the example in Figure 5-9 it can be seen that the "p" entry is 5, which indicates that the field to be sorted begins in column 5 of the input record. The "m" entry is used to specify the length of the control field. In the example in Figure 5-9, the value 3 is used because the field to be sorted, the salesman field, is 3 bytes in length.

The "f" entry is used to specify the format of the data in the field to be sorted. The valid entries in the "f" field are BI for an unsigned binary sort, CH for an unsigned binary (character) field, ZD for a signed zoned-decimal field, PD for a signed packed-decimal field, FI for a signed fixed-integer field, and FL for a signed floating-point field. Note in the example in Figure 5-9 that the entry BI is specified for the salesman number field. The BI indicates that a "logical" comparison is to be performed on the field and the value which is greater in the EBCDIC collating sequence on the System/360 will be considered the greater value. This is illustrated in the following example.

EXAMPLE

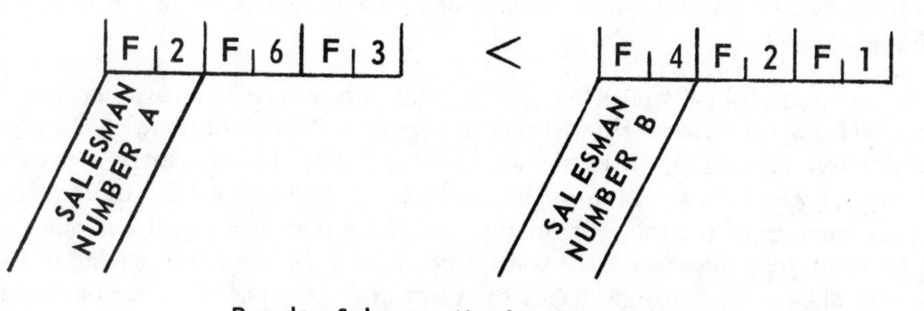

Result: Salesman Number B is considered

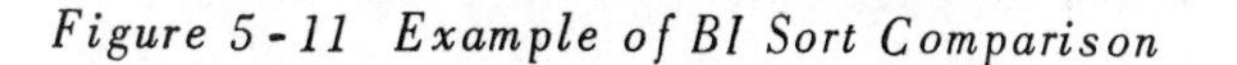

Figure 5-11 Example of BI Sort Comparison

In the example in Figure 5-11, the value in the Salesman Number B field is considered larger than the value in the Salesman Number A field because it has a higher value in the collating sequence on the System/360. Note again that when the BI format is specified, a sign has no meaning. Thus, as illustrated in the following example, a negative field could still be considered greater than a positive field.

EXAMPLE

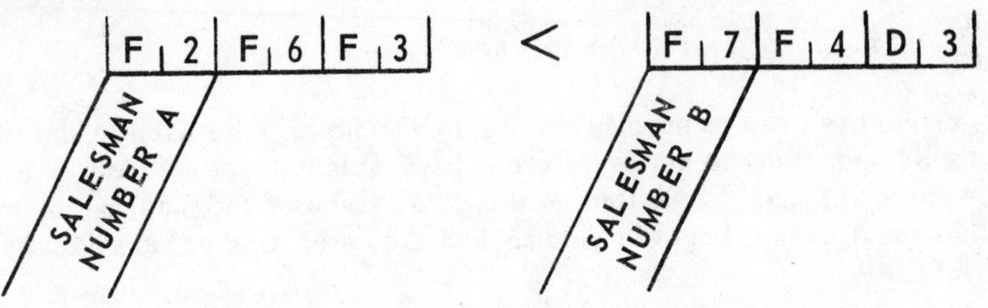

Figure 5-12 Example of Negative Number

Note in the example above that the field Salesman Number B contains a negative sign in the low-order byte which indicates that the value in the field is to be considered negative. When the BI format is specified in the SORT control card, however, the value in Salesman Number B would be considered a greater number than the value in Salesman Number A because the hexadecimal value X'F7' is greater than the hexadecimal value X'F2'. In order for the negative sign to be considered by the sort program, the field must be specified as a ZD (zoned-decimal) field. When zoned-decimal is indicated on the SORT control card, the sign of the field is considered when the sorting of the record takes place. If ZD were specified for the two fields illustrated in Figure 5-12, the value in the field Salesman Number A would be considered greater than the value in the field Salesman Number B.

The entry WORK=n in the Sort control card is used to specify the number of files which are to be used as work areas for the sort program. As noted previously, the sort program utilizes work areas to sort and process the data which is read from the input file before writing it on the output file. The "n" value may be any number 1-8 if disk work areas are to be used or the values 3-9 if tape work areas are to be used (note that the minimum number of tape work areas is 3). In the example in Figure 5-9 it can be seen that one work area is specified in the SORT control card and this corresponds to the job control statements which include one DLBL and one EXTENT card for a work area. In many applications, the use of more than one work area will increase the efficiency of the sort program and the same amount of input data will be sorted in a shorter period of time.

The "FIELDS" entry and the "WORK" entry are the only required entries on the SORT control card. The other entries illustrated in Figure 5-10 are optional and may be included if necessary. The SIZE=n entry is used to specify the number or approximate number of records which will be contained in the input file. The value "n" is assumed to be an estimate of the number of records but for the most efficient sorting, the number specified should be relatively accurate. If this optional entry is used, the estimate for the number of records should be too high rather than not high enough. If this operand is not used on the SORT control card, the sort program will be able to process as many records as can be processed based upon the size of the core storage area available to it and the amount of space which is reserved on the work areas.

The CHKPT entry is used if it is desired that the sort program write checkpoint records while the sort is being processed. Checkpoint records are normally used when the number of records being sorted is quite large. Checkpoint records allow the sort to be restarted at a point other than at the beginning if the processing of the sort is interrupted for some reason, for example, a power failure on the computer.

The FILES=n entry in the SORT control card is used to indicate the number of files which will be input to the sort program. A maximum of 9 separate files may be input to the sort. If this optional entry is omitted, the sort program assumes that there will be one input file. Note that the entry in the SORT control card in Figure 5-9 specifies that one input file is to be processed by the sort program.

The SORT control card is used to specify the information to the sort program which indicates the type of sorting to be performed and the resources that the program has to sort the data. The RECORD control card is used to specify information concerning the input and output records. The general format of the RECORD control card used with fixed-length records is illustrated below.

RECORD TYPE= F ,LENGTH=l_1

Figure 5-13 General Format of RECORD Control Card

In the general format illustrated above, note that the word RECORD must be included to identify the type of control card which is to be processed. The word RECORD must begin to the right of column 1, as must the keyword in all sort control cards. One or more blanks must follow the identifier.

The "TYPE" entry in the RECORD control card is used to specify the format of the input and output records, that is, whether they are fixed-length or variable-length. If the records to be processed by the sort are fixed-length, the entry TYPE=F must be included on the RECORD control card. Note in Figure 5-9 that the records to be processed in the example are fixed-length records. The LENGTH entry is used to specify the length of each logical record which is input to the sort and which is to be output from the sort. The value for l_1 must be the length of the input and output records. In the example in Figure 5-9 note that the length specified is 80, which indicates that the length of each logical record to be processed is 80 bytes.

The INPFIL control statement is used by the sort program to specify the characteristics of the input file. The general format of the INPFIL control statement for disk input is illustrated in Figure 5-14.

Figure 5-14 General Format of INPFIL Control Statement

The statement identifier INPFIL must be included on the card beginning to the right of column 1. It must be followed by one or more blanks. The BLKSIZE=n entry is used to specify the blocksize of the input file. Note that the logical record length is specified on the RECORD control card. The blocksize, or physical record size, must be specified in the "n" operand of this entry. In the example in Figure 5-9, the blocksize of the input file is specified to be 800 bytes. Thus, since the input logical record length is 80, it can be seen that the input file is blocked 10 records per block.

The remaining entries in the INPFIL control statement are optional and if they are not specified, the sort program will utilize default values. The BYPASS entry can be made in the INPFIL control statement if it is desired to bypass any incorrectly read input records or blocks or any wrong-length records which may be contained in the input file. If this operand is included, a brief message is written on SYSLST indicating that an error has occured and the sort continues processing. If it is omitted, a message is written on SYSLST and the sort processing is terminated. In most instances, it is not desirable to sort files which cause I/O errors or which contain wrong-length records. If, however, this is permissable, the BYPASS option may be used to allow the sort to continue processing.

The VOLUME entry is used in the INPFIL statement to indicate the number of volumes on which the input files are stored. If only one input file is to be processed, the volume entry is specified in the format VOLUME=n, where n is the number of volumes on which the single file is stored. If more than one input file is to be sorted, the fomat used is VOLUME=(n_1,n_2, . . .), where each value of "n" specifies the number of volumes for the first, second, etc. input files. For example, the entry VOLUME=(3,2,2) would indicate that the first input file, as identified by the SORTIN1 entry in the label card for the file, would be stored on 3 volumes, the second input file, which would have the filename entry SORTIN2 in its label card, would be stored on 2 volumes, and the third input file would be stored on 2 volumes. It should be noted that if more than one input file is to be processed by the sort program, this must be indicated in the FILES operand in the SORT control card. Thus, for the above example, the entry FILES=3 would be entered in the SORT control card. If the VOLUME operand is omitted from the INPFIL statement, such as in the example in Figure 5-9, the sort program assumes that the input file(s) is stored on a single volume.

The OUTFIL control card is used to specify the characteristics of the output file. When the output file is to be stored on disk, the general format illustrated below is used.

```
OUTFIL BLKSIZE=n
```

Figure 5-15 General Format of OUTFIL Control Card for Disk Output

As with all control cards used with the sort program, the identifier OUTFIL must begin to the right of column 1 and must be spelled as shown. The only entry which is used with the OUTFIL control card when a disk output file is to be created is the BLKSIZE=n entry, where "n" is the size of the block, or physical record, which is to be written by the sort program. Note that the logical record size of the output record is specified in the RECORD control card and must be the same size as the input record unless special processing is performed by a user routine (see the SORT EXITS section in Chapter 6). In Figure 5-9, it can be seen that the output block is to contain 800 bytes.

The OPTION card is used to specify various options which may be incorporated into the sort program for special processing. The general fomat of the Option Statement is illustrated in Figure 5-16.

```
         ┌PRINT=NONE    ┐
OPTION   │PRINT=ALL     │  [,STORAGE=n][,LABEL=(output,input1, . . . ,inputn,work)]
         └PRINT=CRITICAL┘

          [,VERIFY]
```

Figure 5-16 General format of OPTION Control Card

As noted, the Option card is used for special options which may be required or desired for the sort application. It is not required for the operation of the sort. If it is used, it must be included in the control cards as illustrated in Figure 5-9. The key-word OPTION, as with the other sort control cards, must begin to the right of column 1 in the card and be spelled as shown above. One or more blanks must follow the keyword.

The PRINT option is used to determine which messages are to be printed on the device assigned to SYSLST during the portion of the sort program which determines the processing to be accomplished for the particular application. As can be seen from Figure 5-16, one of three options may be selected. The PRINT=NONE option indicates that no messages are to be printed on SYSLST. The PRINT=CRITICAL option specifies that only those messages which are critical to the operation of the sort are to be printed. These messages include error messages that result from conditions which can cause program termination. The PRINT=ALL option indicates that the sort program should print all messages which are generated on the printer. These messages include error and end-of-job messages, control card information, various size calculations pertaining to the size of input, output, and work files, and other informative messages. If the PRINT entry is not included within the sort control cards, the option PRINT=ALL is assumed by the sort program.

The STORAGE=n entry is used on the OPTION card if it is necessary to specify the size of the area in core storage which is available to the sort program. This optional operand is used if it is necessary to restrict the size of core available to the sort to less than the size of the partition in which the sort will be executed. This may be necessary if the sort is to be executed together with another program in the same partition. If this operand is omitted, the sort program assumes that it may use the entire core storage area allocated to the partition in which the sort is to be executed.

The LABEL operand is used to specify the type of labels which are to be used for the input, the output, and the work files. The values which may be placed within the parentheses are S for standard labels, N for nonstandard labels, or U for unlabelled files. The output file is always specified first, followed by the input files, which are followed by the work files. The example in Figure 5-17 illustrates the use of the label operand.

EXAMPLE

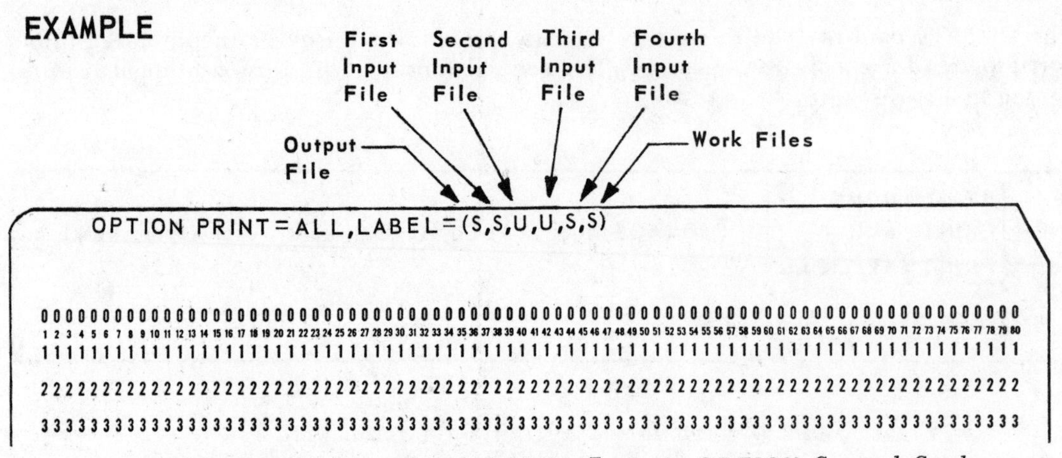

Figure 5-17 Example of LABEL Entry in OPTION Control Card

Note in the above example that the output file is specified first in the list and that it is to have standard labels (entry is "S"). The entries for the input files follow separated only by a comma. The first input file contains standard DOS labels, the second and third are unlabelled files, and the fourth input file contains standard labels. The last entry in the list is for all of the work files. Note that all work files used with the sort program must contain the same type of labels. It should be noted that only tape input and output files can contain labels other than standard DOS labels. If disk input and/or output files are used, these disk files must contain standard labels. If the LABEL operand is omitted from the OPTION Control card, the sort program assumes that all of the labels for the files are standard labels.

The VERIFY option specifies that when a direct-access device is to be used to store the output file, each block which is written on the file will be checked to ensure that it was written correctly. If this operand is omitted, verification will not take place.

The end of the sort control cards is indicated through the use of the END control card. This card must contain the word END beginning to the right of column 1 with no other entries in the card. The END card is a required control card and must be placed following all of the sort control cards.

EXECUTION OF THE SORT PROGRAM

When the job stream in Figure 5-9 is executed, the data records in the input file will be read by the sort program and the output file will consist of sorted records which will be in an ascending sequence based upon the value in the Salesman Number field. The following example illustrates the detailed steps which take place when the sort program is executed.

EXAMPLE

Step 1: The job control program reads the control cards until the sort program is to be loaded into core storage.

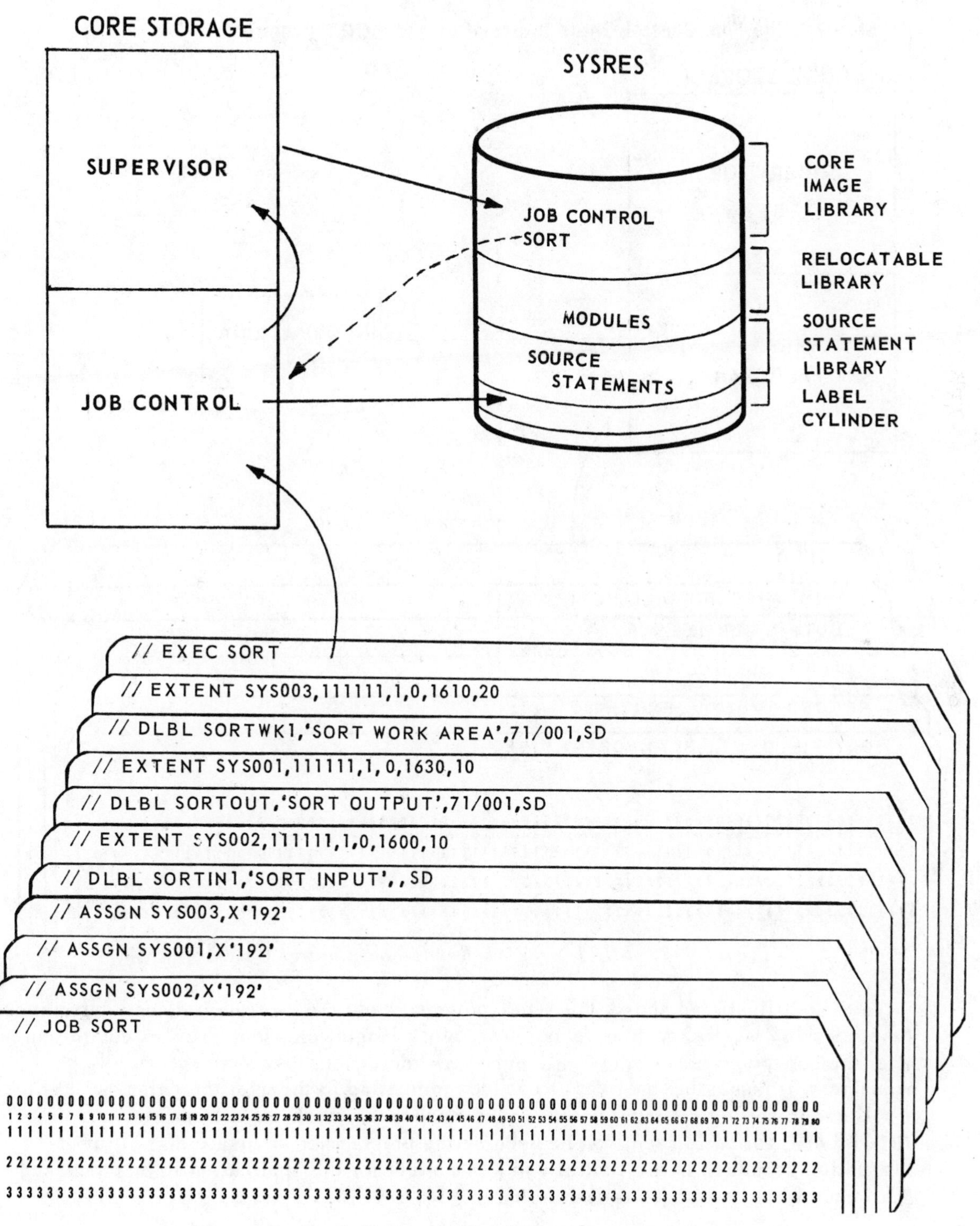

Figure 5-18 Job Control Cards for SORT Program

Note from Figure 5-18 that the job control cards for the sort are read by the job control program and that the labels for the input, output, and work files are stored on the label cylinder of the SYSRES disk pack to be used when the files are opened by the Sort program. When the execute card is read, the SORT program is loaded into core storage by the Supervisor.

Step 2: The Sort Control Cards are read by the SORT Program.

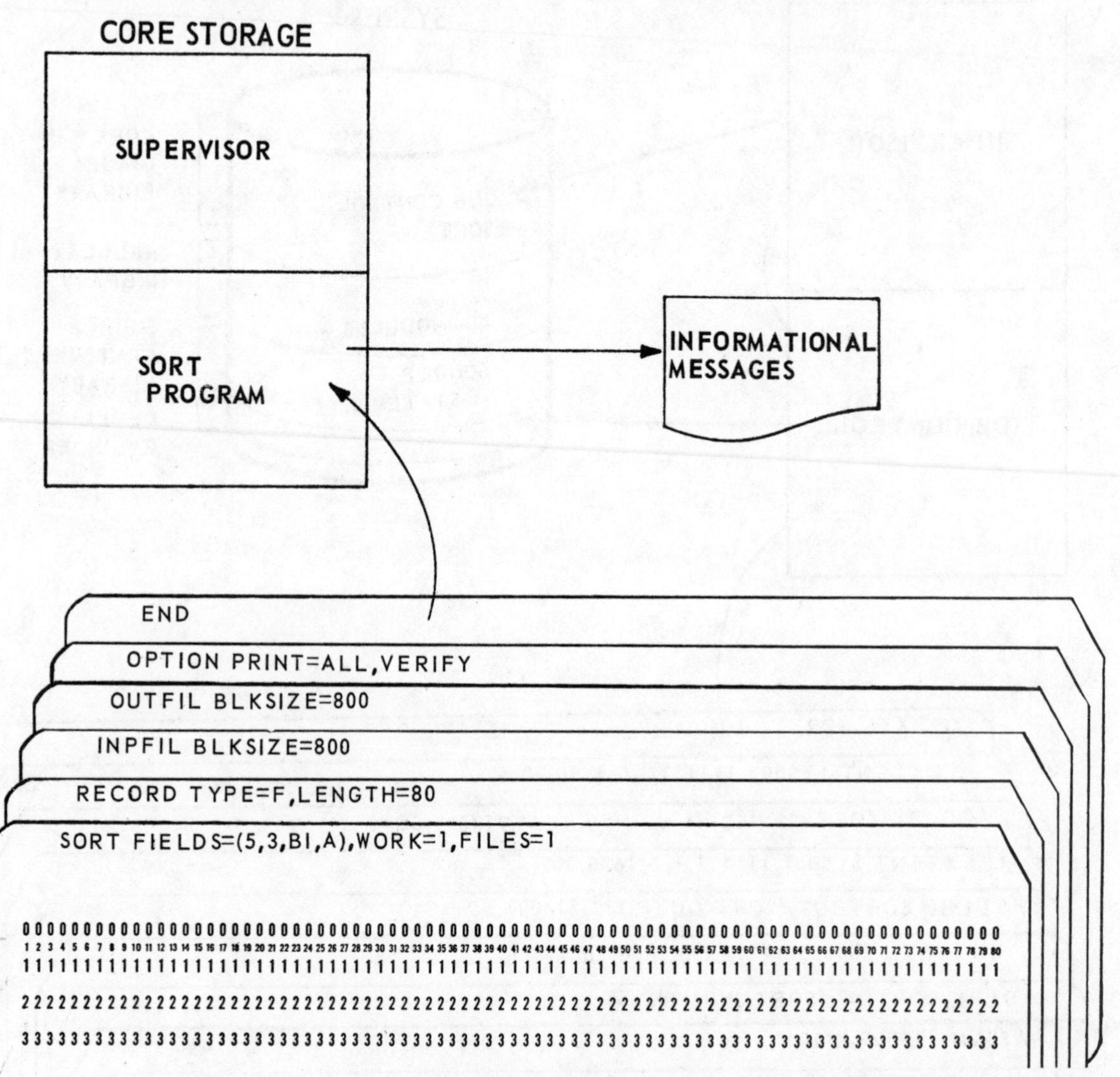

Figure 5-19 SORT Program Reads Sort Control Cards

In Step 2 illustrated above, the SORT program reads the control cards and determines the processing factors, such as record size, block length, etc. from the sort control cards. After the sort program has made the necessary determinations and set all of its parameters, it issues informational messages indicating to the user the parameters which it will use in sorting the input data. These messages, which are only printed when the PRINT=ALL option is used, can be used by the programmer to ensure that all of the information contained in the control cards is correct for the application to be processed. The format of the messages is illustrated in Figure 5-20.

EXAMPLE

```
7000I    SORT FIELDS=(5,3,BI,A),WORK=1,FILES=1
7000I    RECORD TYPE=F,LENGTH=80
7000I     INPFIL BLKSIZE=800
7000I     OUTFIL BLKSIZE=800
7000I     OPTION PRINT=ALL,VERIFY
7000I     END
7050I NMAX = 00000438
7051I B = 00003601
7052I G = 00000409
7001I PHASE 0 END,NO DETECTED ERRORS
```

Figure 5-20 Messages from Sort Program

Note from the example of the messages printed from the Sort program that the sort control cards are printed by the message number 7000I. These messages indicate the parameters requested by the sort control cards. Following these messages are values which are calculated by the sort program based upon the values indicated in the sort control cards. The message 7050I NMAX = 00000438 is used to indicate the estimated maximum number of records which can be sorted based upon the specifications provided in the sort control cards. Thus, as can be seen in the example, the maximum number of records which may be sorted are 438 using the specifications illustrated in the sort control cards.

The message 7051I B = 00003601 indicates the blocking factor used by the sort program for the intermediate work files. This value again depends upon the specifications provided by the sort control cards. The message 7052I G = 00000409 states the number of records that can be contained in the record storage area of the internal sort phase. The record sizes and the amount of core storage available to the sort program are used to determine this value.

After all of the sort control cards have been read by the sort program and have been checked and processed, the message 7001I PHASE 0 END,NO DETECTED ERRORS is printed if all of the control cards are used properly. This indicates to the user that the syntax of the control cards is correct and that the sort should take place properly. If one or more error messages are printed on the report, the sort control cards must be corrected before processing can continue. A complete list of the messages which can be printed from the Sort program is contained in Appendix D.

After the sort control cards have been read and processed by the sort program, the internal sort phase is loaded into core storage to begin actual processing of the files. This is illustrated in Figure 5-21.

Step 3: The SORT Program opens the input file and the work file, reads the INPUT Data, and arranges it on the Work Area.

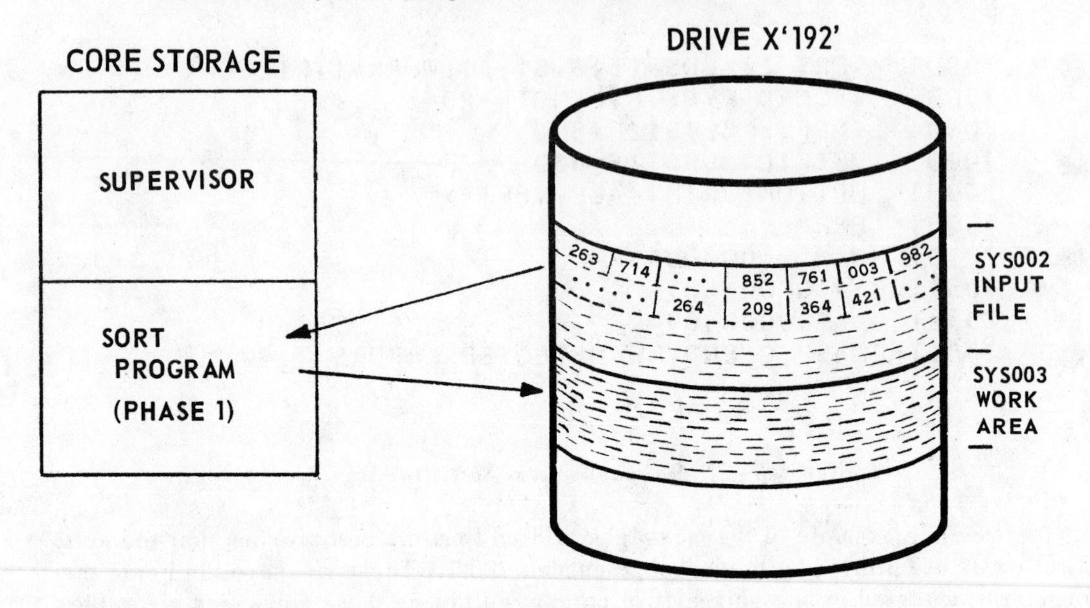

Figure 5-21 SORT Program Reads Input Data and Writes in Work Area

In the example above it can be seen that phase 1 of the SORT program reads the data from the input file after it is opened. The data is read into core storage and is placed in ordered groups of records called SEQUENCES or STRINGS. These Strings are then written on the work file which is used by the sort program. This phase of the sort is called the Internal Sort Phase. Note that the input records are not merged into a final sorted sequence but are merely placed in groups which can be processed by the sort program.

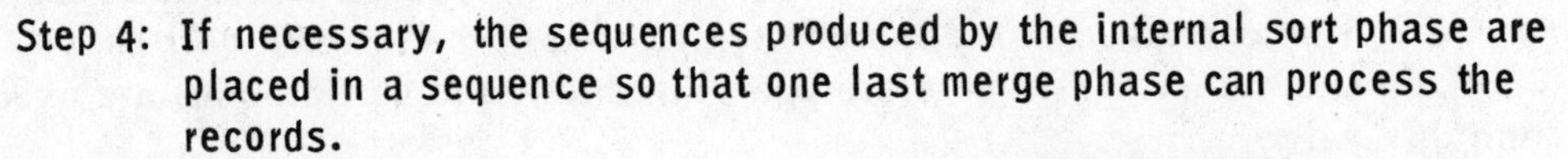

Step 4: If necessary, the sequences produced by the internal sort phase are placed in a sequence so that one last merge phase can process the records.

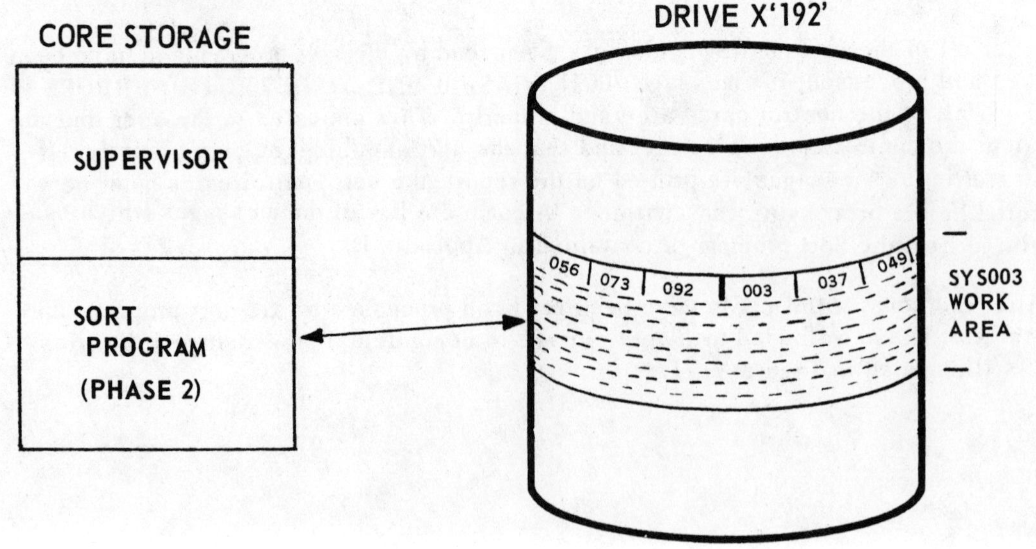

Figure 5-22 Phase 2 of SORT Program

Note from the example in Figure 5-22 that the data stored in the work area by phase 1 of the sort program is processed by phase 2 by reading it into core storage, placing it in the sequence desired, and writing it back onto the work area. This process of doing the sorting of the data will continue for as long as is necessary in order that a single merge of the data will cause it to be in the proper sequence, that is, until the data which is stored in the work area can be merged into the proper sequence. This process may consume a great deal of time or it may be bypassed entirely depending upon the sequence of the data which is written by phase 1 of the sort.

After phase 2, or the external sort phase, of the sort program is complete, the data in the work area is in a sequence which can be merged into a sorted output file. Thus, phase 3, or the final merge phase of the sort, is executed.

Step 5: The data in the work area is merged into the sorted output file.

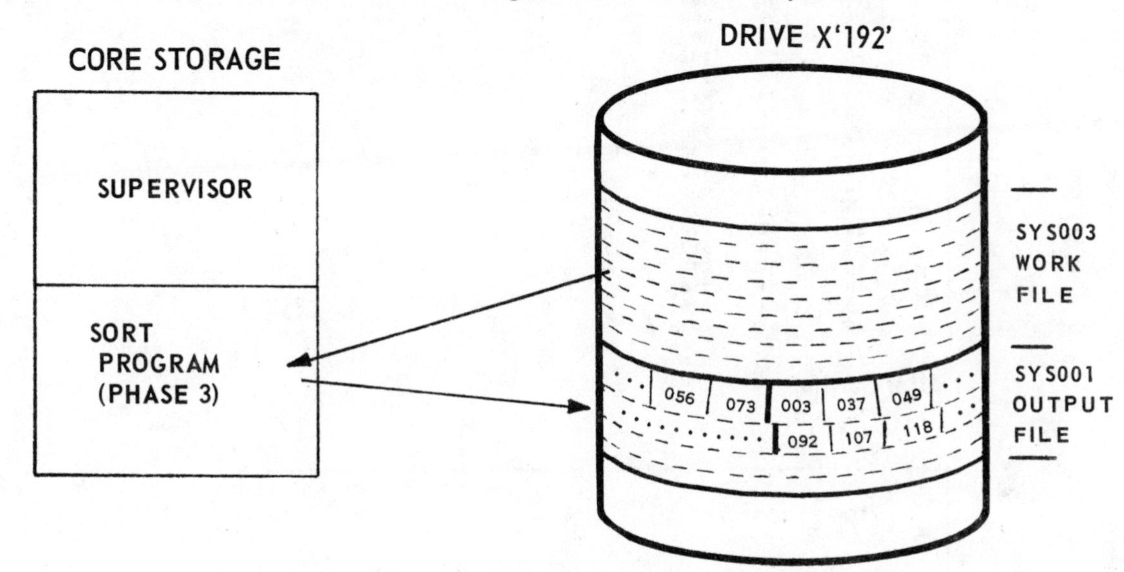

Figure 5-23 Final Merge Phase of Sort Program

In the final merge phase of the sort program, the records which are grouped in a sequence on the work area are read by the sort program and written in the desired sequence in the sort output area. At the conclusion of the final merge phase, the records in the output area are sorted in the sequence specified in the SORT control card. The sorted data is then available for processing by other programs. In the example, the data is sorted on salesman number. The data before and after it is sorted is illustrated in Figure 5-24.

Before Sorting

STORE	DEPT	S/M
20	03	311
10	08	505
20	07	802
20	05	207
20	02	801
20	04	739
10	01	730
20	01	300
10	05	568
10	06	825
10	03	181
20	03	834
10	05	909
10	03	030
10	05	308
20	07	689
10	06	607
20	07	332
10	02	304
10	08	921
10	06	292
20	02	956
20	02	590
10	09	820
20	04	806
10	01	004
20	06	100
20	04	317
10	04	171
10	01	960
10	01	185
20	08	282

After Sorting

STORE	DEPT	S/M
10	01	004
10	04	027
10	03	030
20	06	100
20	08	102
10	09	105
10	02	111
20	05	122
10	04	171
20	06	179
10	03	181
10	01	185
20	05	207
10	07	214
10	09	215
20	08	282
10	06	292
20	01	300
10	02	304
10	05	308
10	07	310
20	03	311
20	09	315
20	04	317
20	03	318
10	08	322
20	01	325
20	07	332
10	06	409
10	03	487
10	08	505
10	05	568
10	09	574

Figure 5-24 Example of data sorted on Salesman Number

As can be seen from Figure 5-24, the data is in no particular order before it has been sorted. The salesman numbers are in a random sequence. After the sort has been performed on the salesman number field, the records are in an ascending sequence by the salesman number. This is the desired result because the records are to be processed based upon the salesman number being in an ascending sequence.

SORTING MORE THAN ONE CONTROL FIELD

As noted earlier in this chapter, it is often desirable to sort records on more than one control field, that is, a major control field and one or more minor control fields. In addition, it may be necessary to sort one field in an ascending sequence and another field in a descending sequence. The following job stream could be used to sort the input records into a sequence where the salesman number is descending within an ascending department number (see Figure 5-4).

EXAMPLE

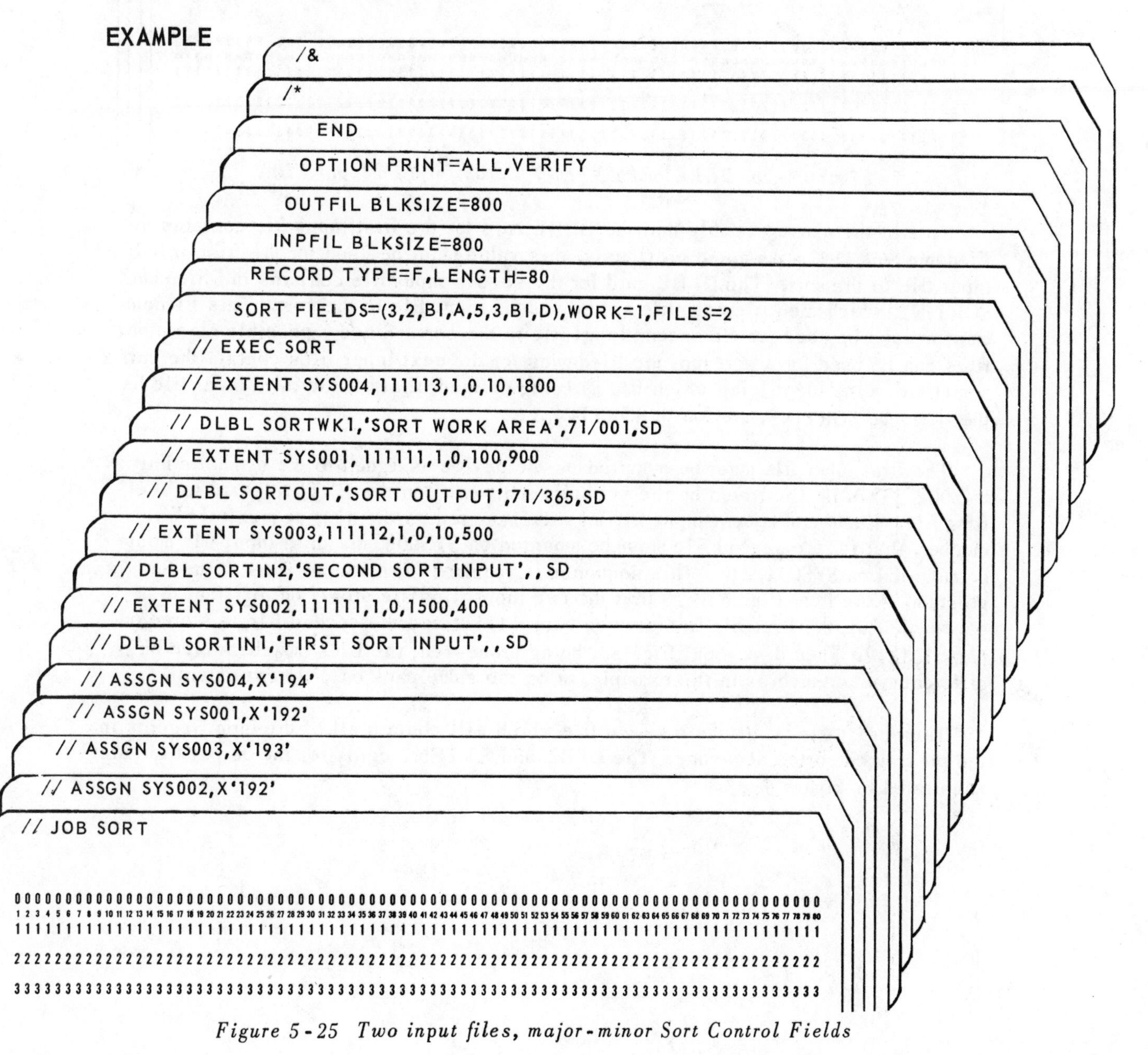

Figure 5-25 Two input files, major-minor Sort Control Fields

In the job stream illustrated in Figure 5-25 it can be seen that two input files are to be sorted by the Sort program and a single, sorted output file is to be created. As noted previously, the Sort program can accept up to 9 input files to be sorted as a single "logical" file and a single output file is created. The DLBL and EXTENT cards for the two input files are again illustrated below.

EXAMPLE

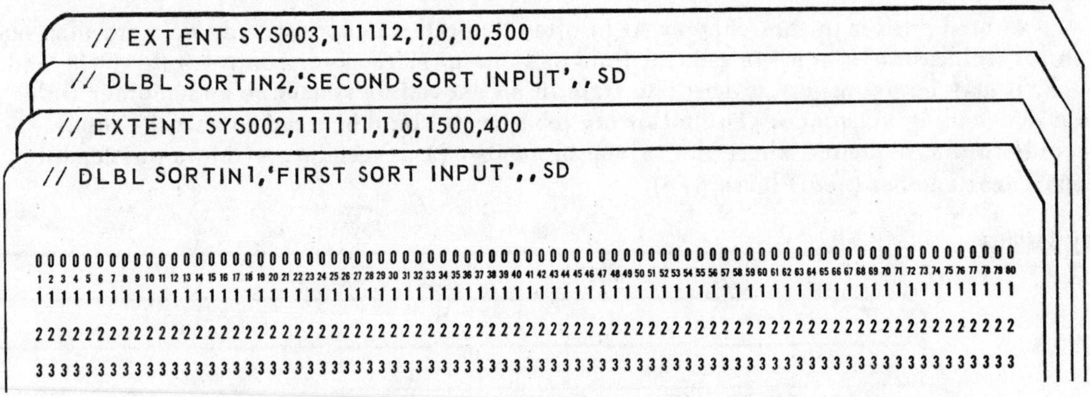

```
// EXTENT SYS003,111112,1,0,10,500
// DLBL SORTIN2,'SECOND SORT INPUT',,SD
// EXTENT SYS002,111111,1,0,1500,400
// DLBL SORTIN1,'FIRST SORT INPUT',,SD

00000000000000000000000000000000000000000000000000000000000000000000000000000000
1 2 3 4 5 6 7 8 9 10 11 12 13 14 15 16 17 18 19 20 21 22 23 24 25 26 27 28 29 30 31 32 33 34 35 36 37 38 39 40 41 42 43 44 45 46 47 48 49 50 51 52 53 54 55 56 57 58 59 60 61 62 63 64 65 66 67 68 69 70 71 72 73 74 75 76 77 78 79 80
11111111111111111111111111111111111111111111111111111111111111111111111111111111

22222222222222222222222222222222222222222222222222222222222222222222222222222222

33333333333333333333333333333333333333333333333333333333333333333333333333333333
```

Figure 5-26 DLBL and EXTENT Cards for Two Input Files

Note in the above example that the DLBL card for the first input file contains the filename SORTIN1. As noted previously, this value must be used for the first or only input file to the sort. The DLBL card for the second input file contains the filename SORTIN2, which identifies the file as the second input file to the sort. This filename must always be used for the second input file to the sort. Since a possible nine input files can be used for a sort run, the filenames for the next input files contain the suffix numerical value identifying which file is being defined, that is, the third input file has the filename SORTIN3, the fourth SORTIN4, etc.

The first input file must be mounted on the device assigned to the symbolic unit SYS002. This is illustrated by the symbolic-unit entry in the EXTENT card for the first file. Each subsequent input file must be mounted on the next sequential SYS number, that is, the second file must be mounted on SYS003, the third input file must be mounted on SYS004, etc. This sequence is required for all input files to the Sort program. Note from Figure 5-26 that the two input files are stored on different disk packs, that is, the first file is stored on pack 111111 and the second file is stored on pack 111112. When disk input files are being processed, the files may be stored on different packs, such as in the example, or on the same pack on different extents.

The output file is always a single file which will contain all of the input records in the prescribed sorted sequence. The DLBL and EXTENT cards for the output file are illustrated in Figure 5-27.

EXAMPLE

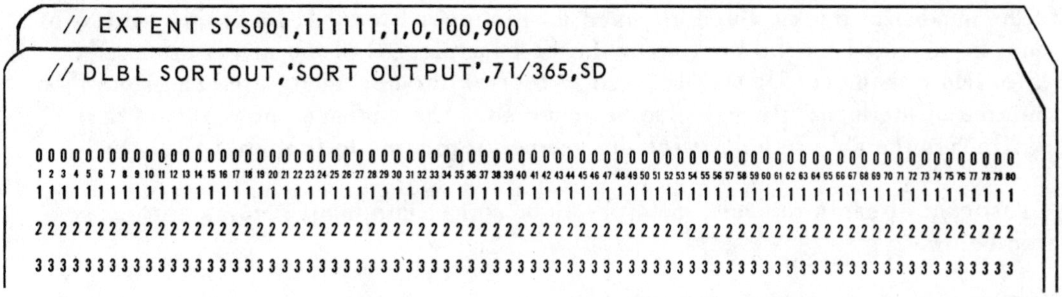

```
// EXTENT SYS001,111111,1,0,100,900
// DLBL SORTOUT,'SORT OUTPUT',71/365,SD

00000000000000000000000000000000000000000000000000000000000000000000000000000000
1 2 3 4 5 6 7 8 9 10 11 12 13 14 15 16 17 18 19 20 21 22 23 24 25 26 27 28 29 30 31 32 33 34 35 36 37 38 39 40 41 42 43 44 45 46 47 48 49 50 51 52 53 54 55 56 57 58 59 60 61 62 63 64 65 66 67 68 69 70 71 72 73 74 75 76 77 78 79 80
11111111111111111111111111111111111111111111111111111111111111111111111111111111
22222222222222222222222222222222222222222222222222222222222222222222222222222222
33333333333333333333333333333333333333333333333333333333333333333333333333333333
```

Figure 5-27 DLBL and EXTENT Cards for Output File

Note from the control cards illustrated in Figure 5-27 that the filename in the DLBL card for the output file must be SORTOUT, which is the name used in the Sort program for the output file. It is always a sequential file, and, as shown by the symbolic-unit entry in the EXTENT card, will always be written on the device assigned to SYS001. In the example above, 900 tracks are reserved for the output file. Note from Figure 5-26 that a total of 900 tracks are used by the input files. The output file must always be large enough to contain all of the records on the input files. In addition, if the blocking factor is changed, that is, if the blocking factor for the output file is less than the blocking factor for the input files, it may be necessary to reserve more space for the output file than is used for the input files.

As noted previously, the work area which is used by the sort program must be able to contain twice the number of records which are stored in the input files. The DLBL and EXTENT cards for the work area are illustrated below.

EXAMPLE

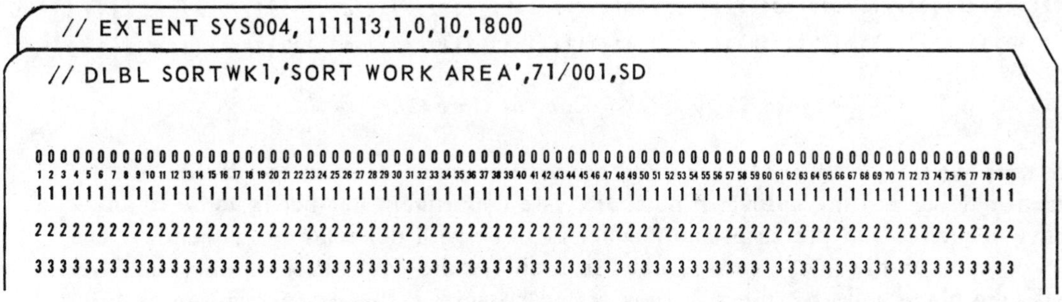

```
// EXTENT SYS004, 111113,1,0,10,1800
// DLBL SORTWK1,'SORT WORK AREA',71/001,SD

00000000000000000000000000000000000000000000000000000000000000000000000000000000
1 2 3 4 5 6 7 8 9 10 11 12 13 14 15 16 17 18 19 20 21 22 23 24 25 26 27 28 29 30 31 32 33 34 35 36 37 38 39 40 41 42 43 44 45 46 47 48 49 50 51 52 53 54 55 56 57 58 59 60 61 62 63 64 65 66 67 68 69 70 71 72 73 74 75 76 77 78 79 80
11111111111111111111111111111111111111111111111111111111111111111111111111111111
22222222222222222222222222222222222222222222222222222222222222222222222222222222
33333333333333333333333333333333333333333333333333333333333333333333333333333333
```

Figure 5-28 DLBL and EXTENT Statements for Work Area

Note that the filename for the work file is SORTWK1. If more than one work file were to be used for this sort, the second file would have the filename SORTWK2, the third would be SORTWK3, etc. A maximum of 8 work files may be defined when the work areas are on disk. The symbolic-unit entry in the EXTENT card specifies that the file will be stored on the device assigned to SYS004. Note that this SYS number is one greater than the last SYS number used for the input files (ie. SYS002 and SYS003 were used for the input files). The symbolic-unit for the work files must always be one greater than the last one used for the input files. If more than one work file is being defined, the SYS numbers must increase by one for each file.

Note also from Figure 5-28 that 1800 tracks are to be used for the work area. This is twice the number of tracks which are used for the input files. At least this amount of space must be reserved for the work area. In the example, all of the area was reserved by one file. More than one file may be used to reserve the necessary area and more than twice the area of the input file may also be reserved. The minimum, however, is that the work file must be able to hold twice the number of records in the input file.

The sort control cards to cause a Major-Minor sort on the input records are illustrated below.

EXAMPLE

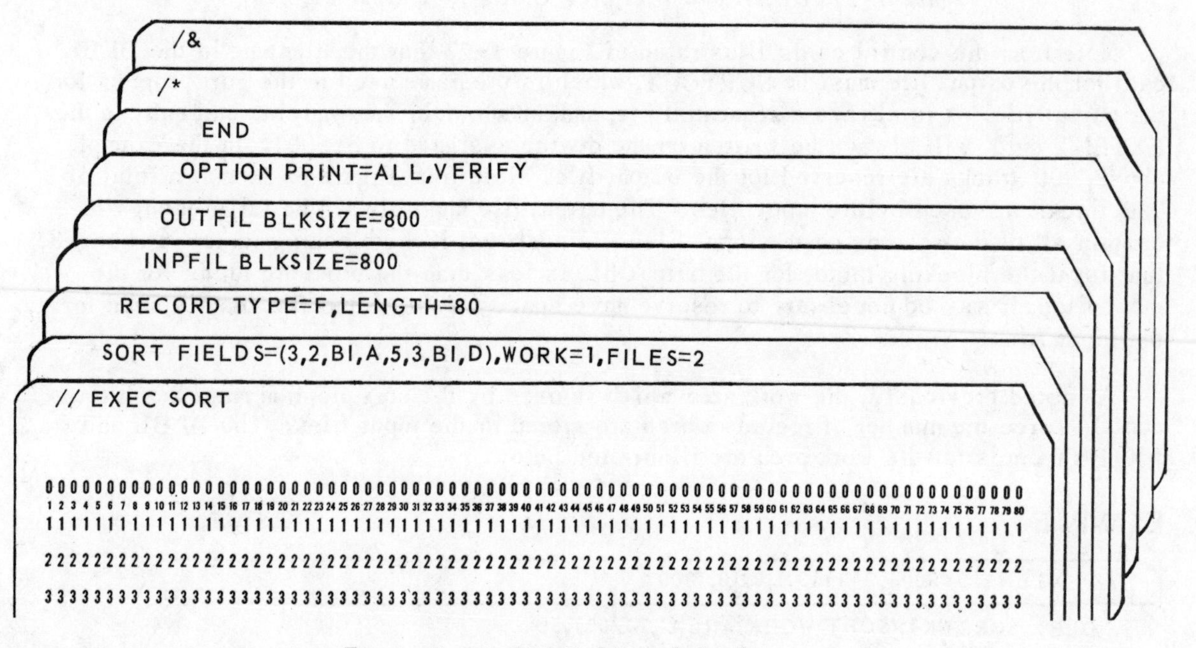

Figure 5-29 SORT Control Cards

The sort control cards above are used to sort two input files on two fields, the department number and the salesman number. The department number is to be in an ascending sequence and the salesman number is to be in a descending sequence. The FIELDS operand on the SORT card is used to indicate the control fields on which the sort will take place and also the way they are to be sorted, that is, ascending or descending. The sort card and the corresponding fields in the records to be sorted are illustrated in Figure 5-30.

EXAMPLE

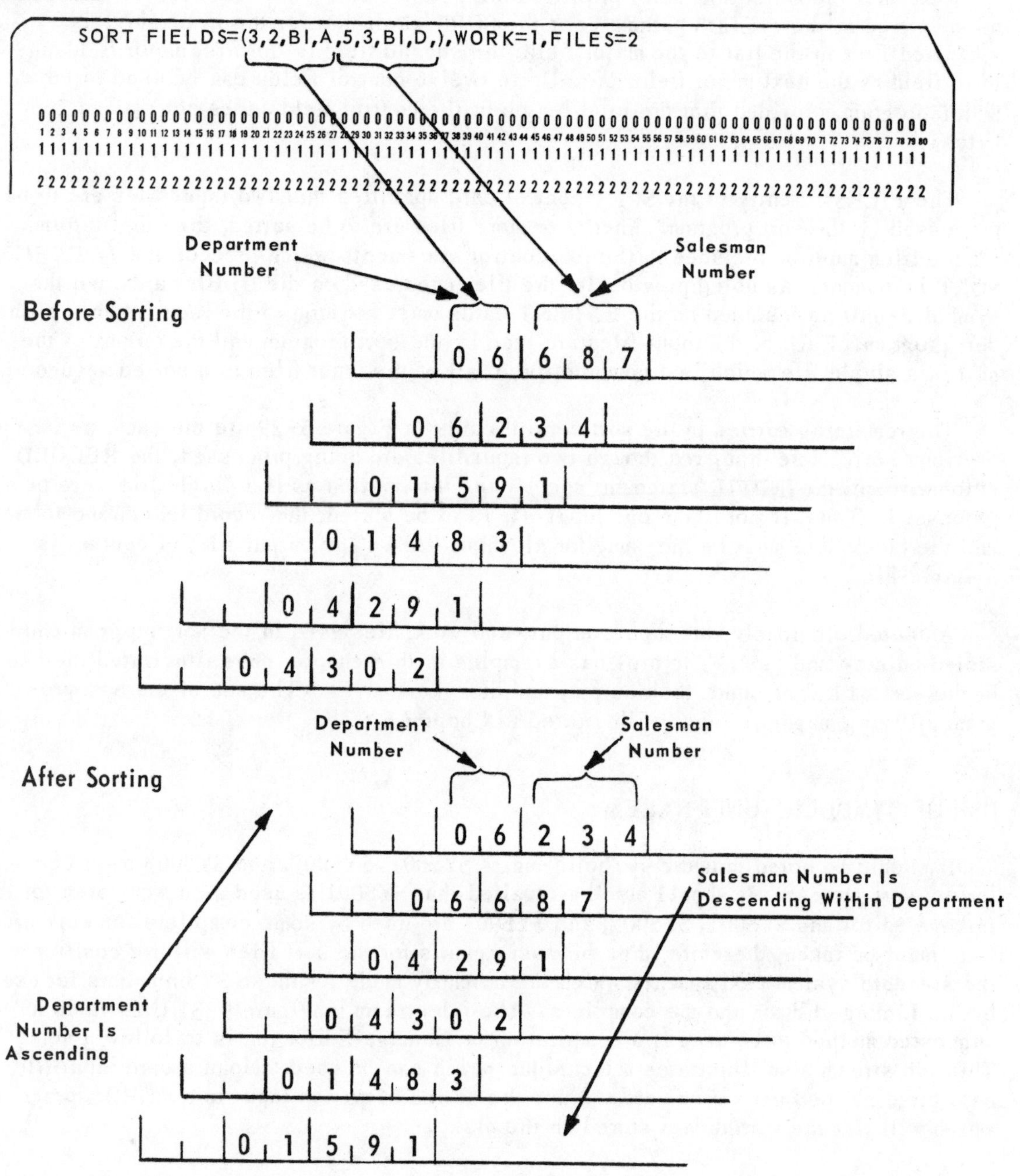

Figure 5-30 SORT Control Card and Records to be Sorted

Note from the example above that the SORT control card specifies that the field which begins in column 3 of the input records (department number) is to be sorted for two bytes in an ascending sequence (indicated by the "A"). The field which begins in column 5 (salesman number) is to be sorted for three bytes in a descending sequence ("D"). Thus, the MAJOR sort is on department number, that is, all of the records will be in an ascending order by department number. The MINOR sort is on salesman number, which means that the salesman numbers will be in a descending sequence within each ascending department number. If more fields were specified in the SORT card, they would be minor to salesman number and the salesman number sequence would have priority over any of the fields minor to it in the sort.

Note that the major and minor priorities are determined by the sequence of the fields as specified in the FIELDS parameter of the SORT card, that is, the field which is specified first in the list is the major field, the second field is the first minor field, the third field is the next minor field, etc. Up to twelve control fields can be used in the SORT program, provided that the total length of the control fields does not exceed 256 bytes.

The FILES=2 entry in the SORT control card specifies that two input files are to be processed by the sort program. When 2 or more files are to be sorted, the label information for the files must be included in the job control statements which precede the // EXEC SORT statement. As noted previously, the filenames used on the DLBL cards and the symbolic-unit names used on the EXTENT cards must conform to the requirements of the sort program. Each of the input files are read by the sort program and the output of the sort is a single file which is a combination of all of the input files in a sorted sequence.

The remaining entries in the sort control cards in Figure 5-29 are the same as for the previous sort. Note that even though two input files are being processed, the RECORD statement and the INPFIL statement specify the information as if a single file were being processed. Thus, if more than one input file is to be sorted, the record length and format and the block size must be the same for all input files. The output file, of course, is only a single file.

As noted previously, the input, output, and work files used in the Sort program can be stored on disk and tape. The previous examples in this chapter have illustrated the use of the sort with disk input, disk output, and disk work areas. The use of the Sort program utilizing magnetic tape is illustrated in Chapter 6.

USE OF SYMBOLIC-UNIT NAMES

It should be noted that the symbolic-units SYS001, SYS002, and SYS003 must be used for the sort program. It should also be recalled that SYS001 is used as a work area for the linkage editor and SYS001, SYS002, and SYS003 are used by some compilers for work areas. Care must be taken, therefore, that the assignments for the sort files will not conflict with the standard system assignments which are normally made for these SYS numbers for use by the Linkage Editor and the compilers. The job stream in Figure 5-31 illustrates a suggested method to be used if a compilation or Linkage Editor run is to follow a sort. This job stream also illustrates a technique which can be used to load a card input file onto the disk, perform a disk sort and use the sorted output as input to a COBOL program which will list the sorted data stored on the disk.

EXAMPLE

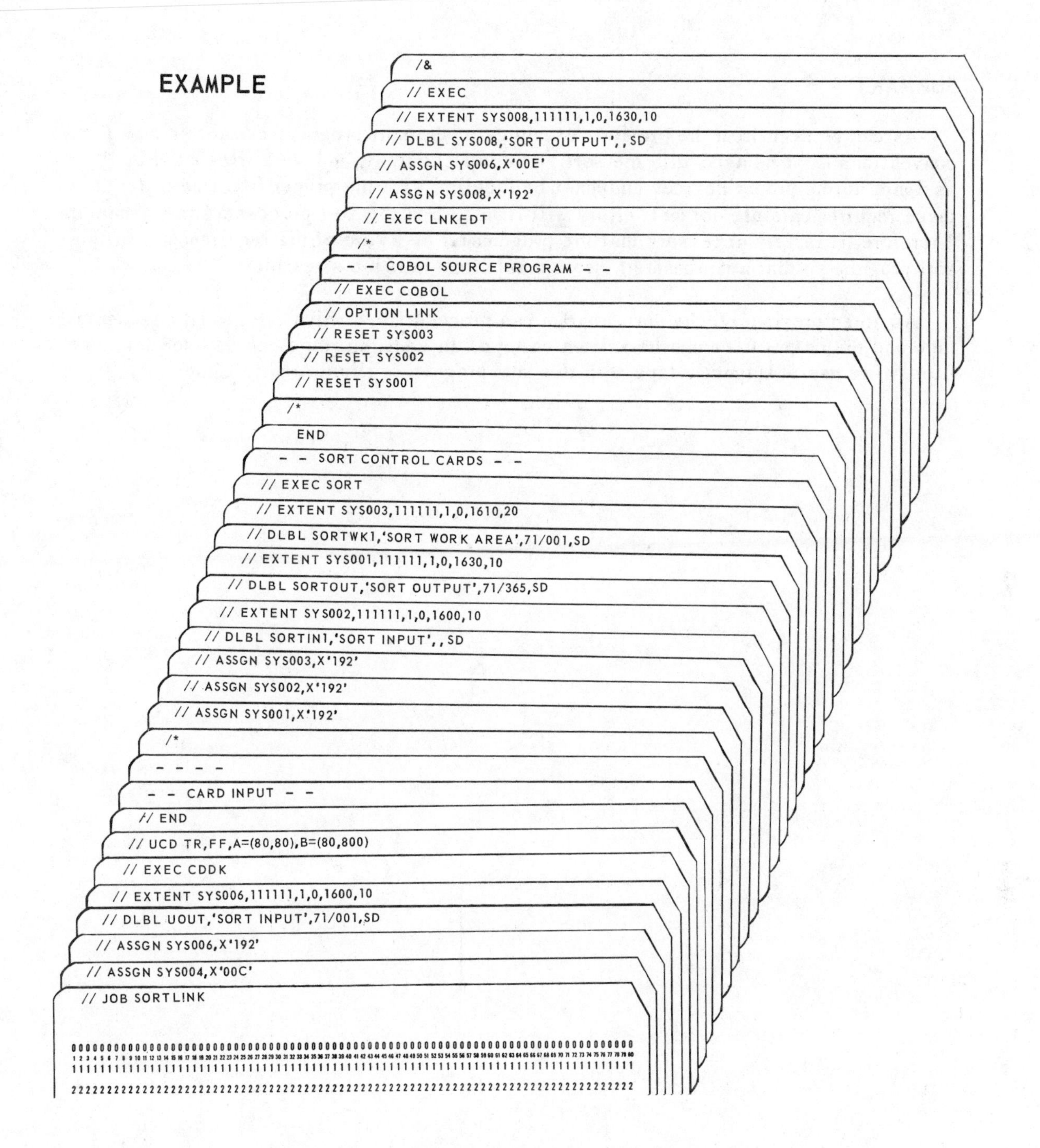

Figure 5-31 Example of RESET Command

Note in the above example that a card file is placed on disk using the card-to-disk utility program and then the data is sorted using SYS001 as the output file, SYS002 as the input file and SYS003 for the work file. Following the job sort step, a COBOL program is to be compiled, link-edited, and executed to read the sorted file and create a report. The // RESET job control card, as illustrated above, is used to reset the assignments of SYS001, SYS002, and SYS003 to the standard assignments for those symbolic units. Thus, when the COBOL compiler references the devices assigned to SYS001, SYS002, and SYS003, it will be referencing the standard assignments for the devices instead of the assignments made by the ASSGN job control cards preceding the sort job step. If the RESET statement had not been used, the compiler would have attempted to use drive X'192' for its work area and the results are unpredictable.

SUMMARY

As can be seen from the previous examples of the sort program, certain standard conventions must be used with the sort. The input, output, and work files must be assigned to the proper devices and must be labelled with the proper filenames. If these requirements are not met, errors will result when the sort processing is attempted. Therefore, it is very necessary that the programmer be aware of the requirements of the sort program so that a minimum of errors result from sort job streams.

As noted previously, the sort program can process files which are stored on magnetic tape and the output files can be written on tape. In addition, tape can be used for work files. The use of magnetic tape with the sort program is illustrated in Chapter 6.

STUDENT ASSIGNMENTS

1. Write the SORT control card to sort a single field which begins in column 24 of the record and is 6 bytes in length. The field is to be sorted in an ascending sequence. One input file is to be processed and one work area is used.

1	2	3	4	5	6	7	8	9	10	11	12	13	14	15	16	17	18	19	20	21	22	23	24	25	26	27	28	29	30	31	32	33	34	35	36	37	38	39	40	41	42	43	44	45	46	47	48	49	50	51	52	53	54	55

2. Write the SORT control card to sort three fields. The major field, which begins in column 55 of the input record, is 43 bytes in length. The intermediate field begins in column 1 and is 3 bytes in length and the minor field begins in column 38 and is 13 bytes in length. The major field is to be in a descending sequence and the other fields are to be in an ascending sequence. Two input files are used and a single work area is used.

1	2	3	4	5	6	7	8	9	10	11	12	13	14	15	16	17	18	19	20	21	22	23	24	25	26	27	28	29	30	31	32	33	34	35	36	37	38	39	40	41	42	43	44	45	46	47	48	49	50	51	52	53	54	55

3. Write the sort control cards which are read by the sort program to sort a 90 byte record on three fields: column 34 for 5 bytes in an ascending sequence (major), column 68 for 10 bytes in a descending sequence (intermediate), and column 90 for one byte in an ascending sequence (minor). The single input file consists of 5 records per block and the records are a fixed-length. The output file should be blocked in the same format as the input file and one work area is to be utilized by the sort. The output file should be checked when it is written and all messages from the Sort program are to be printed on the device assigned to SYSLST.

1	2	3	4	5	6	7	8	9	10	11	12	13	14	15	16	17	18	19	20	21	22	23	24	25	26	27	28	29	30	31	32	33	34	35	36	37	38	39	40	41	42	43	44	45	46	47	48	49	50	51	52	53	54	55

4. Write the control cards read by the sort program to sort an input file consisting of 120 byte records blocked 30 records per block. The fields to be sorted are: Column 45 for 6 bytes in a descending order, Column 68 in an ascending order for 8 bytes, and Column 119 in a descending order for 2 bytes in that major-minor order. The field in column 45 consists of signed numeric values and should be sorted in a descending order, that is, the highest positive number should be first and the lowest negative number should be last. The other two fields are to be sorted based upon their logical sequence. The output file should be blocked in the same manner as the input file and two work files are used.

1	2	3	4	5	6	7	8	9	10	11	12	13	14	15	16	17	18	19	20	21	22	23	24	25	26	27	28	29	30	31	32	33	34	35	36	37	38	39	40	41	42	43	44	45	46	47	48	49	50	51	52	53	54	55

5. Write the complete job stream to sort an input file stored on disk drive X'193'. This file runs from Cylinder 150 Track 0 to Cylinder 190 Track 9 and contains fixed-length records 70 bytes in length. The blocking factor for the input records is 5. The output file should be written on the disk pack stored on disk drive X'192' and the programmer is to choose the extents used. The work area must be stored on the pack on drive X'193'. The programmer must determine the extents for the work area. Each record is to be sorted on two fields: Column 23 for 7 bytes, ascending (major); Column 45 for 15 bytes, descending (minor). The major sort is on a signed zoned-decimal field. All sort messages are to be printed on SYSLST and the output should be verified.

1	2	3	4	5	6	7	8	9	10	11	12	13	14	15	16	17	18	19	20	21	22	23	24	25	26	27	28	29	30	31	32	33	34	35	36	37	38	39	40	41	42	43	44	45	46	47	48	49	50	51	52	53	54	55

6. Write the complete job stream to sort the output of the sort in Problem #5 into the following sequence: Column 2 for 8 bytes, descending; column 55 for 3 bytes, descending. The output of this sort should be written on the same area of disk used for the input to the sort in Problem #5.

1	2	3	4	5	6	7	8	9	10	11	12	13	14	15	16	17	18	19	20	21	22	23	24	25	26	27	28	29	30	31	32	33	34	35	36	37	38	39	40	41	42	43	44	45	46	47	48	49	50	51	52	53	54	55

7. Write the job stream necessary to accomplish the following processing:

   A. The cards illustrated in Appendix E are to be placed on a disk file through the use of the DOS Utility Card-To-Disk program.
   B. The disk file created above is to be sorted on the invoice number (ascending), the customer number (ascending), and the quantity (descending). See the record format in Appendix E.
   C. The sorted output file is to be printed on the printer using the DOS Utility Disk-To-Print program. Only the invoice number, the customer number and the quantity are to be printed. Headings should be included on the report to identify the fields.

The record formats for the above records are fixed in length. The record lengths of the input and output records are to be determined by the programmer. The areas to be used on the disk for output of the card-to-disk utility program, the work areas, and the output area of the sort are to be determined by the programmer. The job stream should be keypunched and executed on a System/360 operating under DOS.

1	2	3	4	5	6	7	8	9	10	11	12	13	14	15	16	17	18	19	20	21	22	23	24	25	26	27	28	29	30	31	32	33	34	35	36	37	38	39	40	41	42	43	44	45	46	47	48	49	50	51	52	53	54	55

1	2	3	4	5	6	7	8	9	10	11	12	13	14	15	16	17	18	19	20	21	22	23	24	25	26	27	28	29	30	31	32	33	34	35	36	37	38	39	40	41	42	43	44	45	46	47	48	49	50	51	52	53	54	55

CHAPTER 6

TAPE AND DISK SORT/MERGE PROGRAM

TAPE FILES

INTRODUCTION

In Chapter 5 it was seen that the Sort Program can utilize disk input and output files and disk work files to sort data into a prescribed sequence. The DOS Tape and Disk Sort/Merge Program also allows tape files to be used for input, output, and work files. The tape files can be used in conjunction with disk files also; that is, the input file may be stored on magnetic tape, the work areas on disk, and the output on tape, or any other combination. The only restriction is that the type of device used for input, work, or output cannot be mixed, that is, a tape and a disk cannot be used for input to the same sort program.

The following sections illustrate the use of the DOS Sort program with magnetic tape and disk storage devices being used.

TAPE INPUT

The following job stream could be used to sort a file of records which are stored on a magnetic tape, with the sorted output stored on disk.

EXAMPLE

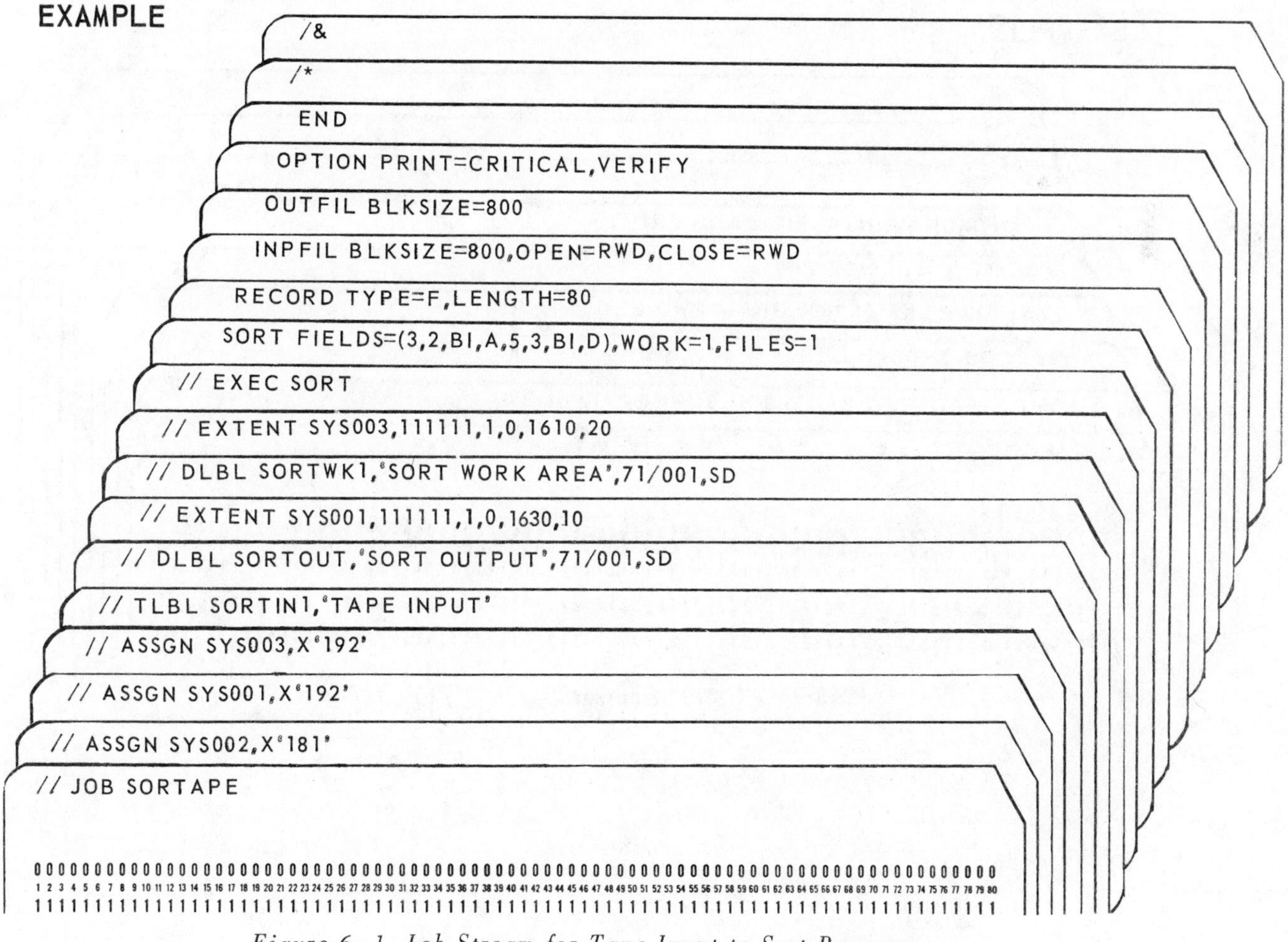

```
/&
/*
END
OPTION PRINT=CRITICAL,VERIFY
OUTFIL BLKSIZE=800
INPFIL BLKSIZE=800,OPEN=RWD,CLOSE=RWD
RECORD TYPE=F,LENGTH=80
SORT FIELDS=(3,2,BI,A,5,3,BI,D),WORK=1,FILES=1
// EXEC SORT
// EXTENT SYS003,111111,1,0,1610,20
// DLBL SORTWK1,'SORT WORK AREA',71/001,SD
// EXTENT SYS001,111111,1,0,1630,10
// DLBL SORTOUT,'SORT OUTPUT',71/001,SD
// TLBL SORTIN1,'TAPE INPUT'
// ASSGN SYS003,X'192'
// ASSGN SYS001,X'192'
// ASSGN SYS002,X'181'
// JOB SORTAPE
```

Figure 6-1 Job Stream for Tape Input to Sort Program

Note in the job stream illustrated in Figure 6-1 that the symbolic-unit SYS002 is assigned to device X'181', which is a magnetic tape drive. As when a disk unit is input to the sort, the magnetic tape device on which the input file is mounted must be assigned to SYS002. If more than one file is to be input to the sort program, SYS003, SYS004, etc. must be used in the same manner as is done with disk drives. Note also that the output file will be written on the device assigned to SYS001 and the work area will be on the device assigned to SYS003, which is one number higher than the last input SYS number. Thus, even though magnetic tape is being used instead of disk storage, the rules concerning symbolic device addresses is still used when processing data with the sort program.

Tape files may contain either standard or no labels when being processed without special user-written routines (see EXITS later in this chapter). If standard labels are to be used, such as in the example in Figure 6-1, a TLBL control card must be included in the job stream to supply tape labelling data. The filename entry for the TLBL card for a tape input file must be SORTIN1 for the first input tape, SORTIN2 for the second input tape, etc. A maximum of 9 input files may be processed, regardless of whether they are stored on disk or tape. Note that the filenames used on the TLBL card for the tape input files are identical to those used for the input files which are stored on disk.

In the example in Figure 6-1, the work file and the output file are stored on disk and the control cards used for them are the same as those used in previous examples. There are, however, several unique entries in the Sort Control cards which are read by the sort program when tape files are being processed. The Sort control cards used in the job stream in Figure 6-1 are again illustrated below.

EXAMPLE

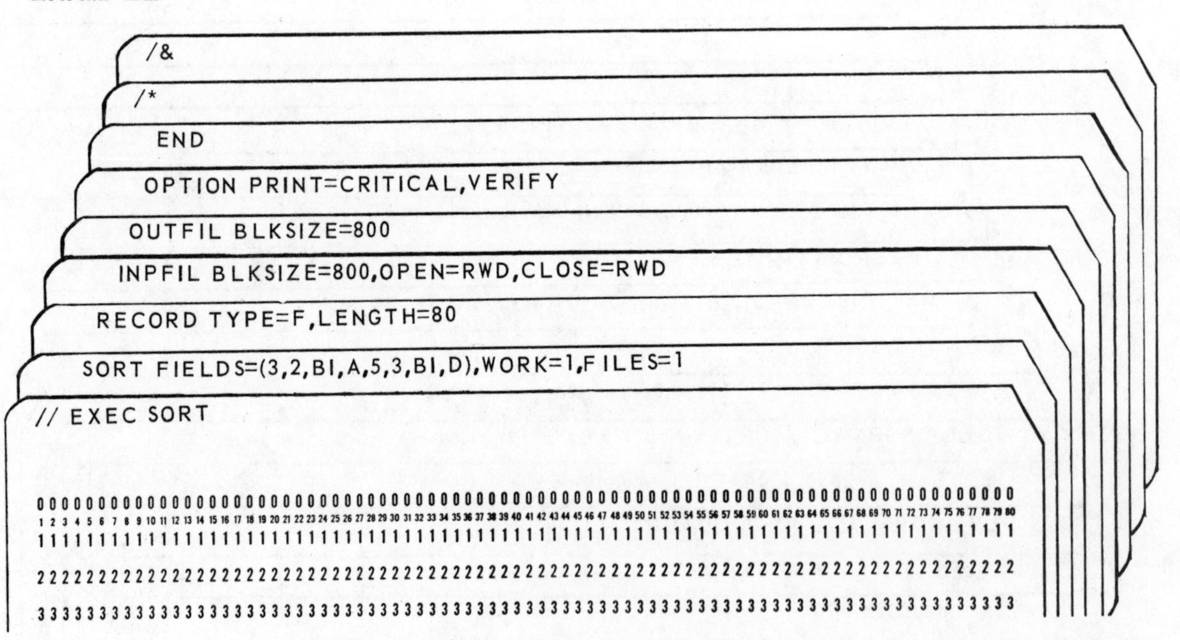

Figure 6-2 Sort Control Cards – Tape Input

In the example in Figure 6-2 it can be seen that the SORT control card and the RECORD control card contain the same information as when a disk input file is used. The INPFIL statement, however, differs from that used with disk because the disposition of the input tape must be specified. The general format of the INPFIL control card when used with tape input is illustrated below.

```
                                                                     [,CLOSE=RWD  ]
INPFIL BLKSIZE=n[,BYPASS] [,VOLUME=n         ] [,OPEN=RWD  ]  [,CLOSE=UNLD ]
                          [,VOLUME=(n1, . . .)] [,OPEN=NORWD]  [,CLOSE=NORWD]
```

Figure 6-3 General Format of INPFIL Control Statement for Tape Input

In the general format illustrated above it can be seen that the BLKSIZE entry, the BYPASS entry and the VOLUME entry are included, if necessary, in the same manner as when a disk input file is to be processed. In addition, two other parameters are included on the INPFIL statement when tape input files are to be processed.

The "OPEN" parameter specifies the tape positioning which should take place when the tape input file is opened by the Sort program. If the entry OPEN=RWD is included on the card, the input tape will be rewound prior to being read as input by the sort program. If the entry OPEN=NORWD is specified, the tape will not be moved prior to being read as input. The OPEN=NORWD option is normally specified when the second or subsequent file on a multi-file volume is to be read as input to the sort program. If this operand is omitted from the INPFIL statement, the tape volume will be rewound before processing begins.

The "CLOSE" operand is used to specify the disposition of the input tape after it has been read as input by the sort program. It may be rewound (CLOSE=RWD), rewound and unloaded (CLOSE=UNLD) or left in the position it occupies when the tape mark which indicates end-of-file is read by the sort program (CLOSE=NORWD). The choice made by the programmer depends upon the use of the tape after it has been read as input to the sort. If the tape is to be used in a subsequent job step, the tape will normally just be rewound. If it is not to be used, it will normally be rewound and unloaded so that the operator may mount another tape on the drive. If another file follows on a multi-file volume, and the file to be processed, it is normally desirable not to rewind the tape so that it will be in a position to process the next file. In the example in Figure 6-2, it can be seen that the input tape will be rewound both before and after processing by the sort program.

Note also in Figure 6-2 the use of the PRINT=CRITICAL operand on the OPTION card. This parameter indicates that the only messages which should be printed on the device assigned to SYSLST are those messages which indicate the cause for an abnormal termination of the sort program. The normal messages as illustrated in Chapter 5 will not be printed.

TAPE INPUT AND OUTPUT

Both the input and the output files of the Sort program can be stored on magnetic tape. The example in Figure 6-4 illustrates the job stream which could be used to execute the sort program with the input file on one tape volume and the output file written on another tape volume.

EXAMPLE

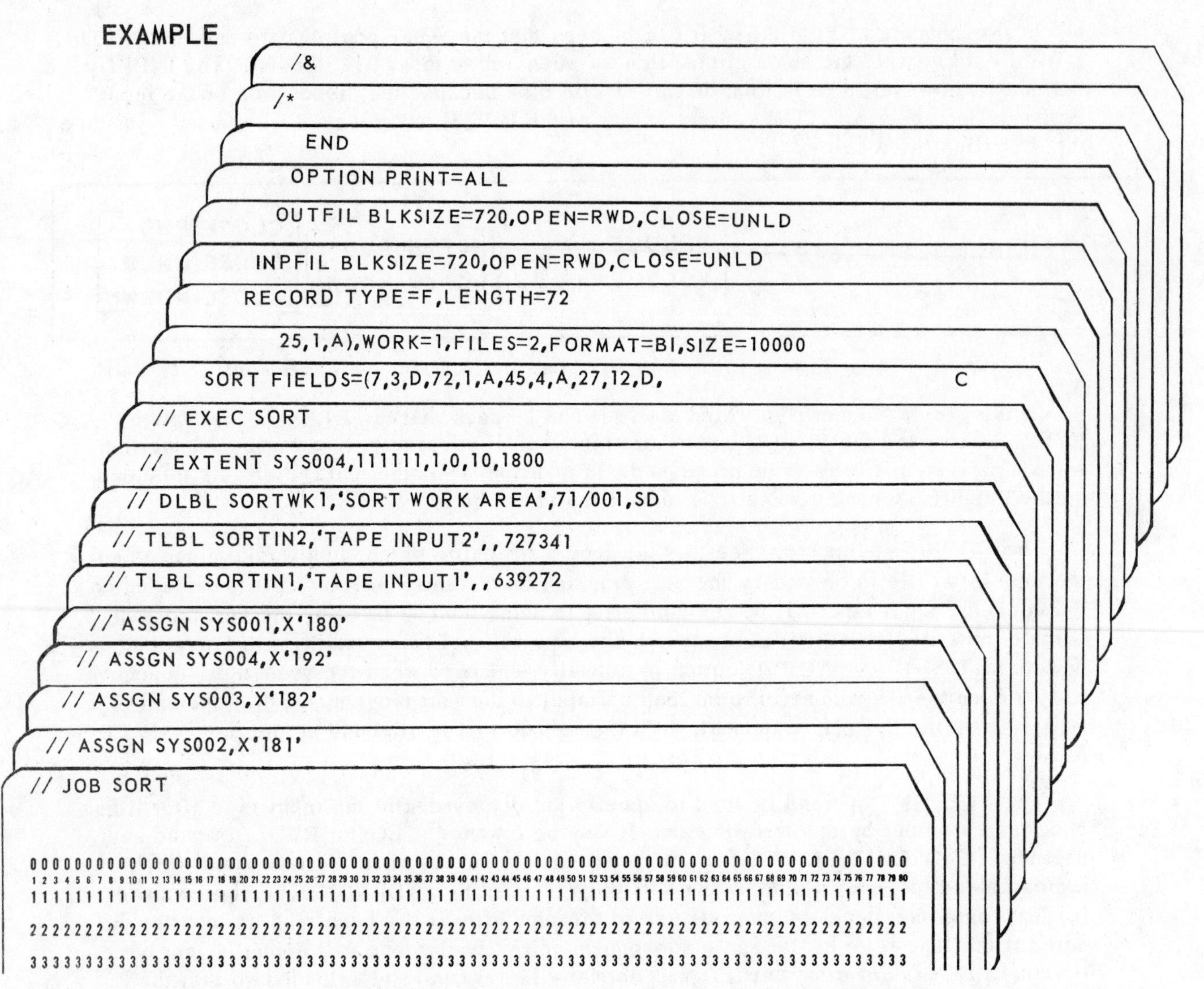
```
/&
/*
 END
 OPTION PRINT=ALL
 OUTFIL BLKSIZE=720,OPEN=RWD,CLOSE=UNLD
 INPFIL BLKSIZE=720,OPEN=RWD,CLOSE=UNLD
 RECORD TYPE=F,LENGTH=72
 25,1,A),WORK=1,FILES=2,FORMAT=BI,SIZE=10000
 SORT FIELDS=(7,3,D,72,1,A,45,4,A,27,12,D,                    C
// EXEC SORT
// EXTENT SYS004,111111,1,0,10,1800
// DLBL SORTWK1,'SORT WORK AREA',71/001,SD
// TLBL SORTIN2,'TAPE INPUT2',,727341
// TLBL SORTIN1,'TAPE INPUT1',,639272
// ASSGN SYS001,X'180'
// ASSGN SYS004,X'192'
// ASSGN SYS003,X'182'
// ASSGN SYS002,X'181'
// JOB SORT
```

Figure 6-4 Example of Tape Input, Tape Output

In the example above a tape input file is to be read from drive X'181', which is assigned to SYS002, as all input files must be. In addition, a second input file will be mounted on drive X'182', which is assigned to SYS003. The output file is written on drive X'180', which is assigned to SYS001. The work file will be stored on disk on drive X'192', which is assigned to SYS004. Note that all of these device assignments correspond to the rules which have been stated for device assignments for the Sort program. Regardless of the type of device used for input, work, and output, whether it is tape or disk, these assignments must be made as shown.

Note in the example above that a TLBL control card is submitted for both of the input files and the output file. This is because these files are to contain standard labels and the label information to be processed when the files are opened must be supplied by the TLBL cards. The DLBL and EXTENT cards are, of course, required for the work file on disk.

The SORT control card in the example in Figure 6-4 is an example of the use of continuation cards when using the sort control cards. If all of the information which must be specified for a particular control card cannot be contained in card columns 2-71, a continuation card must be used. The two cards used for the SORT control statement in Figure 6-4 are illustrated on coding sheets below.

EXAMPLE

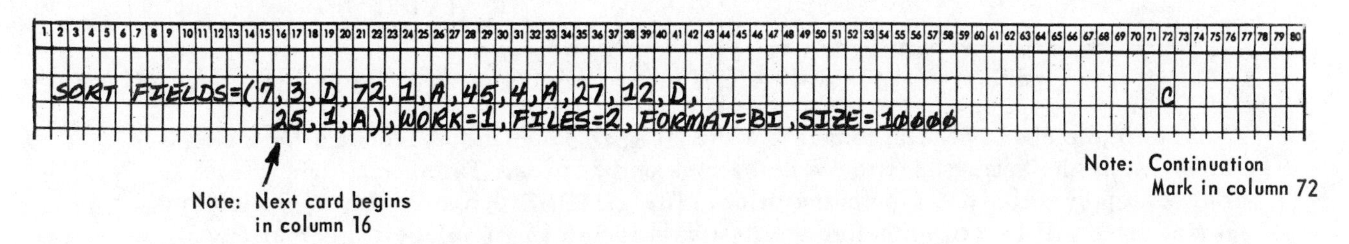

Figure 6-5 Example of CONTINUATION CARD

Note in the example above that two cards are used for the SORT Control Statement. In order to use a continuation card with the sort control cards, two rules must be followed: 1) A continuation punch must be contained in card column 72 of the first card. The value punched in column 72 can be any value other than blank; 2) The information contained on the second card must begin in column 16. If it begins in any other column of the card, the sort program will indicate a control card error and the sort program will be terminated. As many continuation cards as are required may be used; each time the card which is to be continued must contain a non-blank character in column 72 and the information on the next card must begin in column 16.

Note also in the SORT control card that the format of the data to be sorted is not included in the FIELDS parameter. When all of the formats of the data to be sorted are the same, the FORMAT entry may be used. The general format of the SORT control card when using the FORMAT entry is illustrated below.

SORT FIELDS=(p_1,m_1,s_1, . . .),FORMAT=xx,WORK=n[,SIZE=n][,CHKPT][,FILES=n]

Figure 6-6 General Format of SORT Control Statement

In the general format illustrated above it can be seen that the "f" entry in the FIELDS parameter is not included, that is, the "p" entry which specifies the beginning column in the record to be sorted, the "m" entry which indicates the length of the field to be sorted, and the "s" entry which indicates the sequence in which the data is to be sorted, are all included but the "f" entry, which specifies the format of the data, is not included. This may occur only when all of the fields to be sorted contain data in the same format. If this does occur, the FORMAT=xx entry is used to indicate the format of all of the fields. The valid entries for "xx" are the same as the "f" entry in the FIELDS parameter, that is, BI, CH, ZD, PD, FL, or FI.

When magnetic tape is used for the output file of the Sort program, the OUTFIL sort control card contains additional entries to specify the processing of the tape file. The general format of the OUTFIL control card for use with tape output files is illustrated in Figure 6-7.

Figure 6-7 General format of OUTFIL Card for Tape Output

Note in the above general format that the BLKSIZE operand is included for the tape output file in the same manner as with the disk output files. The remaining entries, however, apply strictly to tape output files. The NOTPMK optional entry specifies that no tape mark will be written before the first data record on an unlabelled output tape. This operand applies only to unlabelled output tapes. If it is not specified, a tape mark will be written on the output tape before the first data record is written.

The OPEN and CLOSE parameters are used to specify the positioning of the output tape when the tape is opened and when the file is closed. The output tape will be rewound to the load point when it is opened if the OPEN = RWD option is specified on the OUTFIL control card. The tape will not be repositioned if the OPEN = NORWD option is specified. If neither option is specified, the tape will be rewound to the load point when it is opened. When the file is closed after processing, it will be rewound (CLOSE = RWD), rewound and unloaded (CLOSE = UNLD), or left positioned where it is (CLOSE = NORWD). If one of these three options are not specified, the output tape will be rewound when the file is closed.

The remainder of the sort control cards are the same whether disk output or tape output are to be used. The blocking factors and other criteria which were specified for the disk files in Chapter 5 apply equally to tape input and output files.

TAPE WORK FILES

In the examples presented previously, the work areas used by the Sort program have been on disk storage. It is also possible to have the work areas be stored on magnetic tape. The primary advantage to using tape work areas is the fact that disk storage area need not be used for the temporary work area and this can be quite important when the storage space is limited on disk packs. Thus, tape work areas are normally used only when there is not adequate space available on disk storage. This is true because the use of tape work areas causes the sort to process data more slowly and because a minimum of three tape drives are required when tape is used as the work area.

The job stream illustrated in Figure 6-8 utilizes magnetic tape as the work area for a sort.

EXAMPLE

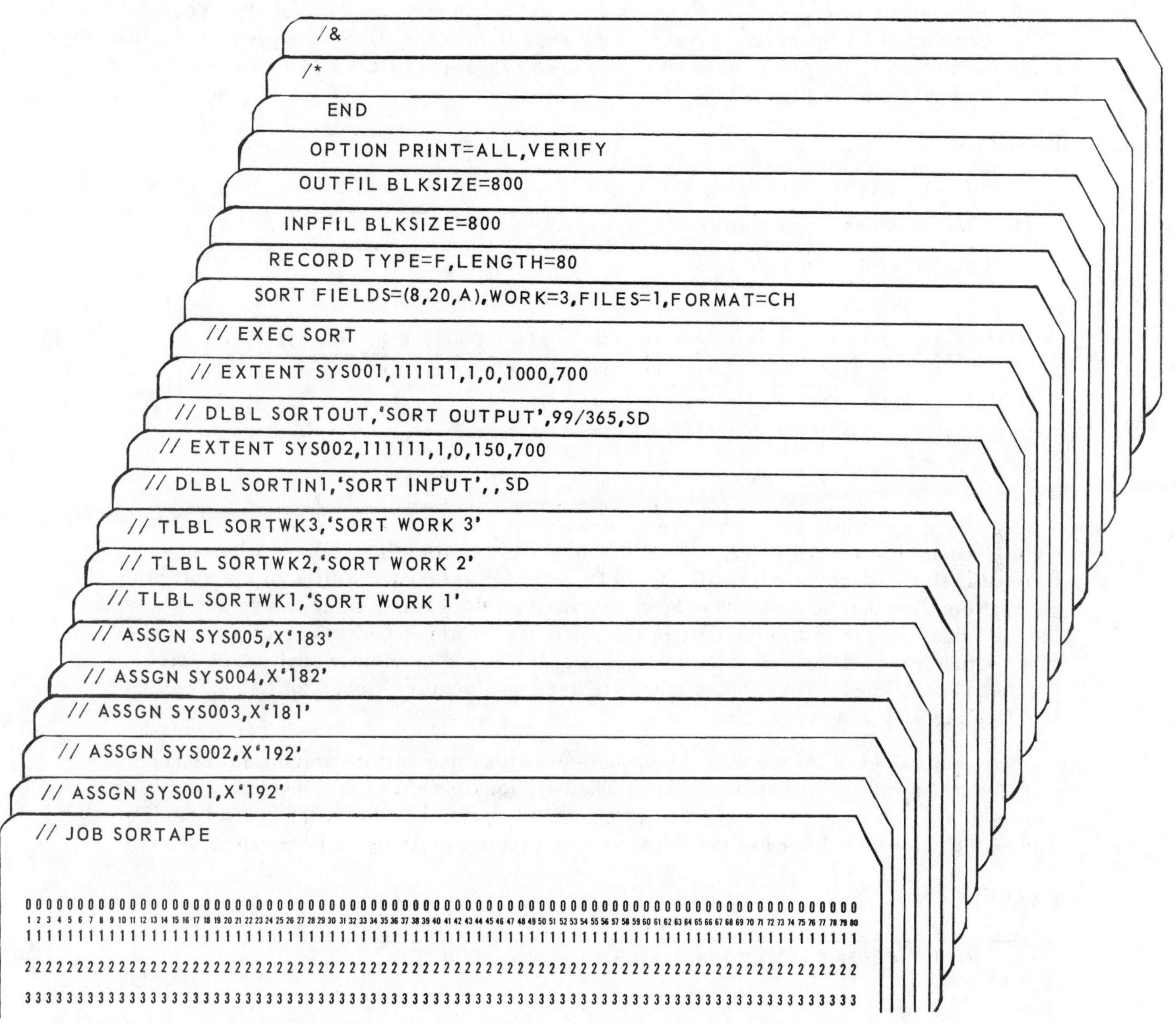

Figure 6-8 Job Stream utilizing Tape Work Files

Note in the above job stream that the input and output files are stored on a disk pack. The work files are to be stored on magnetic tape files. The symbolic-unit addresses which are used for the input, output, and work files must be the same regardless of which medium is used for the files. Thus, the output file will be stored on the device assigned to SYS001, the input file will be read from the device assigned to SYS002, and the work files will begin on the device assigned to SYS003. Since a minimum of three magnetic tape drives must be used for tape workfiles, the symbolic unit addresses SYS003, SYS004, and SYS005 are used for the devices. Note in the job stream that SYS003 is assigned to X'181', SYS004 to X'182', and SYS005 to X'183'. Up to 9 separate tape drives may be used as work areas for the Sort program. The use of more drives increases the efficiency of the sort because the program can use more drives and thus need not write as much information on any single drive.

In the example in Figure 6-8, the work files are to contain standard labels. Thus, TLBL cards must be submitted for each of the work files. The filename entry in the TLBL cards must indicate the proper name of each work file. The TLBL cards for the work files are again illustrated in Figure 6-9.

EXAMPLE

```
// TLBL SORTWK3,'SORT WORK 3'
// TLBL SORTWK2,'SORT WORK 2'
// TLBL SORTWK1,'SORT WORK 1'
```

Figure 6-9 TLBL Cards for Sort Work Files

Note in the example above that the filename entry for the first work file is SORTWK1, for the second file is SORTWK2, and for the third file is SORTWK3. These filenames must be used in order to properly identify the TLBL card for use with the files in the sort program. Note also that no dates have been specified on the cards. Since the default value in tape labels for the expiration date is the same day on which the tape is created, these tapes which are used for work files will be able to be used in another application when the sort is completed. Thus, these tapes are used as "scratch" tapes for use only in an intermediate step within the job.

When tape is used for the work files, there are no unique entries which must be made in the sort control cards to indicate this. The only requirement is that the number of work files which are to be used be indicated in the WORK operand of the SORT control card. The SORT control card used in the job stream in Figure 6-8 is again illustrated below.

EXAMPLE

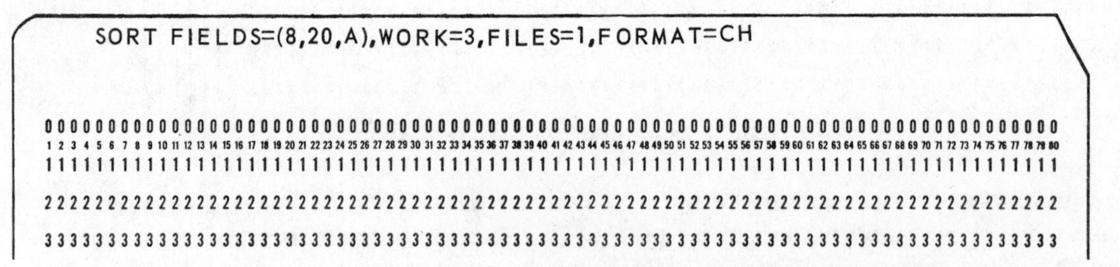

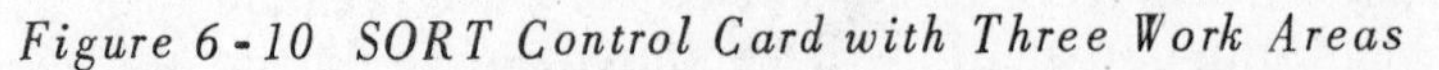
Figure 6-10 SORT Control Card with Three Work Areas

Note in the example above that the entry WORK=3 is used to indicate the number of files which are available for work area. As noted previously, the minimum number of work files which may be used when tape is the medium is 3. Up to 9 different tape files may be used for work areas. Note also that the field to be sorted by the job stream in Figure 6-10 is the employee name field (see Figure 3-14) and that field is a CHaracter field, as indicated by the FORMAT=CH value in the control card. The remaining entries in the job control and sort control cards illustrated in Figure 6-8 are the same as when disk storage is used for the work areas.

UNLABELLED TAPE FILES

When processing magnetic tape files under DOS, the files may contain standard DOS labels, standard labels with user labels, or no labels. If any user labels are to be processed with the sort program, special exit routines must be written to handle them (see EXIT section). The tape input, output, and work files illustrated in previous examples have utilized standard DOS labels. It is possible to process tape files with no labels without writing special exit routines. The job stream illustrated below is an example of the input and work files utilizing non-labelled tapes. The output file is a disk file.

EXAMPLE

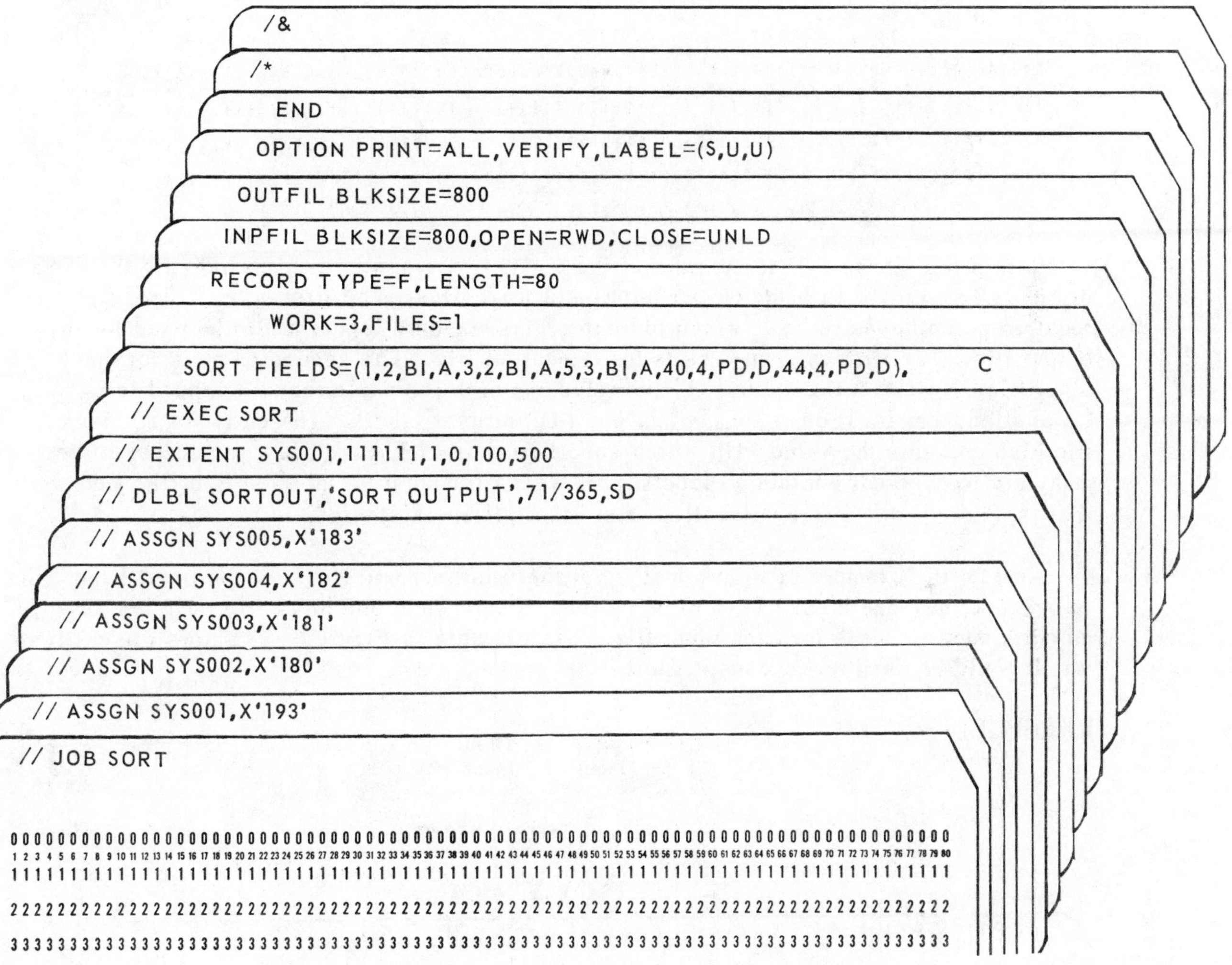

Figure 6-11 Magnetic Tape Files with No Labels

Note in the above example that the device assignments for the input, output, and work files are the same as have been used for all of the previous examples. Again, these assignments are standard for the sort program and must be used as shown. Note also that the only label cards included in the job stream are the DLBL and EXTENT cards for the output file, which is to be stored on disk where standard labels must be used. There are no label cards (TLBL) for either the tape input file or the tape work files.

It was noted previously that the sort program assumes that standard labels are to be used for all of the files which are to be processed. If unlabelled tapes are to be used, the LABEL operand in the OPTION control card must be utilized. The OPTION card used in the job stream for unlabelled tapes in Figure 6-11 is again illustrated below.

EXAMPLE

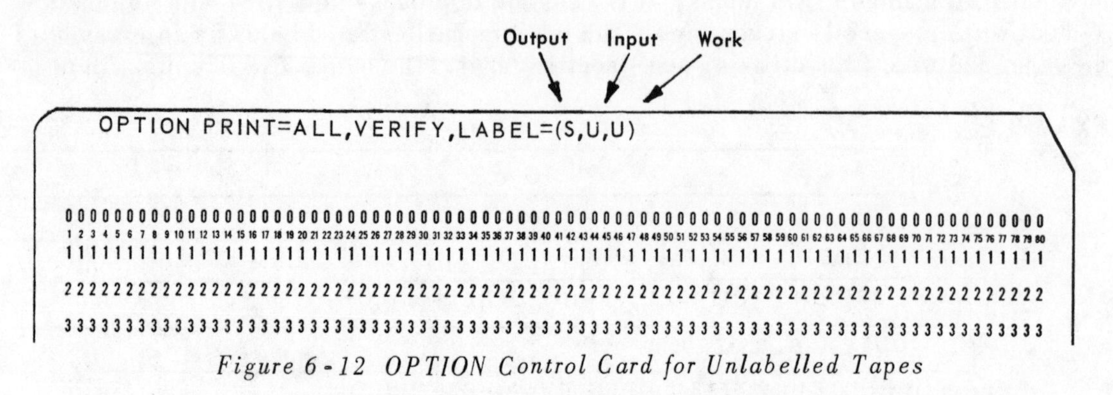

Figure 6-12 OPTION Control Card for Unlabelled Tapes

Note in the above OPTION card that the LABEL operand is used to indicate the type of labels associated with the output, input and work files. The first entry within the required parentheses is "S", which indicates that standard labels are to be used for the output file. The first entry always is for the output file. The second entry is for the input file. Note that the value "U" is included which indicates that the input file is Unlabelled, that is, it does not contain any DOS or user labels. The entry for the work file also contains the value "U" which specifies that ALL work files, regardless of how many are to be used, contain no labels. All work files must be labelled in a like manner, that is, if one work file is unlabelled, they must all be unlabelled.

Note in the example in Figure 6-12 that there is only one entry for the input file. This is because only one input file is to be sorted. If more than one input file is to be sorted, an entry must be made for each input file. The example in Figure 6-13 shows the entries which could be used for four input files.

EXAMPLE

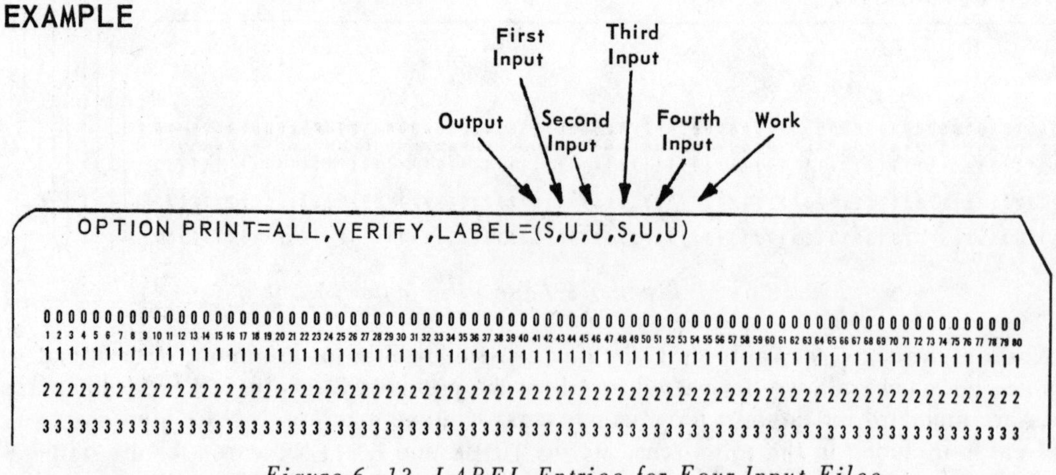

Figure 6-13 LABEL Entries for Four Input Files

In the Option Control Card in Figure 6-13 it can be seen that the LABEL operand contains entries for one output file, four input files, and a single entry for the work files. Note that the first and second input files will be contained on nonlabelled tapes, the third tape file will contain standard labels, and the fourth input file will be unlabelled. As can be seen, each separate input file may contain different types of labels and these types must be indicated by the LABEL entry on the Option Card. There is only one entry for the output file in the LABEL operand because only one output file is created from the sort program regardless of the number of input files. There is only one entry for the work files because all of the work files must contain the same types of labels. The input files, however, can have varying types of labels.

It should be noted that even though the input files can contain different types of labels, they must all be on the same storage medium. That is, they must all be contained on tape or they must all be on disk. Input files to a single sort run cannot be stored on both tape and disk. Neither, it should be noted, can work files.

The SORT control card in Figure 6-11 contains entries in the FIELDS operand to sort four different fields in different formats. This card is illustrated in Figure 6-14.

EXAMPLE

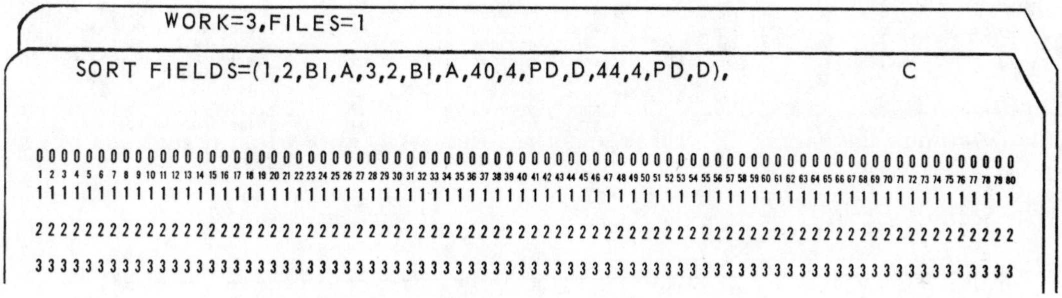

```
            WORK=3,FILES=1
   SORT FIELDS=(1,2,BI,A,3,2,BI,A,40,4,PD,D,44,4,PD,D),                    C
```

Figure 6-14 Example of SORT Control Card

Note in the example above that four different fields are to be sorted by the Sort program. The first two fields, which are the Store Number (Col. 1), and the Department Number (Col. 3) are to be sorted by a straight binary sort. The third and fourth fields, YTD Returns (Col. 40) and Current Returns (Col. 44) are stored in a packed-decimal format on the input file and are to be sorted using the "PD" format, or packed-decimal format. The length of the fields, 4 in both cases, indicates the number of bytes which are to be processed by the sort program, not the number of digits in each field (ie. there could be 7 digits in a 4-byte packed-decimal field). Thus, both of these fields will be sorted in a descending sequence based upon the sign in the field and the numeric value in each field. The input to the sort program and the output after it has been sorted is illustrated in Figure 6-15.

Before Sorting

STORE	DEPT	S/M	EMPLOYEE NAME	YTD	CURR	COMM	MONTHS	YTD RETURN	CURR RETURN
10	01	004	ACHER, WILLIAM C.	0190726	0057524	10	03	0000850F	0000000F
10	04	027	ALHOUER, ELAINE E.	0112573	0022066	12	05	0001880F	0000400F
10	03	030	ALLOREN, RUTH W.	2132020	0000000	15	10	0114000F	0020933F
20	06	100	BATES, TONY F.	0819066	0207645	05	04	0088340F	0005170F
20	08	102	BELLSLEY, ARTHUR A.	0883000	0099000	15	09	0019033F	0002762F
10	09	105	BOYLE, RALPH P.	0878044	0143055	12	06	0016070F	0007787F
10	02	111	CARTOLER, VIOLET B.	0280998	0075006	12	04	0003020D	0001140F
10	04	171	COSTA, NAN S.	0580356	0056002	12	10	0013020D	0000903F
10	09	820	TELLER, STEPHEN U.	1957044	0177009	12	10	0051920F	0032222D
10	01	730	REEDE, OWEN W.	0900555	0105144	10	08	0010520F	0002310F
20	04	739	RIDEL, ROBERT R.	0557504	0082571	12	07	0033303F	0007224F
20	09	740	RIDGEFIELD, SUZY S.	1804178	0190506	12	10	0033024F	0002310F
20	02	801	SCHEIBER, HARRY T.	0095237	0032508	12	03	0000520F	0000520F
20	07	802	SHEA, MICHAEL H.	0642033	0082009	10	08	0012920F	0004680F
20	01	300	FELDMAN, MIKE R.	0250000	0030000	10	09	0000600F	0000000F
10	02	304	FROMM, STEVE V.	0650000	0120000	12	05	0006090F	0001823F
10	05	308	GLEASON, JAMES E.	0145000	0039000	10	04	0000925F	0000000F
10	07	310	GORMALLY, MARIE N.	0389022	0064055	10	06	0007320F	0000866F
20	03	311	GROLER, GRACE B.	2300643	0205420	15	10	0260480F	0048369F
20	09	315	HALE, ALAN A.	1274000	0168000	12	08	0025954F	0003829F
20	04	317	HANBEE, ALETA O.	0385000	0039500	12	10	0007530F	0001180F
20	03	318	HANEY, CAROL S.	0975000	0145000	15	05	0012020F	0006019F
10	08	322	HARLETON, JEAN H.	0780899	0120089	15	06	0010180F	0000000F
20	01	325	HATFIELD, MARK I.	0151122	0020539	10	07	0001840F	0000220F
20	07	332	HELD, ANNA J.	0244000	0029500	10	09	0005115F	0000926F
10	06	409	ICK, MICK W.	0410122	0195080	05	02	0000590F	0000590F
20	07	689	OWNEY, REED M.	0437788	0053066	10	09	0008635F	0001132F
20	04	721	RASSMUSEN, JOHN J.	0810000	0100000	12	08	0017680F	0002346F
10	03	181	DELBERT,EDWARD D.	1250659	0130554	15	09	0037540D	0000000F
20	06	179	DAMSON, ERIC C.	0352502	0180888	05	02	0000223F	0000223D
BLOCK NO. 000004, INPUT AREA UNDERFLOW									
10	01	185	DONNEMAN, THOMAS M.	0650423	0090019	10	07	0006700F	0001020F
20	05	207	EBERHARDT, RON G.	0564007	0094009	10	06	0006300F	0000980F
10	07	214	EDMONSON, RICK T.	0079067	0033057	10	03	0000290F	0000000F
10	09	215	EDSON, WILBUR S.	0607050	0082005	12	07	0011885D	0001575F
20	08	282	ESTABAN, JUAN L.	1984055	0000000	15	10	0040505F	0005050F
10	03	487	KING, MILDRED J.	1850896	0180429	15	10	0038260F	0005322D
10	06	292	EVERLEY, DONNA M.	0332000	0177500	05	02	0000322D	0000322F
10	05	568	LYNNE, GERALD H.	0924487	0133311	10	07	0016530D	0001428F

END OF DATA
PAGE 1

After Sorting

STORE	DEPT	S/M	EMPLOYEE NAME	YTD	CURR	COMM	MONTHS	YTD RETURN	CURR RETURN
10	01	004	ACHER, WILLIAM C.	0190726	0057524	10	03	0000850F	0000000F
10	01	185	DONNEMAN, THOMAS M.	0650423	0090019	10	07	0006700F	0001020F
10	01	730	REEDE, OWEN W.	0900555	0105144	10	08	0010520F	0002310F
10	02	111	CARTOLER, VIOLET B.	0280998	0075006	12	04	0003020D	0001140F
10	02	304	FROMM, STEVE V.	0650000	0120000	12	05	0006090F	0001823F
10	03	030	ALLOREN, RUTH W.	2132020	0000000	15	10	0114000F	0020933F
10	03	181	DELBERT,EDWARD D.	1250659	0130554	15	09	0037540D	0000000F
10	03	487	KING, MILDRED J.	1850896	0180429	15	10	0038260F	0005322D
10	04	027	ALHOUER, ELAINE E.	0112573	0022066	12	05	0001880F	0000400F
10	04	171	COSTA, NAN S.	0580356	0056002	12	10	0013020D	0000903F
10	05	308	GLEASON, JAMES E.	0145000	0039000	10	04	0000925F	0000000F
10	05	568	LYNNE, GERALD H.	0924487	0133311	10	07	0016530D	0001428F
10	06	292	EVERLEY, DONNA M.	0332000	0177500	05	02	0000322D	0000322F
10	06	409	ICK, MICK W.	0410122	0195080	05	02	0000590F	0000590F
10	07	214	EDMONSON, RICK T.	0079067	0033057	10	03	0000290F	0000000F
10	07	310	GORMALLY, MARIE N.	0389022	0064055	10	06	0007320F	0000866F
10	08	322	HARLETON, JEAN H.	0780899	0120089	15	06	0010180F	0000000F
10	09	105	BOYLE, RALPH P.	0878044	0143055	12	06	0016070F	0007787F
10	09	215	EDSON, WILBUR S.	0607050	0082005	12	07	0011885D	0001575F
10	09	820	TELLER, STEPHEN U.	1957044	0177009	12	10	0051920F	0032222D
20	01	300	FELDMAN, MIKE R.	0250000	0030000	10	09	0000600F	0000000F
20	01	325	HATFIELD, MARK I.	0151122	0020539	10	07	0001840F	0000220F
20	02	801	SCHEIBER, HARRY T.	0095237	0032508	12	03	0000520F	0000520F
20	03	311	GROLER, GRACE B.	2300643	0205420	15	10	0260480F	0048369F
20	03	318	HANEY, CAROL S.	0975000	0145000	15	05	0012020F	0006019F
20	04	317	HANBEE, ALETA O.	0385000	0039500	12	10	0007530F	0001180F
20	04	721	RASSMUSEN, JOHN J.	0810000	0100000	12	08	0017680F	0002346F
20	04	739	RIDEL, ROBERT R.	0557504	0082571	12	07	0033303F	0007224F
20	05	207	EBERHARDT, RON G.	0564007	0094009	10	06	0006300F	0000980F
20	06	100	BATES, TONY F.	0819066	0207645	05	04	0088340F	0005170F
BLOCK NO. 000004, INPUT AREA UNDERFLOW									
20	06	179	DAMSON, ERIC C.	0352502	0180888	05	02	0000223F	0000223D
20	07	332	HELD, ANNA J.	0244000	0029500	10	09	0005115F	0000926F
20	07	689	OWNEY, REED M.	0437788	0053066	10	09	0008635F	0001132F
20	07	802	SHEA, MICHAEL H.	0642033	0082009	10	08	0012920F	0004680F
20	08	102	BELLSLEY, ARTHUR A.	0883000	0099000	15	09	0019033F	0002762F
20	08	282	ESTABAN, JUAN L.	1984055	0000000	15	10	0040505F	0005050F
20	09	315	HALE, ALAN A.	1274000	0168000	12	08	0025954F	0003829F
20	09	740	RIDGEFIELD, SUZY S.	1804178	0190506	12	10	0033024F	0002310F

END OF DATA
PAGE 1

Figure 6-15 Example of Sorting of Packed-Decimal field

Note from the input and output file listings in Figure 6-15 that the records have been sorted and also that the YTD Returns and the Current Returns fields contain both positive numbers (Sign is "F") and negative numbers (Sign is "D"). When these fields are sorted in a descending sequence, as specified in the sort control cards in Figure 6-14, the positive numbers will fall before the negative numbers. If they were sorted in an ascending sequence, the negative numbers would be placed before the positive numbers. Thus, it can be seen that when the PD (packed-decimal) format is specified for a field, the sign is considered in the sequence of the fields.

POOLING I/O DEVICES

As has been noted previously, the availability of devices on which to place the input, output, and work files may be a problem when executing the sort program especially if the sizes of the files being sorted are large or tape work areas are being used. Because of this, the Sort program allows a limited amount of "I/O Pooling", that is, using the same hardware device for more than one purpose. For example, the same tape drive may be used for both the input file and the output file. The example below illustrates the use of three tape drives to store three input files, the output file, and the three work files.

EXAMPLE

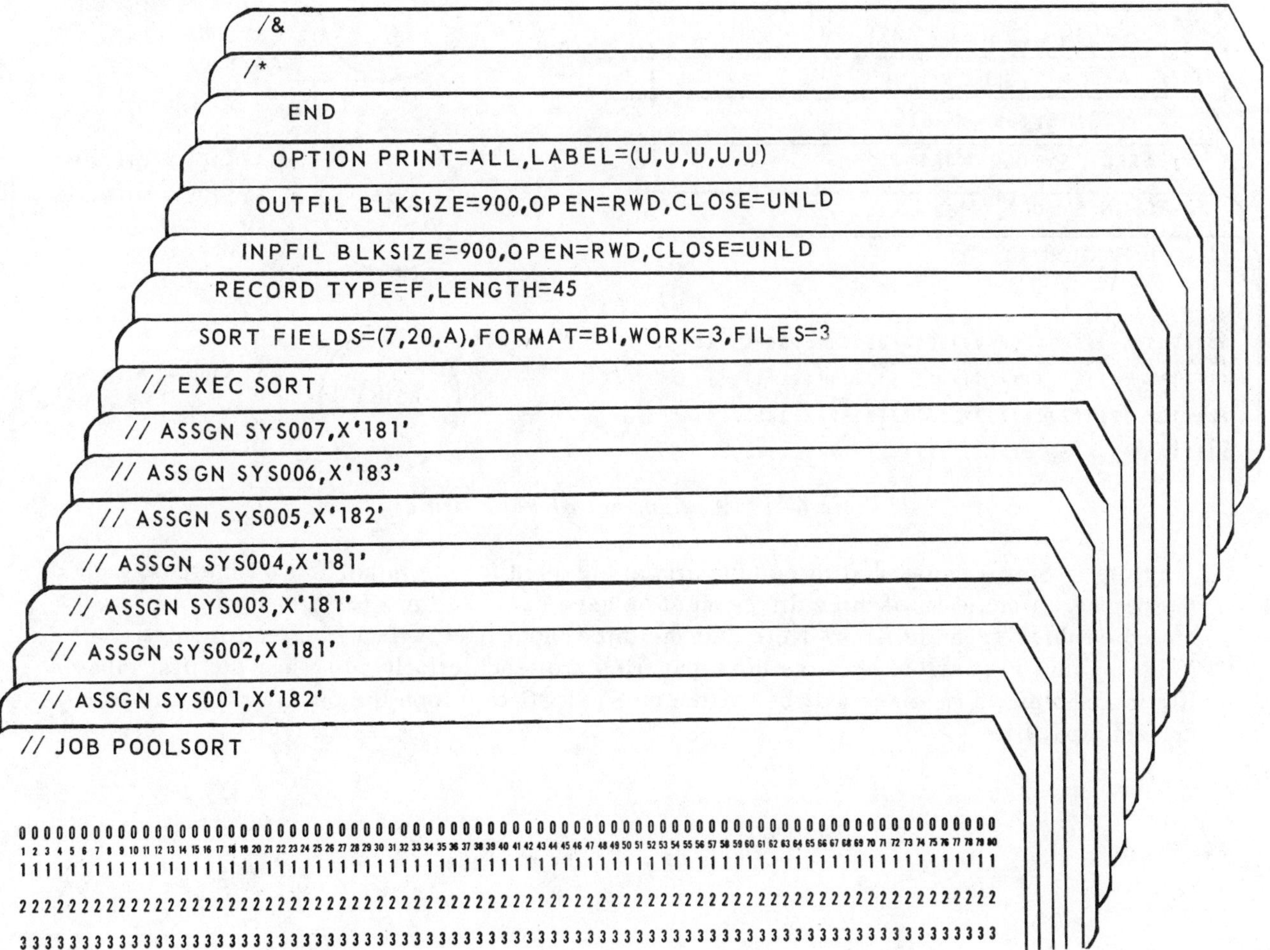

Figure 6-16 I/O Pooling of Tape Drives

In the example in Figure 6-16 it can be seen that three tape drives, X'181', X'182', and X'183' are to be used for the input to the sort program, the work areas, and the output areas. As was noted previously, it is necessary that three separate tape files on three tape drives be utilized when tape work areas are used. Thus, the same drives which are used to store the input and the output files are also used to store the tapes which are used for the work areas. The following example illustrates the steps which are taken by the sort program when I/O pooling such as is illustrated in Figure 6-16 is used.

EXAMPLE

Step 1: The job control cards are read by Job Control and the assignments to the tape drives are made.

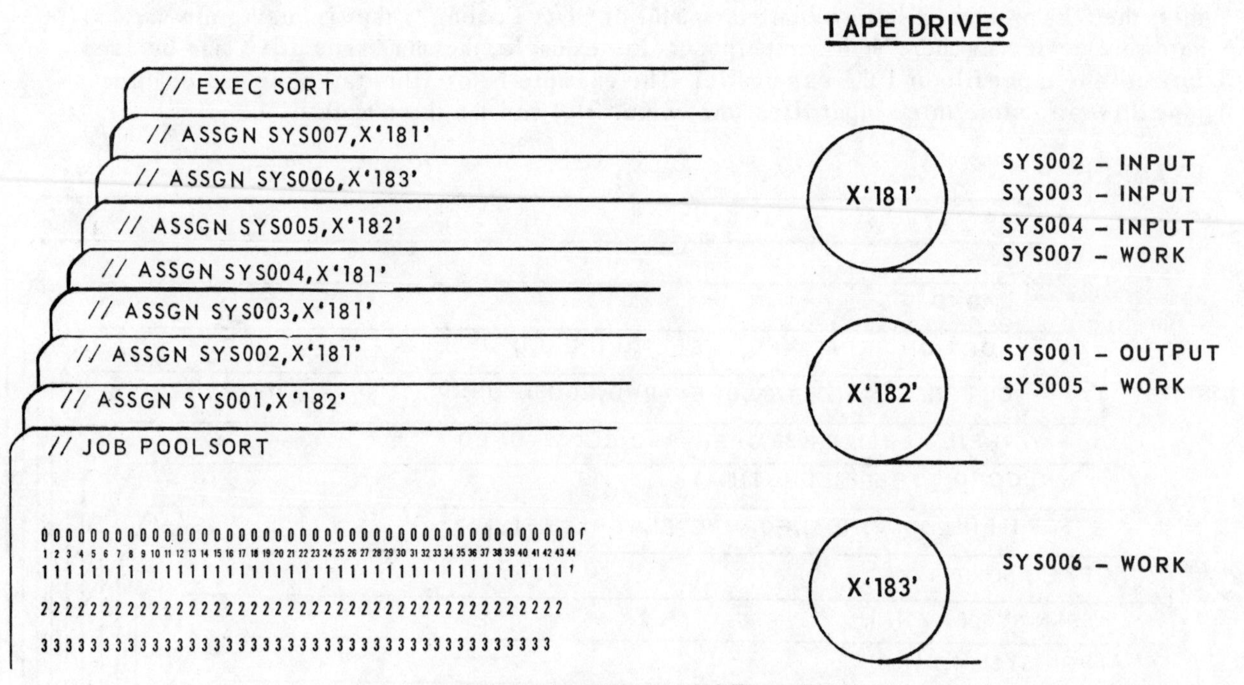

Figure 6-17 Assignment of Tape Drives

Note in Step 1 above that three tape drives are used for the input, the work areas, and the output. A minimum of three drives must be used because the work areas must be stored on three separate files. Note that the three input files will all be read from drive X'181'. This is possible because the input files are read serially and when the first input file is complete, a message will be written on SYSLOG to inform the operator to mount the second input file.

Step 2: The Sort Program is loaded into core storage and it reads the sort control cards.

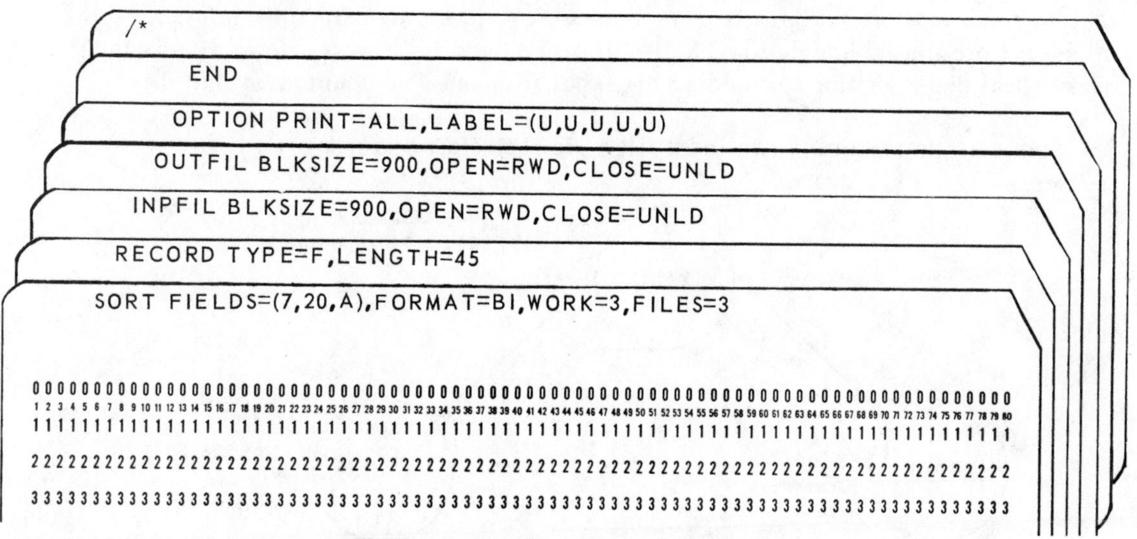

Figure 6-18 Sort Control Cards

Note from the sort control cards illustrated above that the SORT card indicates that there are three work files (WORK = 3) and three input files (FILES = 3). The input files will be read from the tape drive assigned to SYS002, SYS003, and SYS004. The work files will be on the drives assigned to SYS005, SYS006, and SYS007. Note from Figure 6-17 that SYS007 is assigned to drive X'181', which is the same drive assigned to the work files. Whenever the input file(s) is to be on the same drive which is to be used for a work file, the work file must be the last file which is to be used by the sort. Since SYS007 is the third work file of the three to be used, this requirement is satisfied. If five work files were to be used, the fifth work file would be mounted on the same drive which is used for the input files. The input file must always be on the drive which is used for the last work file.

Step 3: The sort program opens the first input file and reads the input data.

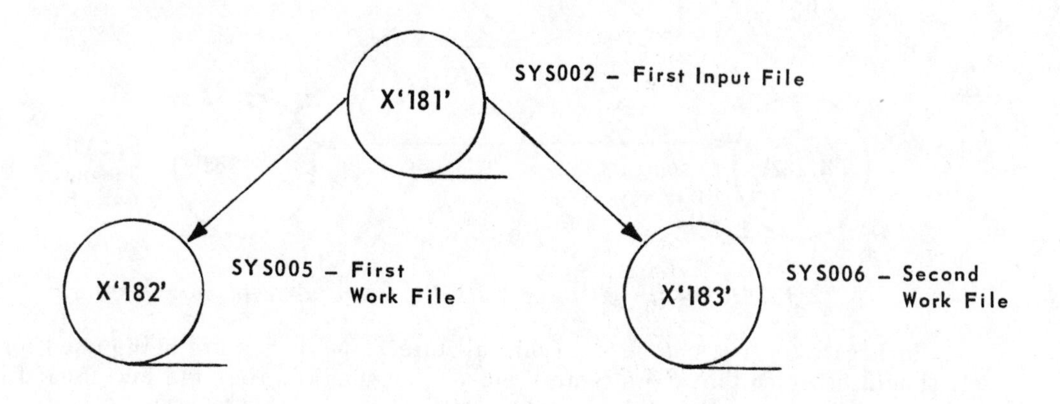

Figure 6-19 First Input File is Read

Note from Figure 6-19 that the first input file is read from the device (X'181') assigned to SYS002 and the first two work files, SYS005 (X'182') and SYS006 (X'183') may be used to store the groups of records which are created from the Internal Sort Phase. Note that although drive X'181' is to be used for a work file, it is not used in this initial phase of the sort and so the input files may be mounted on that drive.

Step 4: The remaining two input files are read from drive X'181'.

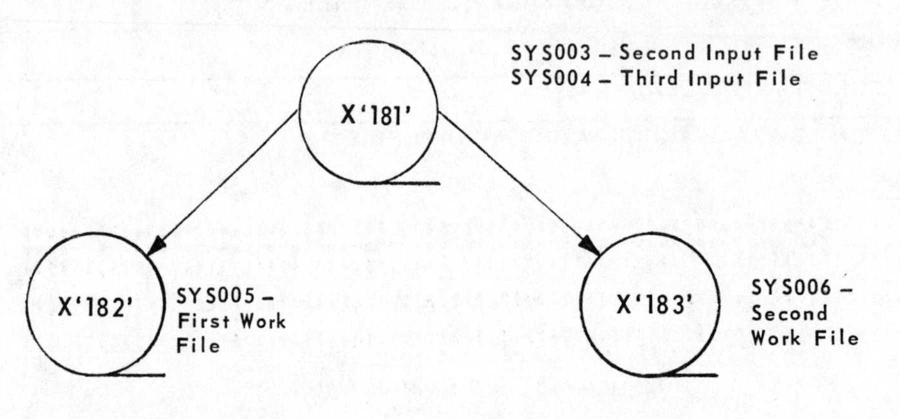

Figure 6-20 Second and Third Input Files are Read

Note from Figure 6-20 that the second and third input files are also read from drive X'181'. This is because the symbolic-units for the second and third files (SYS003 and SYS004) are assigned to X'181'. Note that the operator will have to mount new tapes for the second and third input files. Multi-file volumes cannot be used when I/O Pooling is to be performed. After the third tape has been read, the sort program begins phase 2 of the sort and this phase requires all three work files. Thus, drive X'181' will now be used as a work file.

Step 5: All three drives are used as work files.

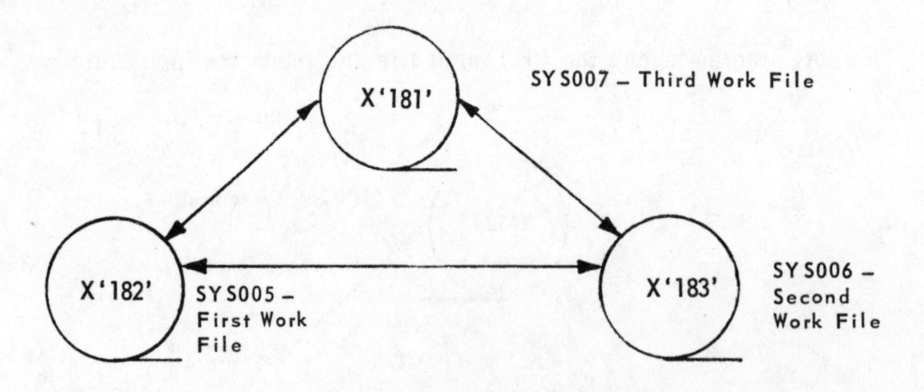

Figure 6-21 All three Work Files are used to Sort the Data

In Figure 6-21 it can be seen that all three tape drives are being used for work files. It should be noted that there is no "pause" between the time the last input file is read and when the sorting begins using drive X'181' as a work file. Thus, if the third input file were merely rewound, without being unloaded, it would be used as a work file and would be destroyed. Therefore, the entry CLOSE=UNLD is entered in the INPFIL control statement (see Figure 6-18) so that the third input file will be unloaded and not used as a work file. This process, of course, takes some time and this will cause the sort to run longer on the computer than if the third file were used as the work file.

Step 6: After the External Phase of the sort is complete, the output file is written on the device assigned to SYS001.

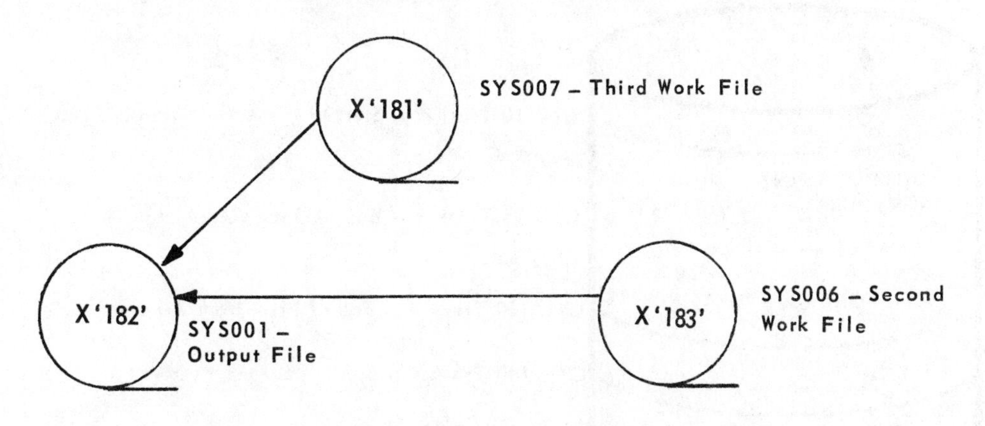

Figure 6-22 The Output File is written on SYS001

As has been noted previously, the device assigned to SYS001 is always used as the output file. From Figure 6-17 it can be seen that SYS001 is assigned to drive X'182'. Thus, drive X'182' is used as the output file. Drive X'182' was also used as the first work file. Whenever the output file is to be written on the same drive as a work file, it must be written on the drive which was used as the first work file. Thus, drive X'182' was assigned to SYS005, which was the first work file, and SYS001, which is the output file.

The example presented above is one example of the use of I/O pooling. Other combinations involving disk and tape may also be used. These combination are explained below.

1. The tape input and the tape output files may be read and written on the same tape drive. This option may be used when either tape or disk are used for the work area. Thus, if disk is being used for the work area, only a single tape drive need be used for both input and output. If this option is used with tape work areas, at least four tape drives must be used because input, output, and work may not all three be on the same tape drive.

2. When a disk work area is to be used, the output may be pooled with a portion of the work area. If this option is used, the work file and the output file must begin at the same cylinder and track address on the same disk drive. In addition, two separate extents must be specified for the disk work area, and the output file must not extend beyond the mid-point track address of the work area. The example in Figure 6-23 illustrates the extents which could be used for this option.

EXAMPLE

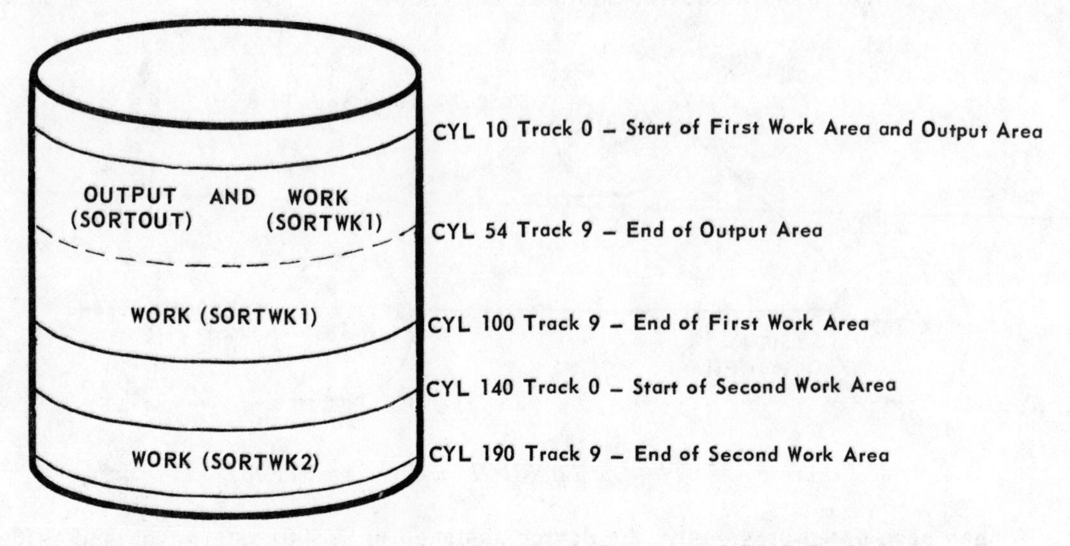

Figure 6-23 Example of pooling Disk Output and Disk Work Areas

In the above example it can be seen that the output file (SORTOUT) and the first extent of the disk work area (SORTWK1) both begin at the same disk location (Cylinder 10 Track 0). This is a requirement if the output and work areas are to be pooled. Note also that the output area ends at Cylinder 54 Track 9, which means that it does not extend past the mid-point track of the work area. This also is a requirement when pooling the output and work areas. A second extent for the work file extends from Cylinder 140 Track 0 to Cylinder 190 Track 9. This second extent must be specified for the disk work area if the output is to be written on the same portion of the disk as a part of the work area.

Note that the above example illustrates the output file being written on the same cylinders and tracks as a portion of the work file. This is the pooling of the I/O. If the output and work areas are contained on the same disk pack but do not use the same cylinders and tracks, no I/O pooling is taking place and the rules for I/O pooling need not be followed.

There are several important restrictions on I/O pooling. First, multiple pooling, that is, placing the input, output, and work on the same tape drive or on the same extents on the disk pack is not valid. If this occurs, the sort program will be abnormally terminated. Another restriction is that a disk input file cannot be pooled with the disk work area. The disk input file and the disk work file(s) must be contained on different extents on the same disk pack or on different disk packs. Multifile input tape volumes cannot be pooled nor can multivolume input tape files. If the tape input file is to be pooled with a tape work file or with the tape output file, the entire tape input file must be contained on a single volume.

VARIABLE-LENGTH RECORDS

In all of the examples presented previously, the records which were sorted by the Sort Program were fixed-length, that is, all of the input and output records contained the same number of bytes in each record. The Sort Program also has the capability to sort VARIABLE-LENGTH records, that is, records which contain a variable number of bytes in each record. The example below illustrates variable-length records.

EXAMPLE

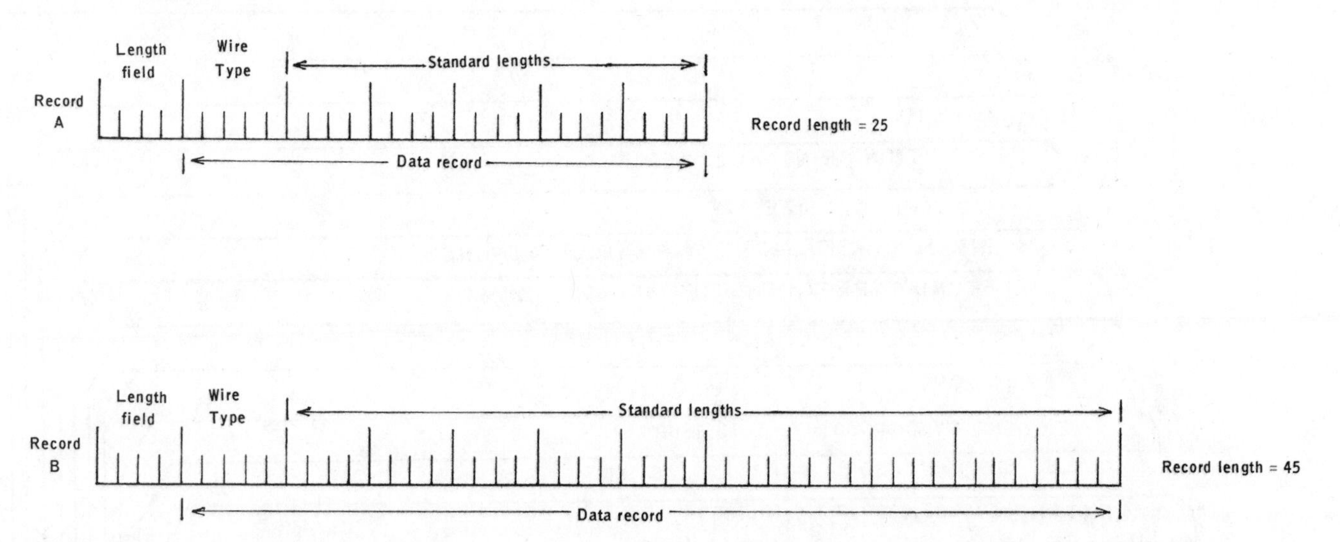

Figure 6-24 Example of VARIABLE-LENGTH RECORDS

In the example above, the records on a file contain a wire-type field, which contains an identifying number for a particular type of wire which is held in the inventory of a company, and the various standard lengths which are used with the wire type. Some wire types have only a few standard lengths while others have many standard lengths. Thus, variable-length records are utilized for the file in order that disk or tape storage space not be wasted by those wire types which have only a few standard lengths.

In addition, the variable length records contain a "length field" which always is stored in the first four bytes of the record and provides information to the Logical IOCS routines which process the record concerning the length of the record. This four-byte field is contained in every variable-length record.

When variable-length records are blocked, an additional length field for the entire block is required. This is illustrated in the following diagram.

EXAMPLE

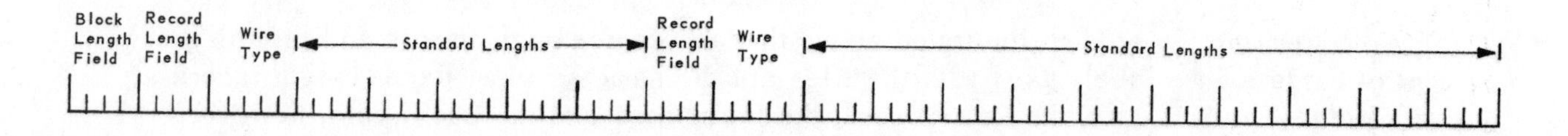

Figure 6-25 Example of BLOCKED Variable-Length Records

Note in Figure 6-25 that a 4-byte block length field precedes the record length field for the first record in the block. In addition, each variable-length record in the block contains the 4-byte record length field. In order to sort variable-length records, the lengths of these fields and blocks must be included in the sort control cards. The job stream below illustrates the necessary control cards to sort a tape input file containing variable-length records and create an output file on disk.

EXAMPLE

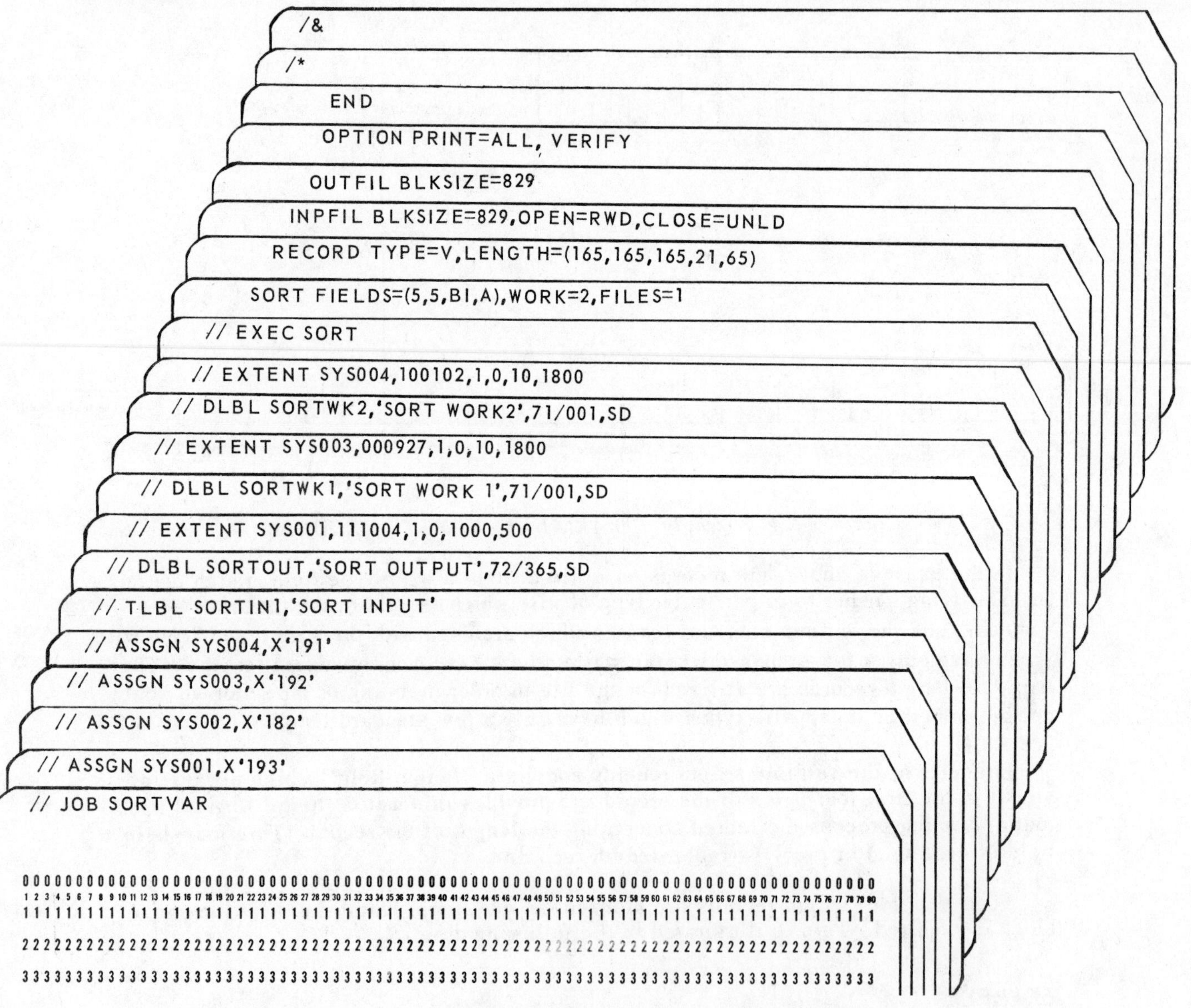

Figure 6-26 Job Stream to Sort Variable-Length Records

Note from the job stream illustrated above that the device assignments and the job control cards for the labels used with the files are the same as when fixed-length records are sorted. The Sort card, the Record card, the Inpfil card and the Outfil card, however, must contain unique entries for variable-length records.

SORT Control Card

As with previous examples, the SORT control card is used to specify which fields in the input record are to act as control fields for the sort. When variable-length records are to be sorted, the length field in each record must be considered when specifying the relative position in the record where the control field is to begin. Thus, in order to sort the variable-length records for the "wire" file illustrated in Figure 6-24, the first byte of the "Wire Type" field, which is to be the control field, is specified as beginning in column 5 of the record. Even though the first byte of the wire-type field is the first actual byte of data in the record, it must be specified as the fifth byte in the record because the length-field is considered the first four bytes in the record. The SORT control card in the job stream in Figure 6-26 illustrates the entries which would be made in order to sort using the "wire-type" field as the control field.

RECORD Control Card

As noted previously, the RECORD control card is used to specify the format of the input and output records and also the record lengths. The general format of the RECORD control statement which is used with variable-length records is illustrated below.

RECORD TYPE=V,LENGTH=(l_1,l_2,l_3,l_4,l_5)

Figure 6-27 General Format of RECORD Control card for Variable-Length Records

Note in the general format above that the TYPE=V operand must be specified for variable-length records. Note also that the LENGTH operand can contain 5 different entries to indicate lengths used for the variable-length records. The "l_1" entry is used to specify the maximum length of a record, that is, the length of the longest record in the file. In the example in Figure 6-26, the value stated in 165, which means the longest "wire" record contains the wire-type plus thirty-nine standard length entries. The length field at the beginning must be included in all of the record lengths used in the LENGTH operand. The "l_2" entry is used to indicate the length of the record to be sorted by the internal sort phase. This value will always be the same as l_1 unless special processing is performed on the input records through the use of an EXIT routine. The l_3 entry specifies the length of the output records to be written on the output file, and again, these records will be the same size as the input records unless special processing is performed. The "l_4" entry specifies the minimum length, or the length of the smallest record, in the input file. Although this entry is not required, it is best to include it in the LENGTH operand because it helps the efficiency of the sort program. In the example in Figure 6-26, the minimum length stated is 21, which means that 3 standard lengths and the "wire type" field are included on the smallest record. The "l_5" entry specifies the record length that occurs most frequently in the input file. This length is referred to as the "modal length". In the example, the modal length of the records is 65 bytes, which means the records contain 14 standard lengths plus the wire type.

The entries within the LENGTH operand, with the exception of l_1, are optional. If an operand is to be omitted but one following it is to be included, the omission must be indicated through the use of a comma. The following control card illustrates the omission of the l_2 and the l_3 entries.

EXAMPLE

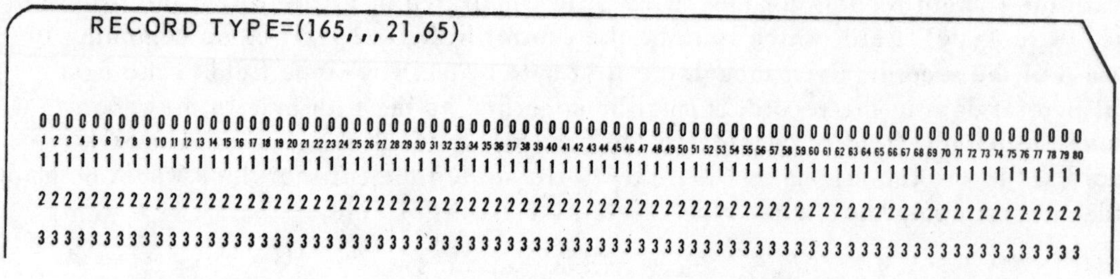

```
RECORD TYPE=(165,,,21,65)

00000000000000000000000000000000000000000000000000000000000000000000000000000000
1 2 3 4 5 6 7 8 9 10 11 12 13 14 15 16 17 18 19 20 21 22 23 24 25 26 27 28 29 30 31 32 33 34 35 36 37 38 39 40 41 42 43 44 45 46 47 48 49 50 51 52 53 54 55 56 57 58 59 60 61 62 63 64 65 66 67 68 69 70 71 72 73 74 75 76 77 78 79 80
11111111111111111111111111111111111111111111111111111111111111111111111111111111
22222222222222222222222222222222222222222222222222222222222222222222222222222222
33333333333333333333333333333333333333333333333333333333333333333333333333333333
```

Figure 6-28 RECORD Control Card with Omitted Entries

Note in the example above that the l_2 and l_3 entries in the LENGTH operand have been omitted and their omission is indicated through the use of commas. When the l_2 and l_3 entries are omitted, the sort program assumes that they have the same value as the l_1 entry, that is, it assumes that the maximum length of the record to be sorted by the internal sort phase and the maximum length of the output are the same as the maximum length of the input record. This is the most common occurrence when variable-length records are sorted because a special EXIT routine must be provided to change the record lengths. The l_4 and l_5 operands are included in the control card and should always be included for maximum sort efficiency when processing variable-length records.

INPFIL Control Card

The INPFIL control card is used to specify the attributes of the input file. Included in these specifications is the block size of the input file. When variable-length records are processed, the block size value must be the length of the largest possible block in the input file. As specified in the INPFIL statement in Figure 6-26, the largest block which will be input to the sort program contains 829 bytes. This block consists of a possible 5 of the largest records ($5 \times 165 = 825$) plus the four bytes contained in the length field at the beginning of the block (see Figure 6-25). Thus, the BLKSIZE=829 entry in the INPFIL control card specifies that the largest block which will be read by the sort program is 829 bytes in length. There may, of course, be shorter blocks because the records are variable in length.

OUTFIL Control Card

The value in the BLKSIZE entry in the OUTFIL control card is used to specify the maximum length of the block of variable-length records which will be written on the output file. The value used in the example in Figure 6-26 is 829, which means that the largest block on the output file, including the four-byte length field at the beginning of the block, will be 829 bytes in length. The sort program will store as many records as possible in the 829 bytes, that is, if all of the records are 165 bytes in length, it will store 5 records in the block but if all of the records are 100 bytes in length, it will store 8 records in the block.

The remaining entries in the sort control cards are the same whether fixed or variable-length records are used. As was noted previously, the decision to use fixed- or variable-length records depends upon the application which is to be processed. The Sort Program, however, is capable of processing both fixed and variable-length records so long as the proper entries are made in the Sort Control cards.

EXITS

In previous explanations of sort processing, reference has been made to EXITS which can be included in the sort processing. These exits are user-written routines which can be used to process data in a special manner dependent upon the needs of the application. These exit routines, which must be written in Assembler Language, are entered at specified points within the sort program processing to allow the user to manipulate the data being processed by the sort program. For example, using the sort exits, the user may add records to the input file, delete records from the input file, change values in the records being sorted, read data from a device or type of file not supported by the sort program (for example, read cards and pass them to the sort program for processing or read an indexed sequential file) or write output records on a device or type of file which is not supported by the sort program (for example, the output of the sort may be printed on a printer by a user routine). The job stream below illustrates the use of an exit from the sort program to read the records to be sorted by the sort program from the card reader and to write the sorted records on the printer.

EXAMPLE

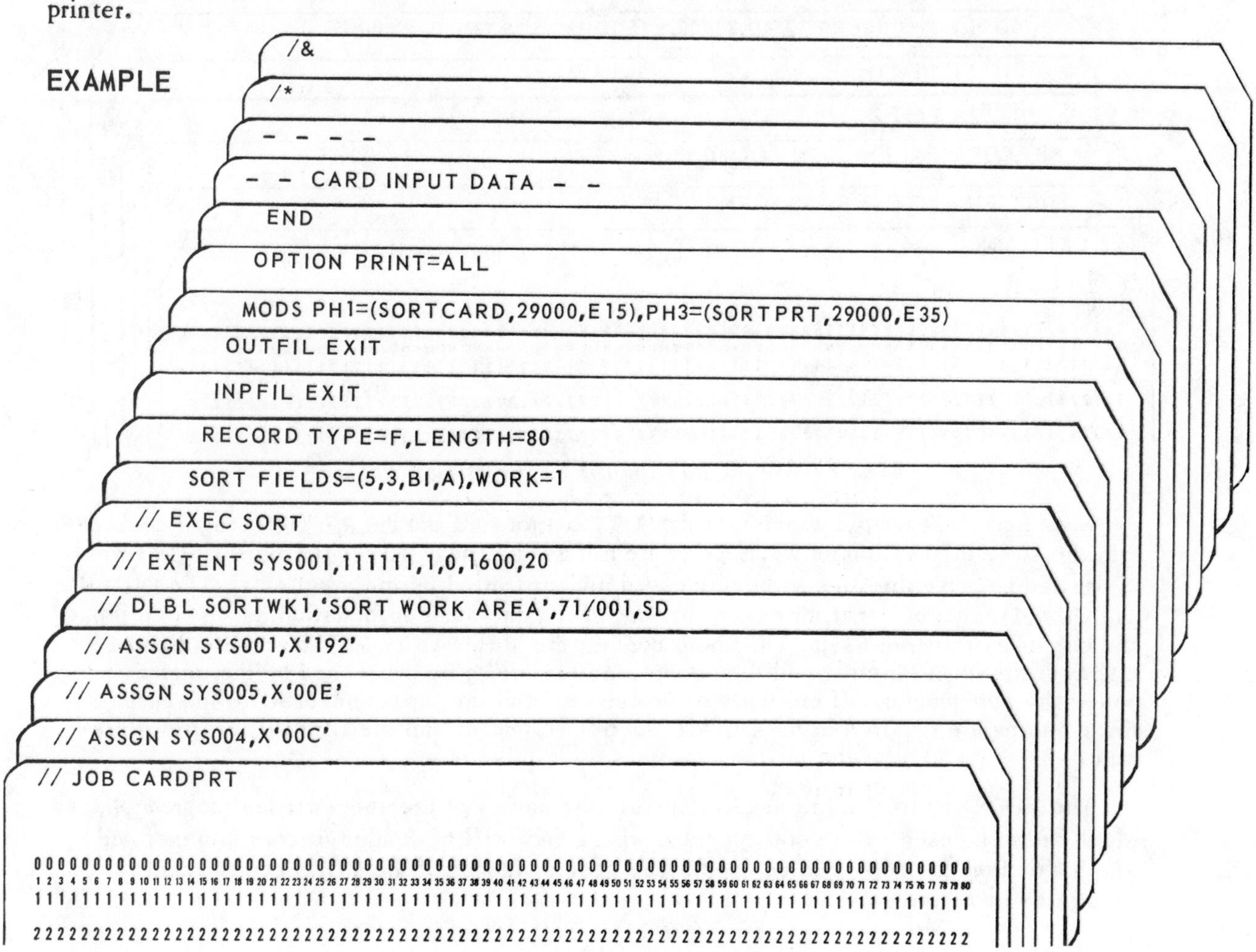

```
/&
/*
- - - -
- - CARD INPUT DATA - -
END
OPTION PRINT=ALL
MODS PH1=(SORTCARD,29000,E15),PH3=(SORTPRT,29000,E35)
OUTFIL EXIT
INPFIL EXIT
RECORD TYPE=F,LENGTH=80
SORT FIELDS=(5,3,BI,A),WORK=1
// EXEC SORT
// EXTENT SYS001,111111,1,0,1600,20
// DLBL SORTWK1,'SORT WORK AREA',71/001,SD
// ASSGN SYS001,X'192'
// ASSGN SYS005,X'00E'
// ASSGN SYS004,X'00C'
// JOB CARDPRT
```

Figure 6-29 Job Stream to Read Cards and Print Sorted Output

The job stream illustrated in Figure 6-29 is to be used to read the data to be sorted from the card reader and to write the sorted output on the printer. Note that the devices supported by the Sort program, that is, tape drives and sequential disk files, are not used for either input or output. Thus, a "user-exit" must be written in order to read the cards and to print the output of the sort. When exits are to be used for both the input and the output, such as in the example, the work files must begin on the device assigned to SYS001. In the example in Figure 6-29, it can be seen that drive X'192' is assigned to SYS001, indicating that the work file will be stored on drive X'192'. The input SYS number and the output SYS number may be any desired by the programmer since it is the user-written exit routines which will read the input and write the output. In the example it can be seen that the card reader is assigned to SYS004 and the printer is assigned to SYS005. Thus, the routines will read the input from device X'00C', which is assigned to SYS004, and the output will be written on the printer, X'00E', which is assigned to SYS005.

In order to utilize exits from the sort program, unique entries must be made in the INPFIL and OUTFIL control cards, and a new control card, the MODS control card, must be used. The Sort control cards utilized in Figure 6-29 are again illustrated below.

EXAMPLE

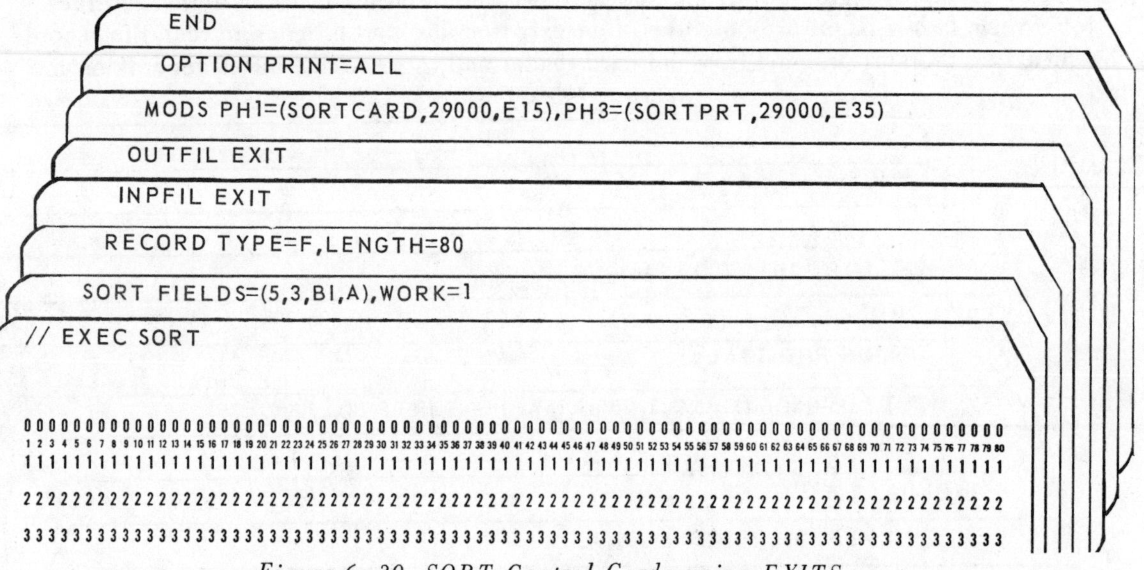

Figure 6-30 SORT Control Cards using EXITS

Note from the example above that the SORT control card and the RECORD control card contain the same information as when Exits are not used. The FILES entry on the SORT card is omitted because the files to be processed are controlled by the input exit. The INPFIL and OUTFIL control cards, however, do not specify the same information as when an input and output exit are not used. They both contain the single entry EXIT, which indicates that both the input function and the output function are to be processed by the user exit, not by the sort program. If exits are to be used to read the input and/or write the output, these values are required in the INPFIL control statement and the OUTFIL control statement.

The MODS control card is used to specify the names of the user-written program phases which are to be used by the sort program, where they will be loaded in core storage, and which exit from the sort is to be used. The general format of the MODS control card is illustrated in Figure 6-31.

General Format

```
MODS PHn=(phasename,loading information,exit, . . .), . . .
```

Figure 6-31 General Format of MODS Control Card

In the general format above it can be seen that the identifier MODS must be written as shown and must, like all sort control identifiers, begin to the right of column 1. One or more blanks must follow the keyword MODS. The PHn value is used to specify which phase of the sort is to have an exit routine used. The valid values are PH1 for the internal sort phase or phase 1 of the sort, PH2 for the external phase or phase 2 of the sort, and PH3 for the final merge phase or phase 3 of the sort program. In the example in Figure 6-30 it can be seen that the values PH1 and PH3 are both used. Thus, there will be an exit in phase 1 of the sort (used to read the cards) and phase 3 of the sort (used to write the sorted records on the printer).

For each phase of the sort which is to process an exit routine, the information within the parentheses must be included. The "phasename" entry is used to indicate the phasename of the routine which is catalogued in the Core Image Library and which will be loaded into core storage to provide the processing at the exit point. Each routine which is to be used for an exit from the sort program must be catalogued in the core image library so that the sort program can load it into core storage during the phase of the sort which is being processed. The "loading information" entry within the parentheses is used to indicate to the sort program either the length of the program phase which is to be loaded into core storage or the address in core storage where the phase will be loaded. If the address where the phase is to be loaded into core storage is indicated, it is specified as shown in Figure 6-30, that is, the decimal address is specified. If the length of the routine is specified, it is stated in the format Lnnnn, where nnnn is the number of bytes within the routine. For example, if the exit routine were 500 bytes in length, the entry in the MODS card would be L500. The determination as to whether the absolute load address or the length of the routine are to be specified is made dependent upon whether the routine is "self-relocating", that is, whether it may be loaded into any portion of core storage for execution. If the program is self-relocating, the length should be stated. If the program is not self-relocating, that is, it has been link-edited with a specific load address, the load address which was used when the program was catalogued should be used as the "loading information" operand.

The "exit" operand is used to indicate which exit is to be used. Each phase of the sort has a number of exits which may be utilized for functions such as reading the input data, checking nonstandard labels, etc. This value must be in the format Enn, where "nn" is the number of the exit. As can be seen from the MODS control card in Figure 6-30, Exit 15 is used to read input from a device not supported by the sort program, and Exit 35 is used to write the output file. A complete list of all of the available exits and the phases in which they are used is illustrated in Figure 6-32.

EXAMPLE

Uses for Exits	Internal Sort Phase (Phase 1)				External Sort Phase (Phase 2)			Final Merge Phase (Phase 3)					
	E11	E15	E17	E18	E21	E25	E27	E31	E32	E35	E37	E38	E39
Assignment (Initialization)	X				X			X					
Taking Checkpoints	X				X			X					
Processing Labels	X		X		X		X	X			X		
Opening Files	X				X			X					
Closing Files	X		X				X	X			X		
Reading Input File		X											
Counting Input Records		X											
Inserting/Deleting Records		X				X				X			
Lengthening/ Shortening Records		X								X			
Modifying Data Contents of Record*		X							X	X			
Modifying Control Fields*		X							X	X			
Summarizing Records						X							
Substituting Records (Merge)									X				
Suppressing Sequence Checking										X			
Writing Output File										X			
Processing Read Errors				X								X	
Processing Direct Access Write Errors													X

*Does not include lengthening or shortening of control fields or data records.

Figure 6-32 List of Available EXITS from Sort Program

Note from the list illustrated above that there are four exits for phase 1 of the sort, three for phase 2, and six for phase 3. Each of the exits can be used for one or more functions as listed in the "Uses for Exits" column. For a detailed explanation of each of the sort exits and their uses, see IBM SRL GC28-6676, IBM System/360 Disk Operating System Tape and Disk Sort/Merge Program.

As noted previously, the exit routines are processed at specified portions of the sort program in order to perform the function designated. The following example illustrates the processing steps that occur when the job stream in Figure 6-29 is executed.

EXAMPLE

Step 1: The job control cards are read and the sort program is loaded into core storage.

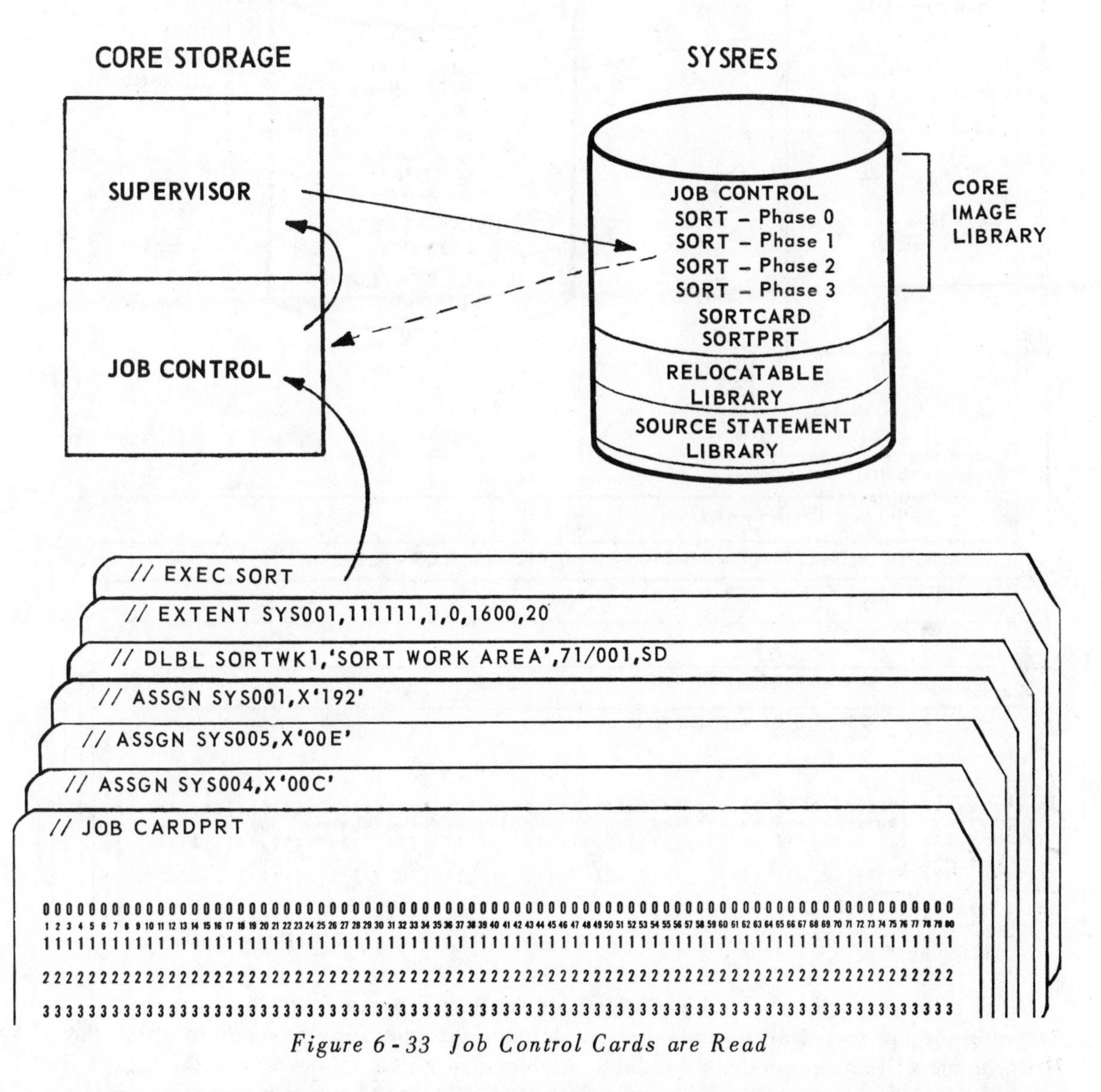

Figure 6-33 Job Control Cards are Read

In the example above, it can be seen that the job control cards are read and processed by the Job Control Program. Note that, since both the input and the output files are to be processed by sort exit routines, the work file is stored on the device assigned to SYS001. The card input file will be stored on device X'00C', which is assigned to SYS004 and the output will be written on the printer, X'00E', which is assigned to SYS005. These assignments must correspond to the devices used in the sort exit routines and are not required for the sort program.

Step 2: The Sort program (Phase 0) reads the sort control cards and establishes the processing parameters for the sort program.

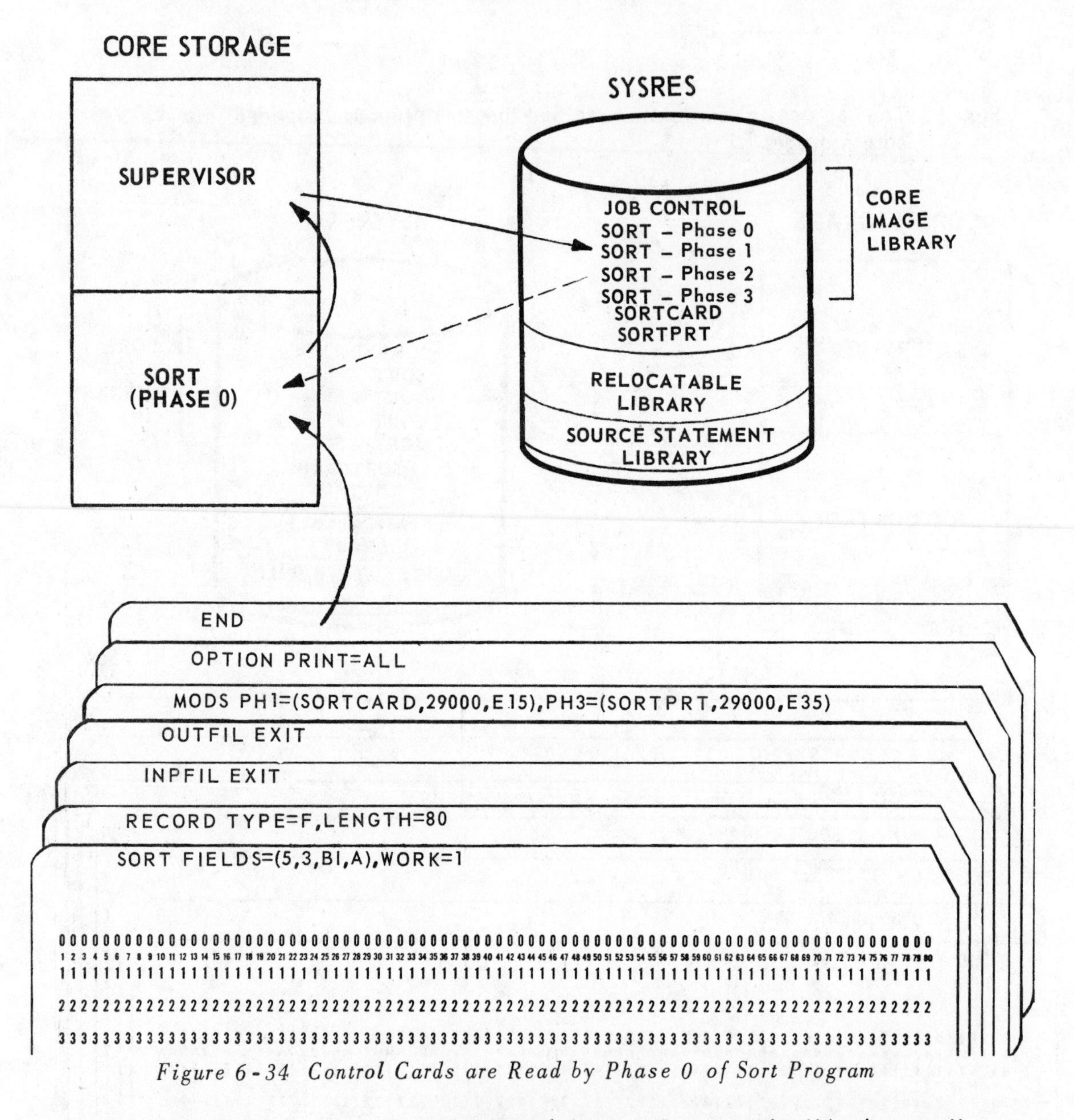

Figure 6-34 Control Cards are Read by Phase 0 of Sort Program

Note in the example above that Phase 0 of the Sort Program (the "Assignment" Phase) reads the sort control cards and establishes the operating parameters which the sort program will use to process the data. Included in these parameters is the fact that a user exit will be used to read the input data (INPFIL EXIT) and the output file will be written by a user exit (OUTFIL EXIT). When the assignment phase has completed processing, it returns control to the Supervisor so that Phase 1 of the sort program can be loaded into core storage.

Step 3: After Phase 1 of the sort is loaded into core storage, it loads the user exit routine, SORTCARD, from the core image library into the core storage address specified on the MODS Control card (29000).

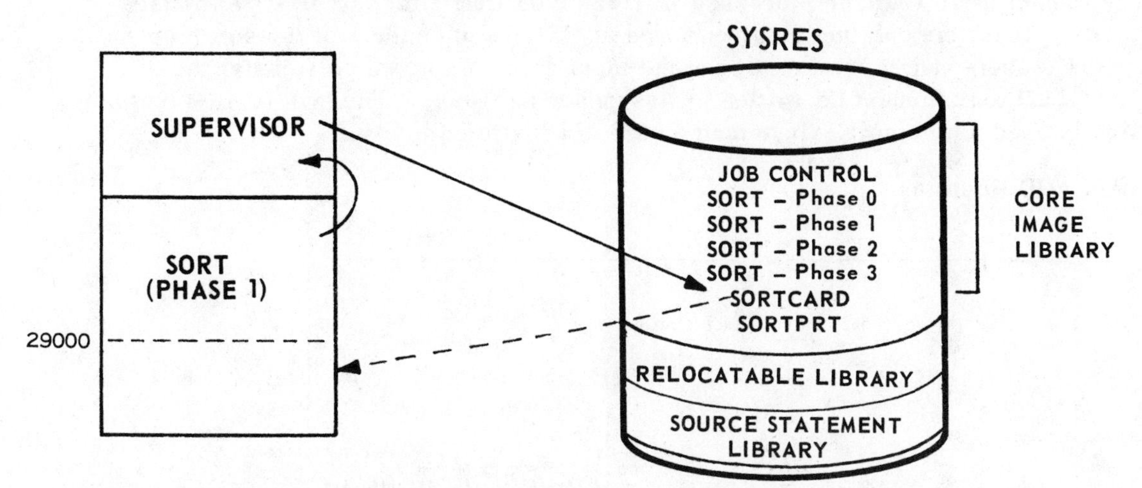

Figure 6-35 SORTCARD Phase is loaded into Core Storage

As can be seen from the example above, when phase 1 (Internal Sort) of the sort program gains control, it determines that a user exit routine is to read the input file. Thus, it must load the user routine into core storage from the Core Image Library. Since the entry in the MODS control card specified that the routine should be loaded into core storage at decimal location 29000, this is where the phase SORTCARD is loaded. It should be noted that the user routine SORTCARD must either be self-relocating or must be link-edited to be loaded at core address 29000 in order for processing to be performed properly.

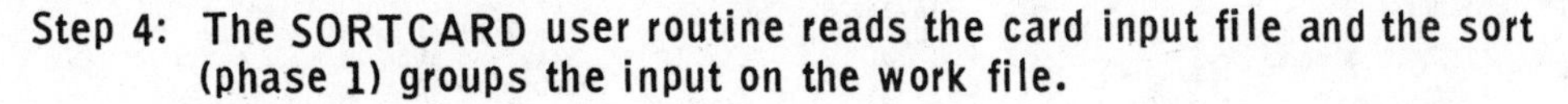

Step 4: The SORTCARD user routine reads the card input file and the sort (phase 1) groups the input on the work file.

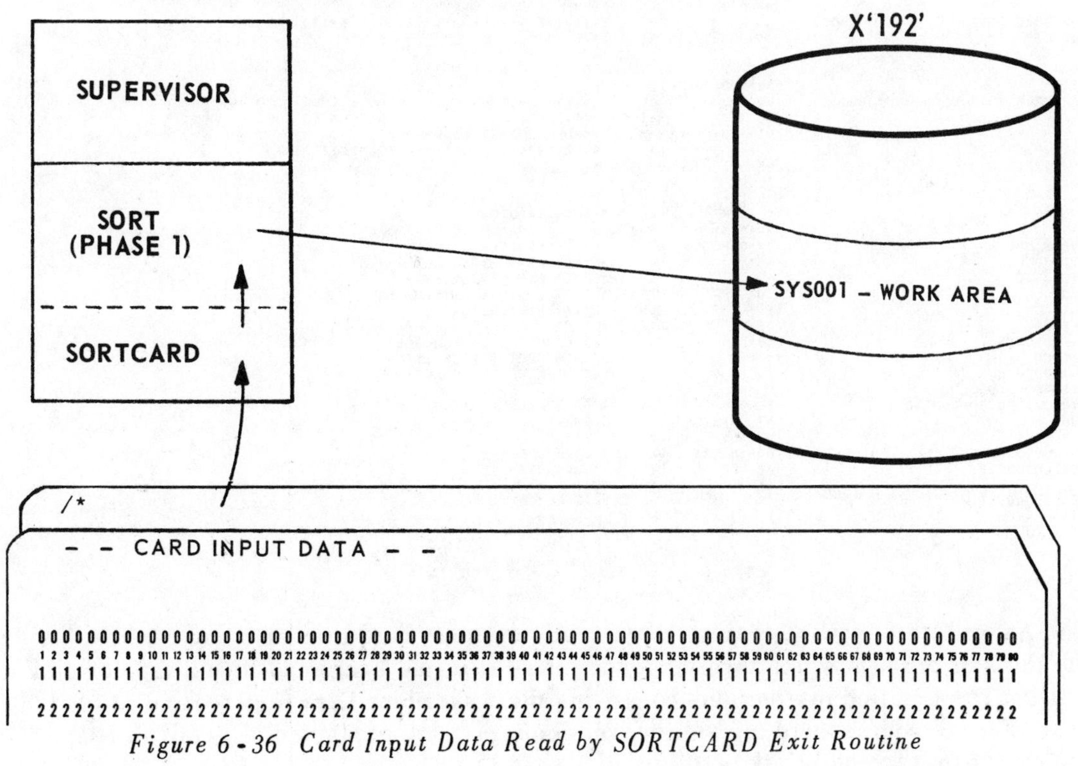

Figure 6-36 Card Input Data Read by SORTCARD Exit Routine

Note in Figure 6-36 that the card input file is read by the SORTCARD exit routine. The card is then passed to phase 1 of the sort program which processes the data in the same manner as it would be processed if it was read from a tape or disk file by the sort program. Thus, the only function performed at EXit 15 of phase 1 of the sort program is to allow a user-written routine to read the input data. As noted previously, the SORTCARD routine must be written in Assembler Language. The SORTCARD routine which is used as the user exit to read a card is illustrated below.

SORTCARD Routine

```
                                                                                                          PAGE    1

  LOC  OBJECT CODE    ADDR1 ADDR2  STMT   SOURCE STATEMENT                                       FDOS CL3-7 11/08/71

000000                                1 SORTREAD START 0
000000                                2          USING *,15
000000 00000000                       3          DC    A(0)
000004 47F0 F010            00010     4          B     E15                         GO TO EXIT 15 ROUTINE
000008 00000000                       5          DC    A(0)
00000C 00000000                       6          DC    A(0)
                                      7 E15      SAVE  (14,12)                     SAVE THE REGISTERS
                                      8+* 360N-CL-453 SAVE      CHANGE LEVEL 3-8                                 3-8
000010 90EC D00C            0000C     9+E15      STM   14,12,12+4*(14+2-(14+2)/16*16)(13)
000014 05C0                          10          BALR  12,0                        ESTABLISH REG 12 AS BASE REG
                                     11          DROP  15
000016                               12          USING *,12
000016 1841                          13          LR    4,1                         SAVE ADDRESS OF TABLE
000018 5821 0004            00004    14          L     2,4(1)                      OBTAIN ACTION WORD

                                     16          EXCP  CARDCCB                     READ AN INPUT CARD
                                     17+* 360N-CL-453 EXCP      CHANGE LEVEL 3-0
00001C 5810 C0C2            000D8    18+         L     1,=A(CARDCCB)
000020 0A00                          19+         SVC   0
                                     20          WAIT  CARDCCB                     WAIT FOR I/O
                                     21+* 360N-CL-453 WAIT      CHANGE LEVEL 3-0
000022 5810 C0C2            000D8    22+         L     1,=A(CARDCCB)
000026 9180 1002      00002          23+         TM    2(1),X'80'
00002A 4710 C01A            00030    24+         BO    *+6
00002E 0A07                          25+         SVC   7

000030 D501 C062 C0B2 00078 000C8    27          CLC   CARDINPT(2),SLSHAST         IS IT END OF DATA
000036 4780 C038            0004E    28          BE    NODATA                      YES, GO TO SET INDICATOR
                                     29 *                                          FOR NO SORT RETURN
00003A D203 2000 C0B4 00000 000CA    30          MVC   0(4,2),INSERT               MOVE INSERT CODE TO SORT AREA
000040 4130 C062            00078    31          LA    3,CARDINPT                  LOAD ADDRESS OF INPUT AREA
000044 5034 0000            00000    32          ST    3,0(4)                      STORE I/O ADDRESS IN TABLE
                                     33          RETURN (14,12)                    RETURN TO SORT
                                     34+* 360N-CL-453 RETURN    CHANGE LEVEL 3-0
000048 98EC D00C            0000C    35+         LM    14,12,12+4*(14+2-(14+2)/16*16)(13)
00004C 07FE                          36+         BR    14

00004E                               38 NODATA   EQU   *
00004E D203 2000 C0B8 00000 000CE    39          MVC   0(4,2),NORETURN             MOVE CODE 8 TO INDICATOR
                                     40          RETURN (14,12)                    RETURN TO SORT
                                     41+* 360N-CL-453 RETURN    CHANGE LEVEL 3-0
000054 98EC D00C            0000C    42+         LM    14,12,12+4*(14+2-(14+2)/16*16)(13)
000058 07FE                          43+         BR    14

                                     45 CARDCCB  CCB   SYS004,CARDCCW
                                     46+* 360N-CL-453 CCB       CHANGE LEVEL 3-5                                 3-5
00005A 0000                          47+CARDCCB  DC    XL2'0' RESIDUAL COUNT
00005C 0000                          48+         DC    XL2'0' COMMUNICATIONS BYTES
00005E 0000                          49+         DC    XL2'0' CSW STATUS BYTES
000060 01                            50+         DC    AL1(1) LOGICAL UNIT CLASS
000061 04                            51+         DC    AL1(4) LOGICAL UNIT
000062 00                            52+         DC    XL1'0'
000063 000070                        53+         DC    AL3(CARDCCW) CCW ADDRESS
000066 00                            54+         DC    B'00000000' STATUS BYTE
000067 000000                        55+         DC    AL3(0) CSW CCW ADDRESS
00006A 000000000000
000070 0200007800000050              56 CARDCCW  CCW   X'02',CARDINPT,X'00',80
000078                               57 CARDINPT DS    CL80
0000C8 615C                          58 SLSHAST  DC    C'/*'
0000CA 0000000C                      59 INSERT   DC    X'0000000C'
0000CE 00000008                      60 NORETURN DC    X'00000008'
                                     61          END
0000D8 0000005A                      62                =A(CARDCCB)
```

Figure 6-37 SORTCARD Routine

The Assembler Language routine above reads a single card at a time and passes this card to the sort program which then processes it as if it had been read from a tape or disk input file. The detailed instructions on writing the Assembler Language routine to read an input file, the address and register conventions, etc. are explained in detail in IBM SRL GC28-6676 Tape and Disk Sort/Merge.

Step 5: When the input file has been read, phase 2 of the sort program is loaded into core storage and the External Sort phase sorts the records on the work file.

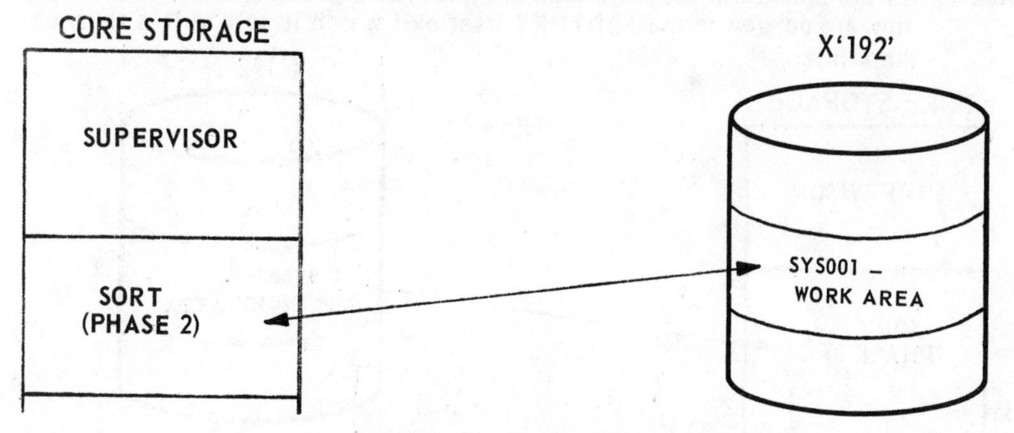

Figure 6-38 Phase 2 of Sort Program

In the above example it can be seen that the input records stored in the work area on disk are processed by Phase 2 of the Sort Program until the records are in a sequence which can be merged by a single pass of the final merge phase. Phase 3 of the Sort program is then loaded into core storage for the final merge pass of the sorted records.

Step 6: After the merge phase (phase 3) of the sort is loaded into core storage, it determines that the output is to be written by an exit routine (OUTFIL EXIT). Thus, it must load the proper program phase from the core image library into core storage

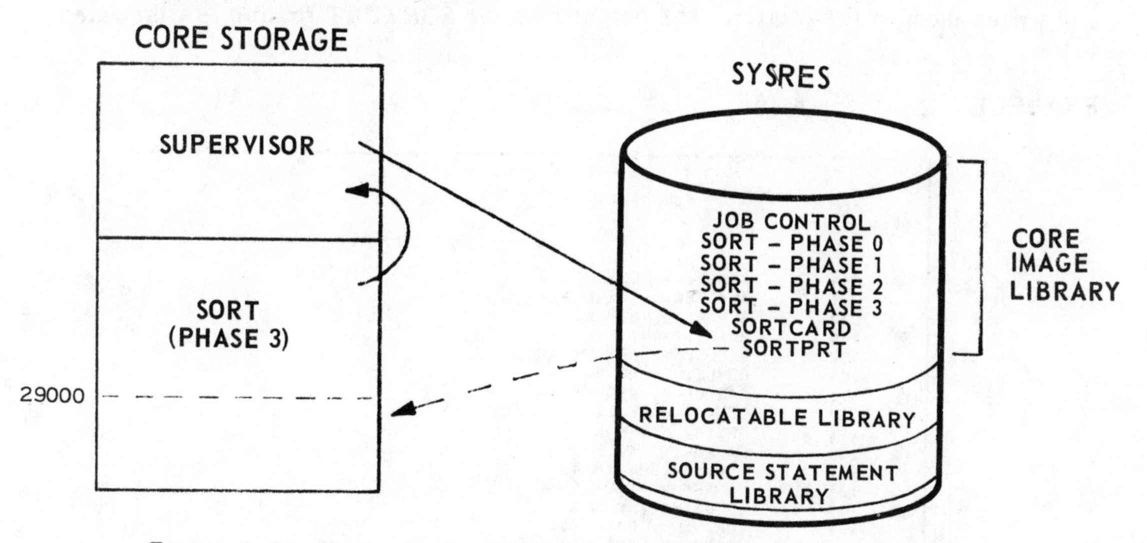

Figure 6-39 User-routine SORTPRT is loaded into Core Storage

Note in the above diagram that the user-written phase SORTPRT is loaded into core storage by Phase 3 of the Sort Program. The routine SORTPRT is to be used to write the sorted output records on the printer. Note also that it is loaded into core storage location Dec 29000, as specified on the MODS Control Statement.

Step 7: **As the records in the work area are merged by phase 3 of the Sort program, they are passed to the SORTPRT user exit which in turn writes them on the printer.**

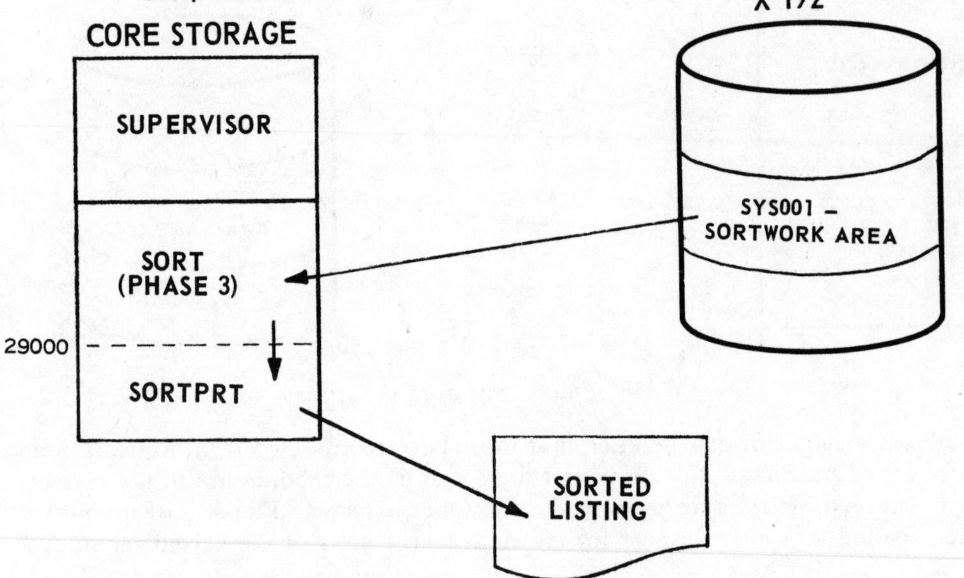

Figure 6-40 SORTPRT Exit Routine Writes sorted record on Printer

As can be seen from the diagram above, the SORTPRT exit routine, which was loaded into core storage by Phase 3 of the Sort program, obtains the records from the sort phase and writes them on the printer. The output from the SORTPRT routine is illustrated below.

EXAMPLE

S/M	EMPLOYEE
004	ACHER, WILLIAM C.
027	ALHOUER, ELAINE E.
030	ALLOREN, RUTH W.
100	BATES, TONY F.
102	BELLSLEY, ARTHUR A.
105	BOYLE, RALPH P.
111	CARTOLER, VIOLET B.
122	CENNA, DICK L.
171	COSTA, NAN S.
179	DAMSON, ERIC C.
181	DELBERT,EDWARD D.
185	DONNEMAN, THOMAS M.
207	EBERHARDT, RON G.
214	EDMONSON, RICK T.
215	EDSON, WILBUR S.
282	ESTABAN, JUAN L.
292	EVERLEY, DONNA M.
300	FELDMAN, MIKE R.
304	FROMM, STEVE V.
308	GLEASON, JAMES E.
310	GORMALLY, MARIE N.
311	GROLER, GRACE B.
315	HALE, ALAN A.
317	HANBEE, ALETA O.
318	HANEY, CAROL S.
322	HARLETON, JEAN H.
325	HATFIELD, MARK I.
332	HELD, ANNA J.
409	ICK, MICK W.

Figure 6-41 Output from SORTPRT Exit Routine

Note that the report illustrated in Figure 6-41 is written directly from the sort program. It is written by the exit routine SORTPRT. In previous examples of the sort, it was necessary to use the Disk-To-Print Utility program in order to print the contents of the output file. In this example, the report was created directly out of the sort program. The detailed information concerning the use of the exit routines may be found in IBM SRL GC28-6676 Tape and Disk Sort/Merge Program.

In addition to reading input files and writing output files, exits from the sort program can perform such functions as checking and processing user-labelled files, altering records which have been sorted, inserting or deleting sorted records, taking checkpoints, and handling Input/Output errors.

MISCELLANEOUS SORT OPTIONS

As has been illustrated in the previous two chapters, the sort program is used primarily to sort data records into a prescribed sequence for processing by other programs. The devices used for input, output, and work files are normally tape and/or disk, but may be any devices provided a user-written exit is used. In addition to sorting data records, the sort may, for particular applications, be used for several other functions. These are described below.

Address Out Option

In previous examples it was seen how disk output records are sorted for processing by other programs. In some applications, it is desirable to not store the output records in the output file, but only a disk address referencing the proper sequence in the input file, that is, the output file consists only of disk addresses which reference the input file and which are in sequence based upon the control fields in the input records. This option is requested by the ADDROUT=A entry in the OPTION control card. If this type of processing is requested, the input and work files must both be stored on disk devices.

If it is desired to have the output file consist of both the disk address of the input file and the control field associated with each record, the entry ADDROUT=D is included in the OPTION control card. This will cause both the address of the sorted input records and the control field of each record to be stored in the output file.

CALCAREA Option

This option may be specified if it is desired to determine the work area required on a direct-access device for the associated sorting operation. When the CALCAREA entry is placed in the OPTION Control Card, only the "assignment phase" (phase 0) of the Sort Program is executed. The sort/merge program prints on SYSLST the minimum number of tracks of work area which is required for the sort, based upon the input record size and block length, and the number of records specified in the SIZE entry in the SORT control card. It also determines the number of work area tracks which are required for the sort to give its best performance, that is, the number of tracks to be used for maximum efficiency.

If the CALCAREA option is used, no sorting will take place and the sort program is terminated after the assignment phase. This option may be used before the actual sorting operation takes place so that the proper amount of work space may be allocated for maximum efficiency when the input file is actually sorted. Although no job control label cards need be submitted when the CALCAREA option is specified, all of the sort control cards must be submitted so that the Sort program can make the determination for the area required for the work file.

ALTWK Option

This option may be used in the OPTION Control card if it is desired to double the size of the input file which may be processed when tape is being used for work files. This option is never used for disk work files. If ALTWK is included on the OPTION control card, a TLBL card should be used if the file is labelled and should contain the filename entry SORTALT. The tape volume must be mounted on a drive assigned to SYSn + 1, where "n" is the last number used for a work file. For example, if SYS003, SYS004, and SYS005 were used for tape work files, the ASSGN and TLBL cards illustrated below would be used for the alternate work file.

EXAMPLE

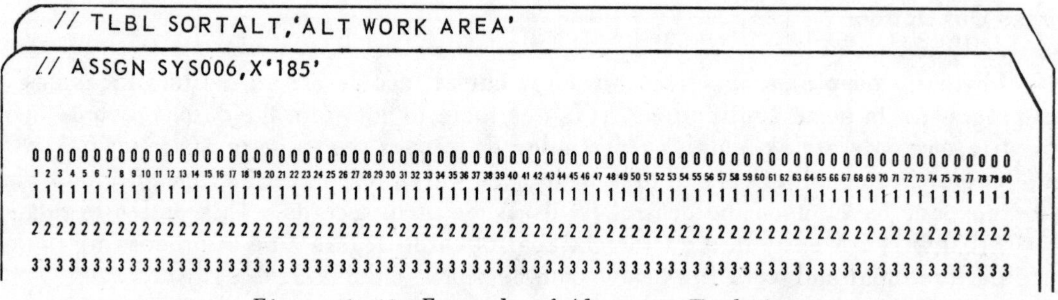

Figure 6-42 Example of Alternate Work Area

Note in the example above, that since the last work file is to be mounted on the device assigned to SYS005, the alternate work area must be assigned to SYS006. Note also the filename entry SORTALT in the TLBL card. The use of the ALTWK option allows two full reels of tape to be used for the input file(s) instead of one as when an alternate work area is not used.

STUDENT ASSIGNMENTS

1. Write the job stream to sort a file of records stored on magnetic tape. The file contains 120 byte records blocked 5 records per block. The records should be sorted on the first 22 columns in an ascending sequence (binary sort), columns 56-62 in a descending sequence (zoned-decimal), and columns 25-27 in an ascending sequence (packed-decimal field). The work area and the output file should be stored on a disk file. The use of the devices to sort the file and the dating criteria should be determined by the programmer.

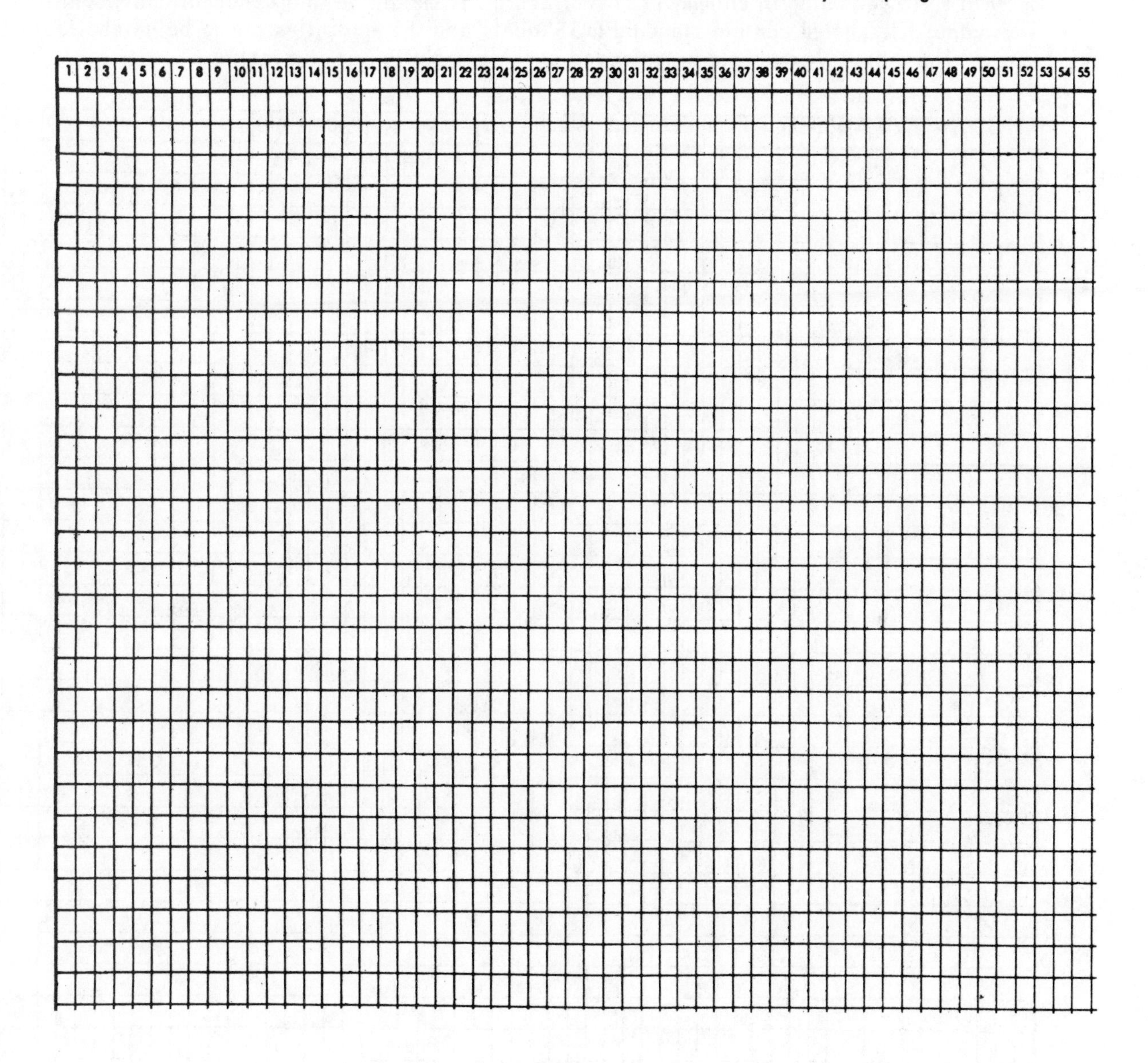

2. Write the job stream to sort two input files which are stored on disk pack 444321. The first file, which has the identification CUSTOMER MASTER, is stored on Cylinder 3 Track 0 through Cylinder 98 Track 9. The second input file, named CUSTOMER MASTER 2, is stored on Cylinder 102 Track 0 through Cylinder 151 Track 9. The work areas for the sort are to be on tape drives X'180', X'181', and X'184'. The output file should be written on tape drive X'185'. The major sort control field is the customer number, which is 7 bytes in length and is stored beginning in column 32 of the 95 byte record. The input records are blocked 10 records per block. The minor field for the sort is the 4 byte packed decimal field beginning in column 19 of the record. It should be in a descending sequence. The output file should contain standard DOS labels and the work files are to be unlabelled tapes. The output tape should be unloaded at the completion of processing.

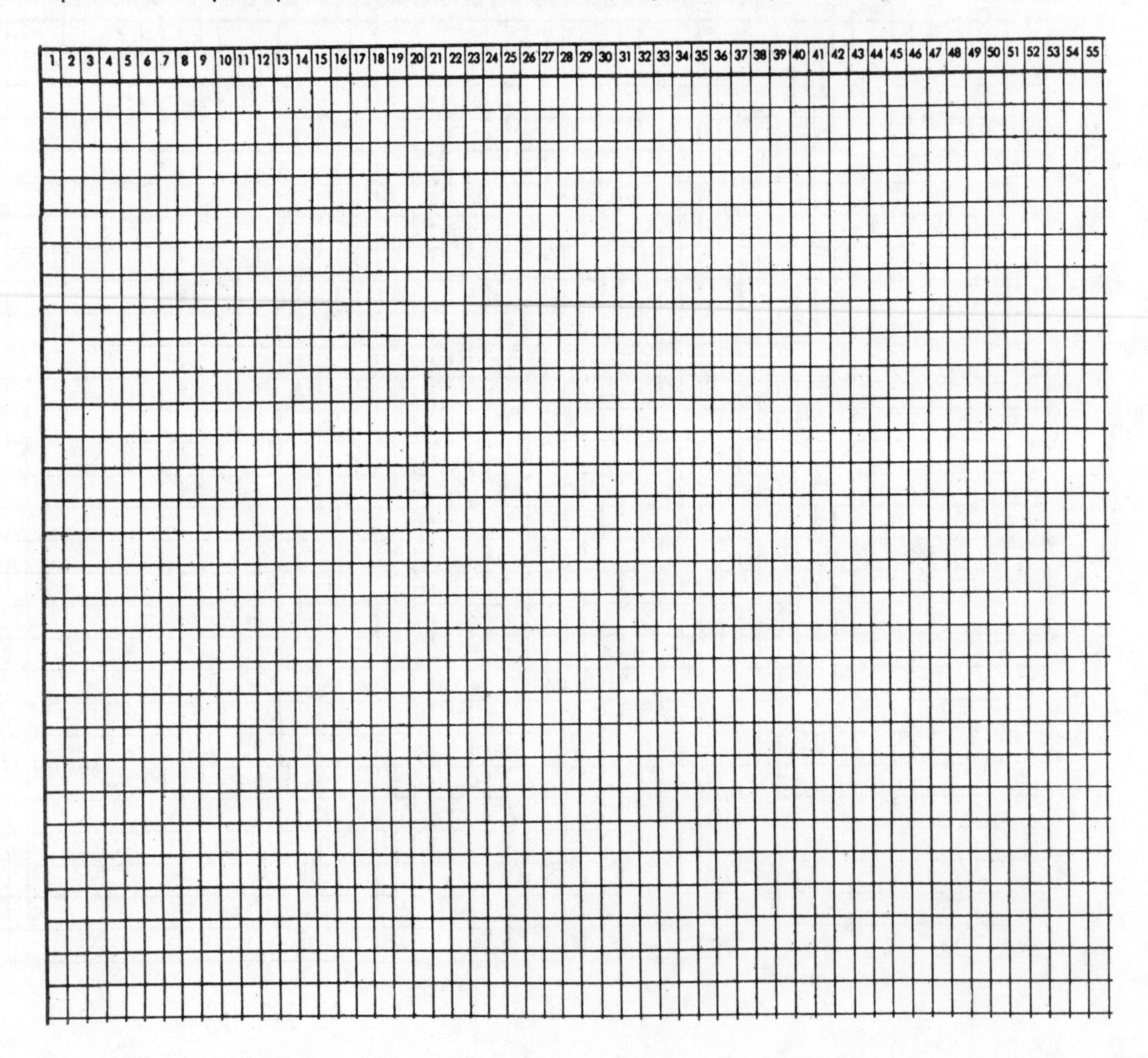

3. Three tape input files are to be sorted on the following control fields.

A. Column 43 for 7 bytes, ascending.
B. Column 68 for 5 bytes, descending.
C. Column 79 for 2 bytes, ascending.
D. Column 3 for 20 bytes, ascending.
E. Column 32 for 1 byte, descending.
F. Column 25 for 2 bytes, ascending.
G. Column 58 for 3 bytes, ascending.
H. Column 56 for 1 byte, descending.
I. Column 50 for 2 bytes, descending.
J. Column 75 for 1 bytes, descending.

The input records are 80 bytes in length, blocked 10 records per block. All of the fields in the record are to be sorted in a binary collating sequence. Each of the input files is to be mounted on drive X'184'. The output file should be written on drive X'185' and the tape work files, which are unlabelled, should be mounted on drives X'182', X'186', and X'184'. The approximate number of input records to be sorted is 20,000 records.

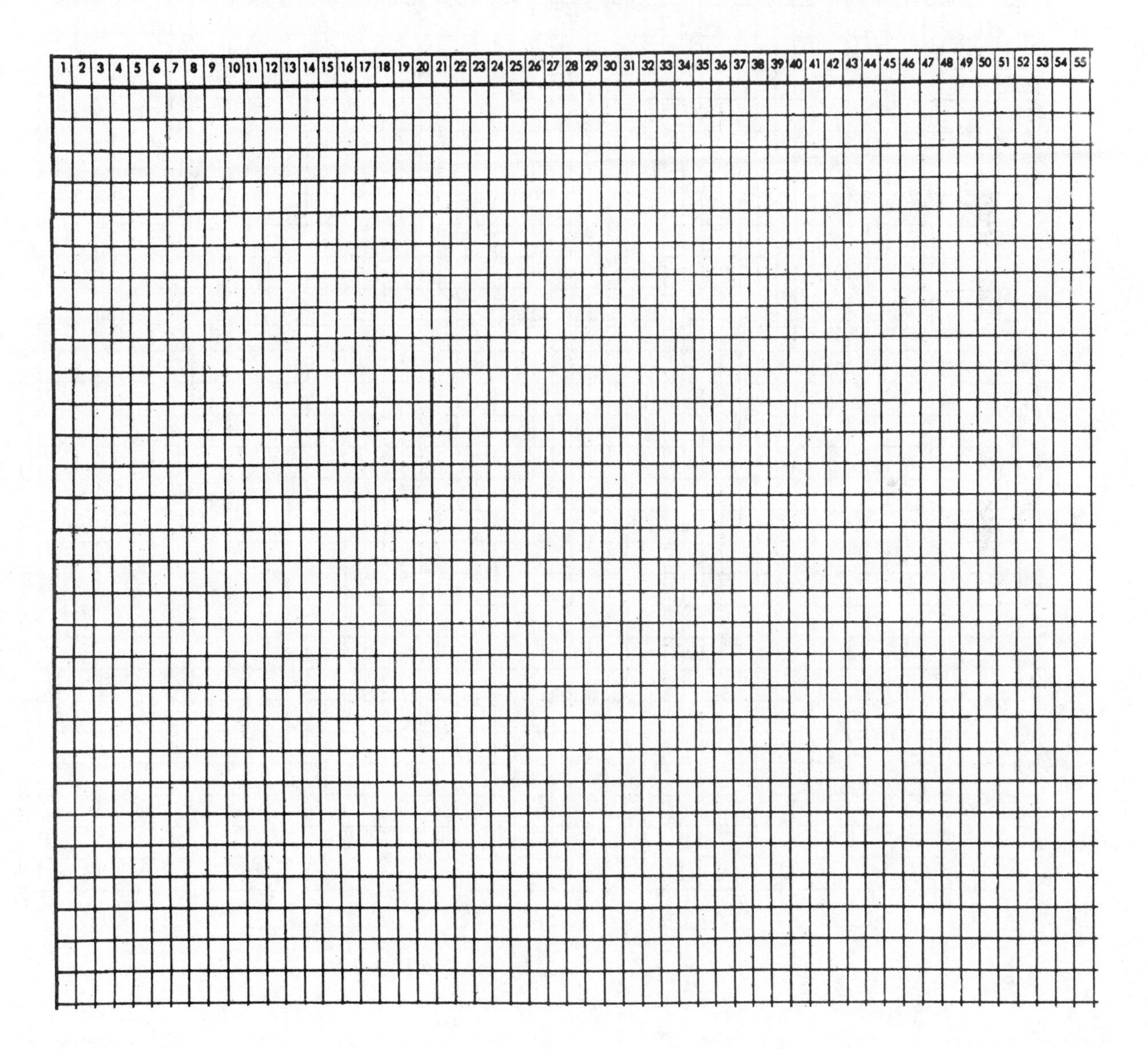

4. An input file containing variable-length records with a maximum length of 280 bytes is to be sorted in an ascending sequence on column 3 of the data record for 5 bytes. The data is stored in a character format. The shortest record in the file is 32 bytes in length and the average length is 120 bytes. The records are unblocked and stored on a magnetic tape volume. The approximate number of input records is 25000 and the work areas should be on magnetic tape. The output file should also be written on tape. The input file is an unlabelled tape, but the work files and the output file should contain standard DOS labels. The devices to be used and the label information is to be determined by the programmer.

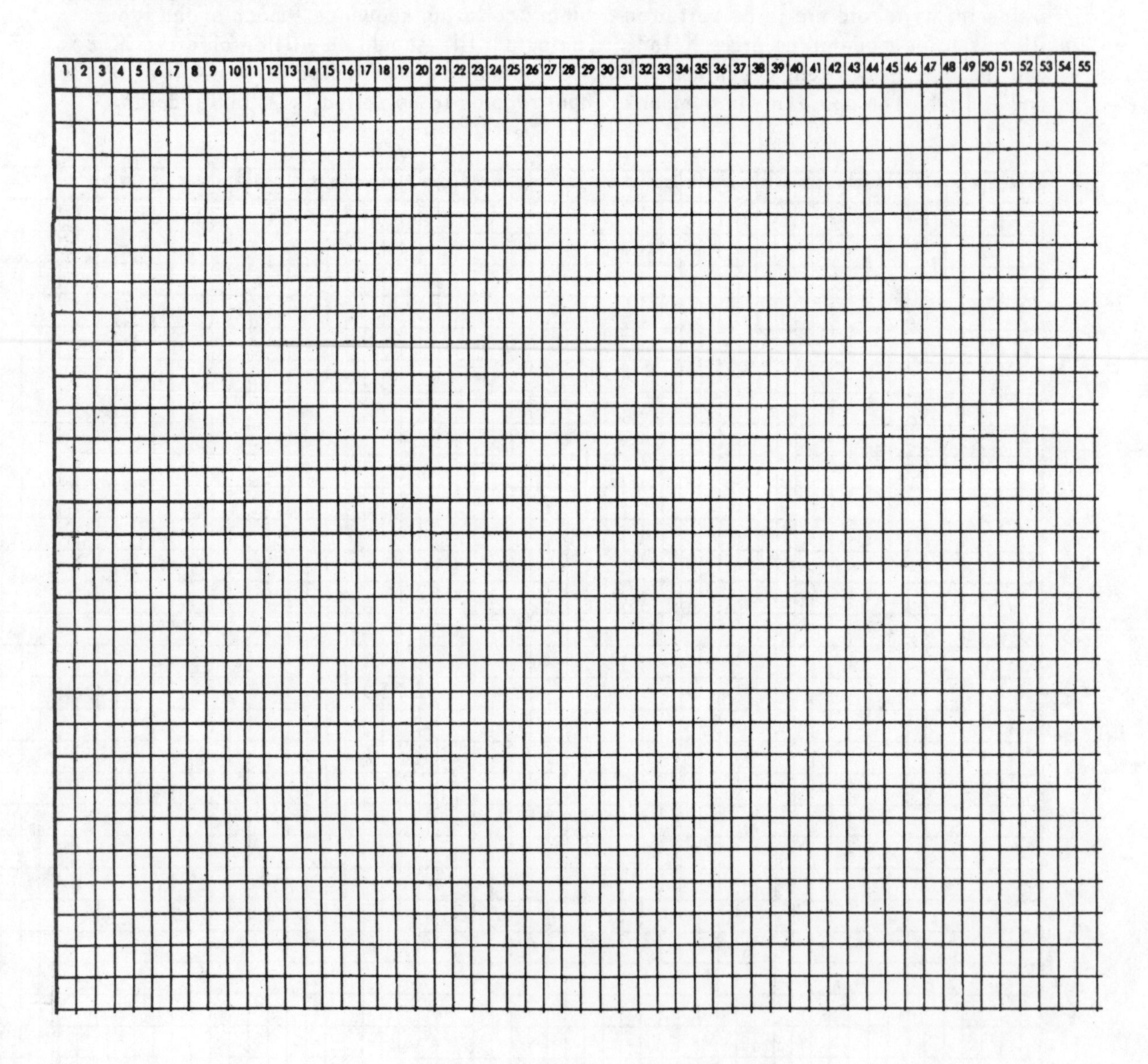

5. A file of cards are to be read by the sort exit CARDINPT (500 bytes in length) to be sorted on columns 25 through 29 in an ascending sequence. The output file is to be stored on tape drive X'183' and is to be an unlabelled file. The work area will be on tape drive X'180', X'182', and X'183'. The file should be blocked 10 records per block.

1	2	3	4	5	6	7	8	9	10	11	12	13	14	15	16	17	18	19	20	21	22	23	24	25	26	27	28	29	30	31	32	33	34	35	36	37	38	39	40	41	42	43	44	45	46	47	48	49	50	51	52	53	54	55

6. Write the job stream to load the data cards illustrated in Appendix E onto a tape file using the DOS Utility Program Card-To-Tape. This tape file should then be used as the input file to the sort program to sort the cards on invoice number, customer number, and description, all in an ascending sequence. The work files should be on magnetic tape and should be unlabelled files. The devices to be used and the labelling information is to be determined by the programmer. Keypunch the job stream and execute the jobs on a System/360

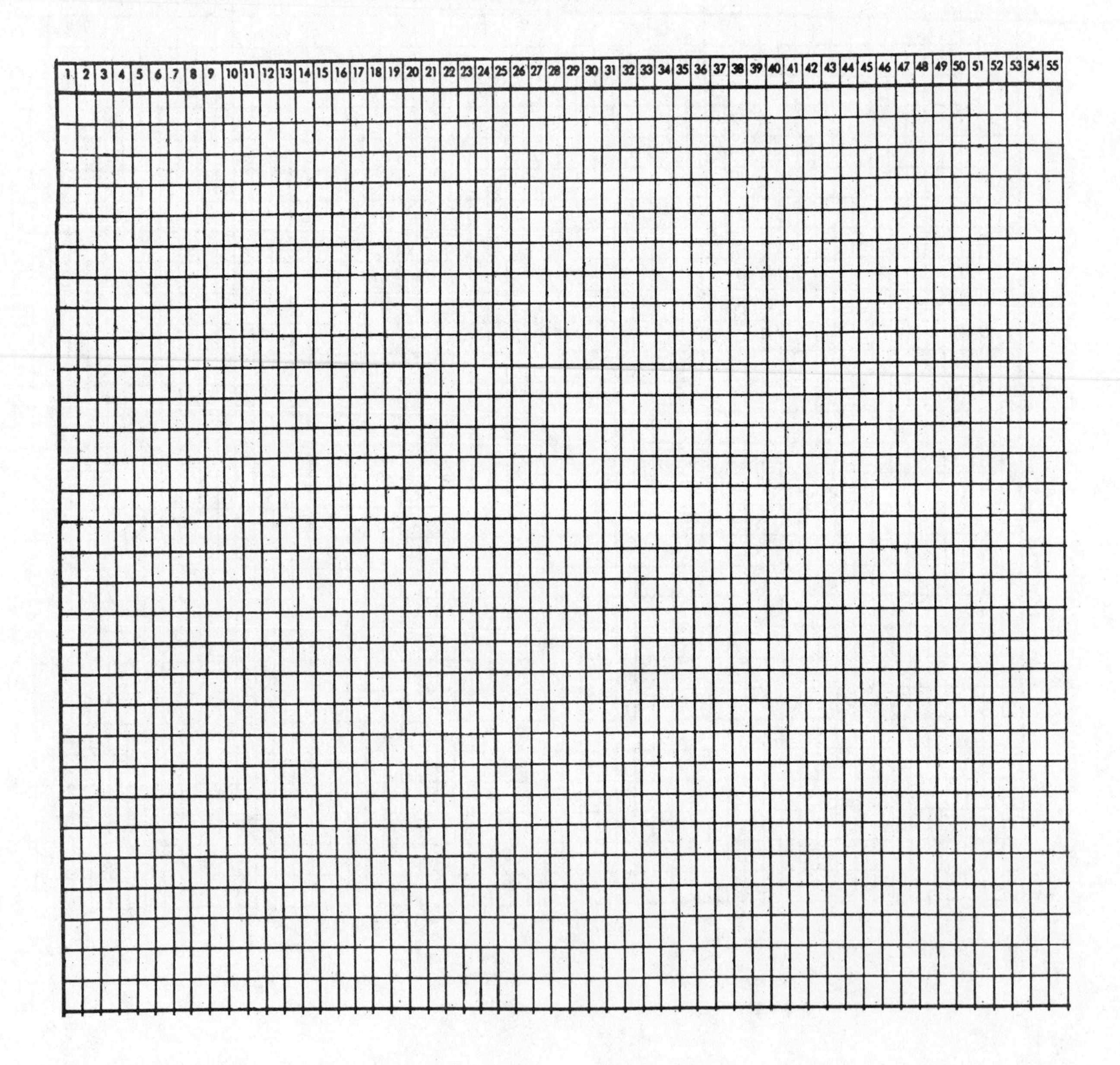

CHAPTER 7

TAPE AND DISK SORT/MERGE PROGRAM

MERGING

INTRODUCTION

In Chapter 5 and Chapter 6 the process of sorting records which were in no ordered sequence was illustrated. In many applications, it is necessary merely to create a single ordered output file from two or more files which by themselves are in the proper order. The process of combining two or more sorted files into a single sorted file is called MERGING. The DOS Tape and Disk Sort/Merge Program is normally used in order to merge two or more files into a single sorted file.

As noted, the input to a merge operation must consist of two or more files which are in the proper sequence prior to the merge operation. The diagram below illustrates three input files which are sorted on salesman number and which are to be merged together to form one file which is sorted on salesman number.

EXAMPLE

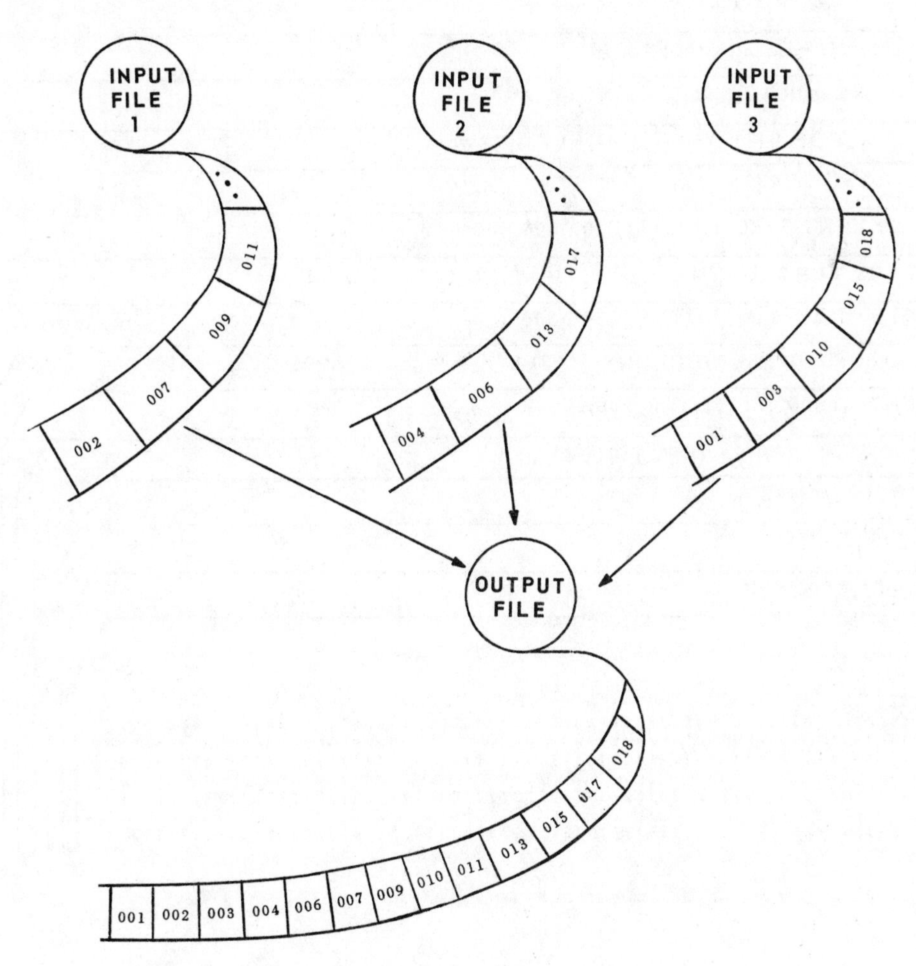

Figure 7-1 Example of MERGE Operation

Note from the drawing in Figure 7-1 that the three input files which are to be merged each contain records which are sorted in the proper sequence, that is, on each of the input files the salesman numbers are in an ascending sequence. After the input files are merged into one output file, the salesman numbers are still in an ascending order but the output file contains all of the salesman numbers from each of the three input files. Thus, the three input files have been merged into one output file.

SORT/MERGE PROGRAM

The DOS Tape and Disk Sort/Merge Program is used to merge up to eight input files into one sorted output file. The job stream illustrated below could be used to merge two disk files into a single disk output file.

EXAMPLE

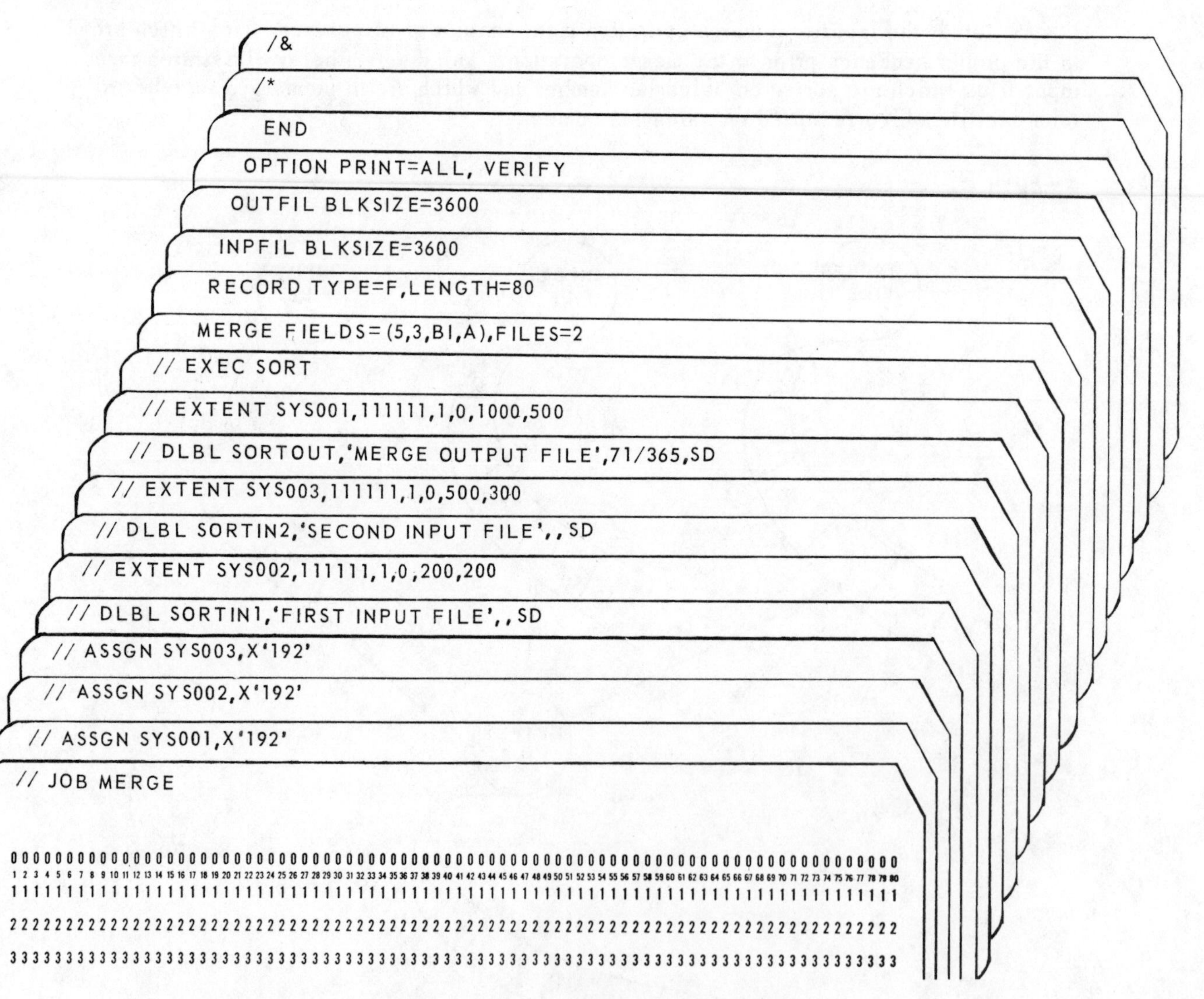

Figure 7-2 Job Stream to MERGE Two Input Files

It can be seen from the job stream illustrated in Figure 7-2 that the job control and merge control cards used for a merge operation are quite similar to those used with the sort program. There are, however, several significant differences and these will be discussed below.

As noted previously, two to eight input files may be merged by the Sort/Merge program. These input files may be stored on disk, tape, or on both disk and tape. The first input file, as with the sort program, must be stored on the device assigned to SYS002. The second input file must be on SYS003, the third on SYS004, etc. up to SYS009, which will store the eighth input file. The output file, as with the sort program, must always be mounted on the device assigned to SYS001. Thus, in the job stream illustrated in Figure 7-2 it can be seen that the output file will be written on the disk pack mounted on drive X'192'. The two input files (SYS002 and SYS003) will also be stored on drive X'192'. The filename entries in the DLBL cards for the input and output files are the same as those used for the sort program, that is, the output file filename is always SORTOUT and the input files are named SORTIN1, SORTIN2, etc. Note from the job stream that no work files are specified. This is because the merge program does not use work files. The data is read directly from the input files, merged, and written on the output file with no intermediate storage being used.

The program name used for the merge operation is SORT, the same phase name which is used for the sort operation. This is because the same program is used for both the sort operation and the merge operation. It is through the control cards which are read by the program that the sort or merge operation is requested. The MERGE control card is the first control card which is read by the sort/merge program and the general format of this control card is illustrated in Figure 7-3.

```
        { FIELDS=(p1,m1,f1,s1, . . . ,p12,m12,f12,s12)           }   { FILES=n }
MERGE   {                                                        } , {         }
        { FIELDS=(p1,m1,s1, . . . ,p12,m12,s12),FORMAT=xx        }   { ORDER=n }
```

Figure 7-3 General Format of MERGE Control Statement

In the above example, the word MERGE is the operation definer and must begin to the right of column 1. The FIELDS entry, which is required, may be specified in one of two formats in a manner similar to that used for the Sort operation. In both formats, the "p" entry specifies the beginning byte in the record, relative to one, where the control field begins. The "m" entry specifies the length of the control field. The "f" entry, if it is used, specifies the format of the data in the control field and the method which is to be used to merge it. The valid entries for the "f" entry are BI for unsigned binary, CH for character, ZD for zoned-decimal numeric data (signed), PD for packed-decimal, FL for floating-point, or FI for fixed-integer or signed binary. These entries are not required if all of the control fields contain the same type of data. Instead , the "f" field is omitted from the FIELDS operand and the FORMAT=xx operand is used to specify the format of all of the fields to be merged. The "s" entry is required and specifies whether the data is to be merged in an ascending sequence (A) or a descending sequence (D).

It should be noted that there is a maximum of 12 control fields which may be used in a merge operation. In addition, the maximum number of bytes which can be contained in the "control word", that is, the sum of all of the "m" entries in the FIELDS operand, must not exceed 256 bytes.

The FILES entry or the ORDER entry are required in the MERGE Control Card. Either the word FILES or the word ORDER must be used and they are equivalent to one another. The "n" entry following the equal sign is used to specify the number of input files which are to be merged by the sort/merge program. Thus, "n" can be any decimal value 2 through 8.

Note that the FIELDS operand, the FILES or ORDER operand, and optionally, the FORMAT operand, are the only valid entries in the MERGE control card. Thus, the entries CHKPT, SIZE, and WORK which are used in the SORT control card are not used in the MERGE control card. The remainder of the control cards used for the Merge operation are almost identical to those used for the Sort operation. The Merge control cards used in the job stream in Figure 7-2 are again illustrated below.

EXAMPLE

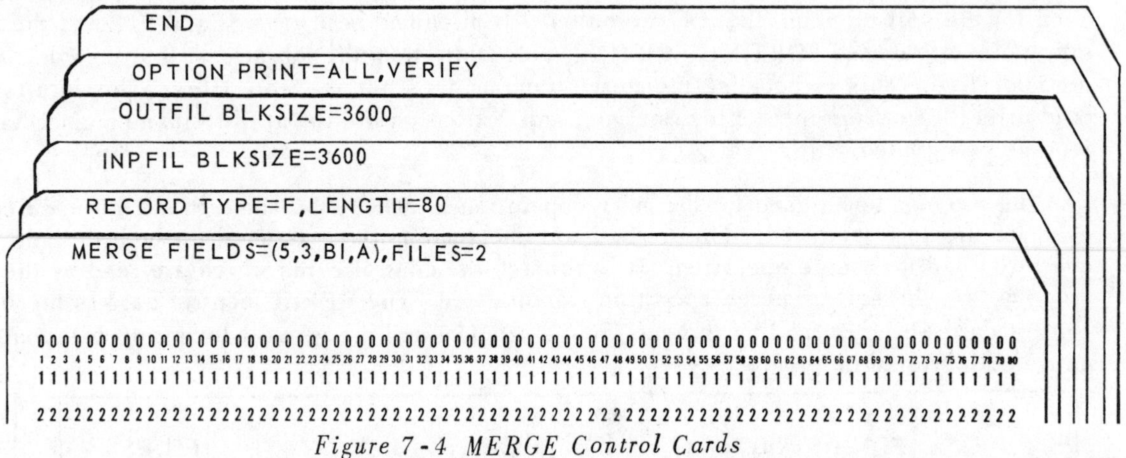

Figure 7-4 MERGE Control Cards

As can be seen from the control cards above, the RECORD control card, the INPFIL control card, the OUTFIL control card, the OPTION control card, and the END control card are utilized for the same purposes in a merge operation as when the sort/merge program is used to sort records. The RECORD control card is used to specify the type of records to be processed (fixed or variable-length) and the record sizes to be processed. The general fomat of the RECORD control statement when used with the merge portion of the sort/merge program is illustrated in Figure 7-5.

Figure 7-5 General format of RECORD Control Statement

Note in the general format above that the type of record (Fixed or Variable) must be specified by the TYPE operand. The LENGTH operand is used to specify the length of the records to be processed. The "l_1" entry is used to specify the length of the input records which will be read by the merge program. The "l_2" entry which is used by the sort portion of the program is not used by the merge portion. Thus, its omission must be indicated through the use of two commas following the value specified for the "l_1" entry.

The "l_3" entry must be included in the LENGTH operand if the length of the output record is to be changed by a user-written exit routine in the merge phase of the program. If the output record length is to be the same as the input record length, the "l_3" operand may be omitted. There is no "l_4" operand for the merge operantion and the "l_5" operand is used only if variable-length records are to be merged. It specifies the "modal length", or the most common length, of the input records. If fixed-length records are to be processed, the "l_5" entry may be omitted. Note in the RECORD control card illustrated in Figure 7-4 that the entry is LENGTH=80. If fixed-length records are to be processed by the Merge program and the record length is to be the same in the input file and in the output file, the "l_1" entry is the only one required and it need not be enclosed within parentheses.

The general format of the INPFIL Control Card when used with the merge is illustrated below.

```
                                                                    [,CLOSE=RWD  ]
                                  [,VOLUME=n         ] [,OPEN=RWD  ] [,CLOSE=UNLD ]
INPFIL BLKSIZE=n[,BYPASS]         [,VOLUME=(n1,...)  ] [,OPEN=NORWD] [,CLOSE=NORWD]
```

Figure 7-6 General Format of INPFIL Statement with MERGE Operation

Note from the general format illustrated above that the INPFIL control statement contains the same entries as for the sort program. The major difference in the use of the entries is in the BLKSIZE entry. When the merge function is to be performed, input files with different blocking factors can be processed, that is, the record lengths for fixed-length records must all be the same but one input file may contain 5 records per block and another input file may contain 10 records per block. The value of "n" in the BLKSIZE operand must be the maximum block length which is to be used for any of the input files. Thus, if one input file contained 80 byte records blocked 5 records to the block and a second input file contained 80 byte records blocked 10 records to the block, the entry BLKSIZE=800 would be used because this is the maximum block size which is to be processed in the merge program. The remaining entries are used in the same manner as with the sort program. The BYPASS entry specifies that I/O errors are to be bypassed, the VOLUME entry specifies the number of volumes for each of the input files, the OPEN operand specifies the positioning of the tape input file prior to opening it and the CLOSE entry specifies the positioning of the input file after closing the file. Note that the EXIT operand is not used when merging input files. This is because the user-written routine to the read data from a device other than tape or sequential disk is processed in phase 1 of the sort/merge program and phase 1 is not used for the merge operation. Only phase 0 (assignment phase) and phase 3 (final merge phase) are used.

The general format of the OUTFIL control card when used with the merge is illustrated below.

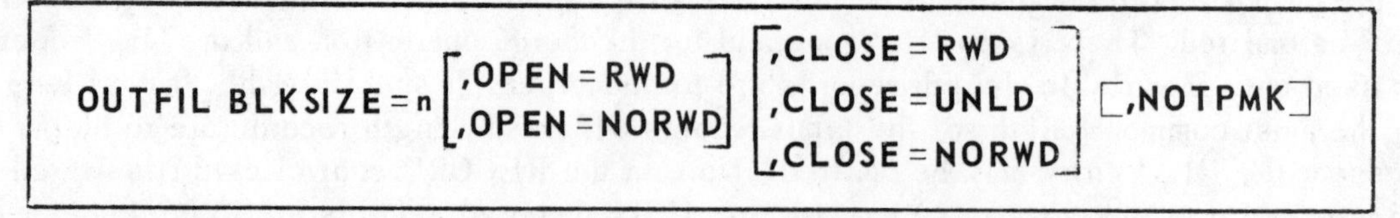

Figure 7-7 General Format of OUTFIL Control Statement

All of the entries in the OUTFIL statement used with the Merge program are the same as the entries when used with the Sort program. The BLKSIZE entry is used to specify the number of bytes which will be contained in the output block. If variable-length records are to be processed, this value must be the maximum block length which is to be written by the Merge program. The OPEN entry specifies the positioning of an output tape that is to occur prior to opening the file and the CLOSE entry specifies the positioning of the output tape after the tape file is closed. The NOTPMK entry requests that no tape mark be written as the first record on an unlabelled output tape. If this entry is omitted and the output tape is unlabelled, a tape mark will be the first record on the tape.

The OPTION control card is used to specify some of the same options which are used with the sort program. The general format of the OPTION statement when used with the merge operation is illustrated in Figure 7-8.

```
        [PRINT=NONE    ]
OPTION  [PRINT=ALL     ] [,STORAGE=n][,LABEL=(output, input1, . . ,input8)][,VERIFY]
        [PRINT=CRITICAL]
```

Figure 7-8 General Format of OPTION Control Statement

In the general format illustrated above, it can be seen that the entries which are available when a merge is to be performed are almost the same as when a sort is to be executed. The PRINT operand is used to specify that all messages from the merge program are to be printed (PRINT= ALL), none of the messages are to be printed (PRINT= NONE) or only the messages relating to abnormal termination of the merge function are to be printed (PRINT=CRITICAL). If the PRINT operand is not included in the OPTION card, PRINT= ALL is assumed by the program. The STORAGE operand may be used to specify the amount of core storage which is available to the merge program. If it is not used, the merge program assumes it can use all of the core storage allocated to the partition in which it will be executed. The LABEL operand is used to specify the type of labels which are to be used with the output file and each of the input files. A single entry is made for the output file and then entries are made for each of the input files. The entries may be S for standard labels, U for unlabelled tapes, and N for non-standard labels. If the N operand is used, a label processing exit routine must be specified in order to process the nonstandard labels. Note that, unlike the sort program, there is no entry for a work file in the LABEL parameter. This is because work files are not used with the merge program. The VERIFY parameter can be included if a disk check is to be performed on a disk output file.

The END control statement must be included in the job stream of the merge control cards to indicate the end of the control statements. The general format of the END control statement is illustrated below.

Figure 7-9 General format of END Control Card

Note that the only entry which is specified in the card is the word END, which must begin to the right of column 1 in the card. No other operands are used on the END control statement.

TAPE INPUT/OUTPUT

The merge portion of the DOS Sort/Merge program supports both tape input and tape output files. The job stream illustrated in Figure 7-10 contains the entries which could be used for four tape input files and a tape output file.

EXAMPLE

```
/&
/*
END
OPTION PRINT=ALL,LABEL=(S,S,U,S,U)
OUTFIL BLKSIZE=1500,OPEN=RWD,CLOSE=RWD
INPFIL BLKSIZE=1500,OPEN=RWD,CLOSE=UNLD
RECORD TYPE=F,LENGTH=75
MERGE FIELDS=(7,4,CH,A,23,4,PD,D),FILES=4
// EXEC SORT
// TLBL SORTIN3,'TAPE INPUT3'
// TLBL SORTIN1,'TAPE INPUT 1'
// TLBL SORTOUT,'SORT OUTPUT'
// ASSGN SYS005,X'184'
// ASSGN SYS004,X'183'
// ASSGN SYS003,X'182'
// ASSGN SYS002,X'181'
// ASSGN SYS001,X'180'
// JOB MERGTAPE
```

Figure 7-10 Job Stream to Merge Four Input Files

Note from the job stream in Figure 7-10 that four input files are to be read by the merge program and these files will be mounted on the devices assigned to SYS002, SYS003, SYS004, and SYS005. Input files to the merge program cannot be pooled either with each other or with the output file, that is, device X'181' cannot be used to read the first and the third input file. Each input file must be contained on a separate drive. In addition, the output file must also be mounted on a separate drive. This is because the four input files are read at one time by the merge program and the output file is created as the input files are being read. No data is stored on a work file such as with the sort program.

Note also from the job stream that TLBL job control cards are included for the output file and for the first and third input files. The second and fourth input files do not have any TLBL cards. Thus, they are to be unlabelled tapes and this must be indicated in the OPTION merge control card. The OPTION card used in the job stream is again illustrated below.

EXAMPLE

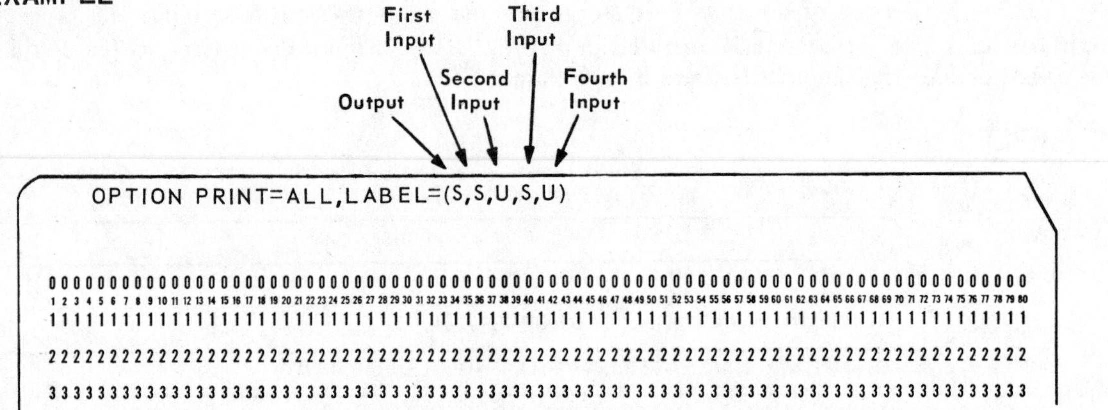

Figure 7-11 OPTION Merge Control Card

Note from the above OPTION card that the LABEL operand is used to specify the type of labelling which will be used for both the output file and the input files. The first entry within the required parentheses specifies the type of labelling for the output file. It will contain standard labels since the entry S is included. This means that a TLBL card must be included in the job stream in order to supply the label information for the output file. The first and third input files are to contain standard DOS tape labels and TLBL control cards will be in the job stream for those files. The second and fourth files are to be unlabelled. Note from the job stream in Figure 7-10 that the filename entries determine which of the files the TLBL cards pertain to, that is, the filename entry SORTIN1 indicates that label information for the first input file is provided and the filename entry SORTIN3 indicates that label information for the third input file is provided. Note that the TLBL card for the output file contains the filename SORTOUT, which must be used for output files.

Note also from the job stream in Figure 7-10 that the OPEN and CLOSE operands are used on the INPFIL and OUTFIL control cards. They are used to specify the tape file positioning for the input and output files. If they are not specified, the input files and the output file are rewound at the conclusion of processing by the merge program.

BLOCKING FACTORS

As was mentioned in the discussion of the INPFIL control statement, input files to the merge program must contain the same record lengths but may be blocked in different sizes. The following job stream contains the statements to merge input files which are stored on both disk and tape and which contain different blocking factors.

EXAMPLE

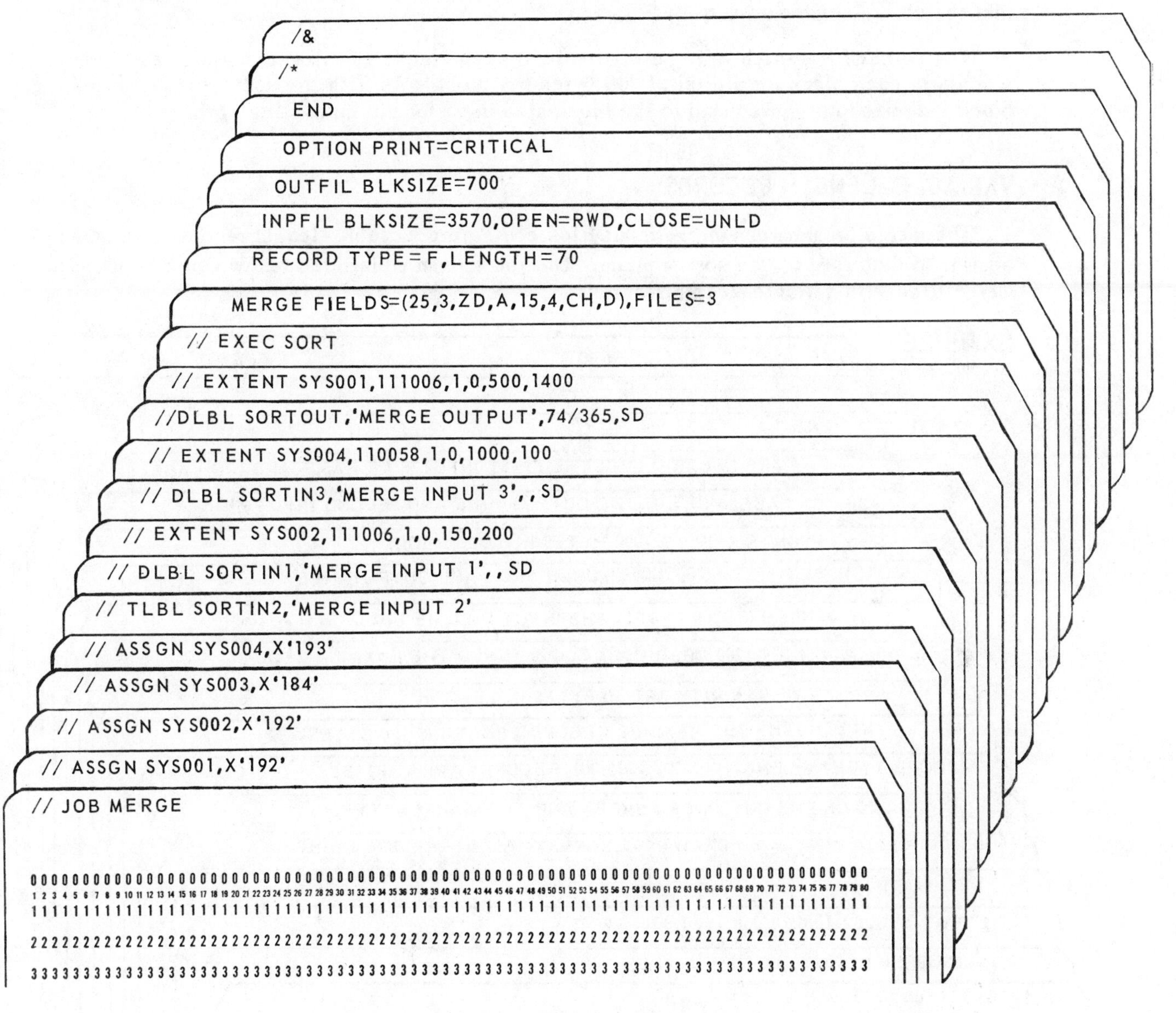

Figure 7-12 MERGE Job Stream

In the example above, the input files are stored on both disk and tape. The first input file is stored on disk device X'192', the second input file is stored on tape device X'184', and the third input file is stored on disk device X'193'. This mixing of device types for input is valid for the merge operation of the sort/merge program but is not valid sort operation. Note that the corresponding label cards in the job stream are used for the two disk input files and the single tape input file. The filenames and symbolic-units used must conform to the rules for merge input files.

The blocking factors for the input files may be different. The record length for all of the input records, as specified in the RECORD control card, must be the same. The blocksizes, however, may differ. The entry which is placed in the BLKSIZE operand of the INPFIL Control card must be the maximum length of any of the input blocks which will be read by the merge program. Thus, in the example in Figure 7-12 it can be seen that the maximum block size which will be encountered by the merge program is 3570 bytes in length, which would indicate that there are fifty-one 70-byte records in one of the input files. Note again that the length of each record in the input files must be the same but that the block length may vary.

The output file, which is to be written on disk device X'192', is to contain ten 70-byte records in each block for a total of 700 bytes per block. As with the sort program, this block size need not correspond to the block size used for the input files.

VARIABLE-LENGTH RECORDS

The merge program can merge input files containing variable-length records in a manner similar to that used by the sort program. The job stream illustrated below could be used to merge files with variable-length records.

EXAMPLE

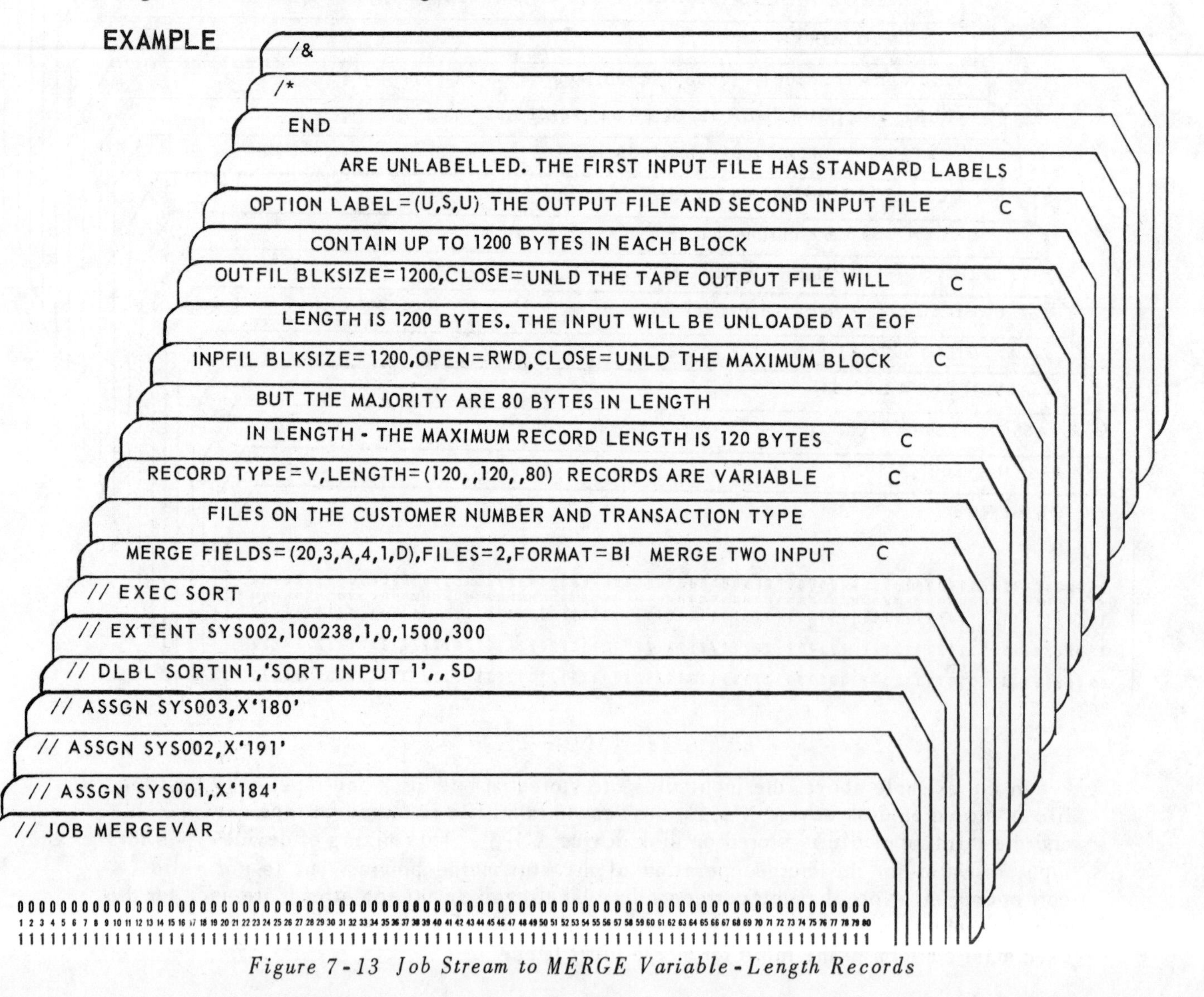

Figure 7-13 Job Stream to MERGE Variable-Length Records

In the job stream in Figure 7-13 it can be seen that two input files, one on tape drive X'180', and one on disk drive X'191', are input to the merge program. These input files contain variable-length records. As specified in the RECORD control card, the maximum length of an input record is 120 bytes and the "modal length" of the input records is 80 bytes. The maximum input block length is 1200 bytes, as indicated by the BLKSIZE entry in the INPFIL control card, and the maximum output block length will also be 1200 bytes. The disk input file is labelled and the tape input file and tape output file are unlabelled.

Note the use of comments on the merge control cards. Comments may be included on each merge control card following the last control entry in the card. One or more blanks must separate the last control entry from the first letter of the comment. The general format of the MERGE control card utilizing comments is illustrated below.

```
MERGE FIELDS=(p1,m1,f1,s1, . . .),FILES=n COMMENT BEGINS HERE
```

Figure 7-14 General Format of MERGE Control card with Comments

Note in the example above that the first letter of the comment begins following one or more blanks after the last entry in the control card. Note also from the example in Figure 7-13 that continuation cards may be used with the comments going from the first control card to the continuation card. The only requirement is that the rules for continuation cards be followed, that is, a continuation punch must be contained in column 72 of the card to be continued and the first entry in the continuation card must begin in column 16. It should be noted that comments are valid in the sort control cards as well as the merge control cards.

MERGE EXITS

The user-written exits available to the programmer when merging data are not as extensive as those available when sorting data because only the assignment phase (phase 0) and the merge phase (phase 3) of the program are executed to merge data. Thus, exits are available to process user-labels on tape files, to process Input/Output errors on either input or output files, and to modify data within an input record. Note that a user-written routine such as illustrated in Chapter 5 to read an input file is not available to the programmer when the merge operation is to take place because phase 1 of the sort/merge program is not utilized. A special merge-only exit (Exit 32) is available to modify data contained in an input record but the input to the merge program must be on tape or disk.

OTHER IBM SORT/MERGE PROGRAMS

The information in Chapter 5 and in this chapter has dealt with the IBM program Disk Operating System Tape and Disk Sort/Merge Program (Program Number 360N - SM - 483). There are three other sort/merge programs which are currently available from IBM for operation with the Disk Operating System. These are the Disk and Tape Operating System Tape Sort/Merge Program (Number 360N - SM - 400), the Disk Operating System Disk Sort/Merge Program (Number 360N - SM450) and the IBM System/360 Disk Operating System Sort/Merge Program (Number 5743 - SM1). The first two sorts, as well as the "483" sort, are available from IBM free of charge. The "SM1" sort is an IBM "program product" and may be leased from IBM.

Each of the available sort/merge programs processes data in approximately the same manner. Differences occur in the type of devices supported by the Sort/Merge program and in the types of special processing which may occur, for example, the user - exit routines which may be used. There are also differences, of course, in some of the control cards which are used with the sort/merge programs. For the most part, however, the types of information which must be indicated for the "483" sort are also necessary for the other sort/merge programs and an understanding of the Tape and Disk Sort/Merge program will be sufficient to reference the appropriate SRL to use the other sort/merge programs.

STUDENT ASSIGNMENTS

1. Write the job stream to merge two input files which are stored on disk pack 123471. The first input file is on Cylinder 5 Track 0 through Cylinder 43 Track 9 and the second file is on Cylinder 145 Track 0 through Cylinder 159 Track 9. The first input file is named CUSTOMER FILE and the second is named SALES FILE. Both files are to be merged on the item number field which begins in column 23 and is 6 bytes in length. The output file, which is to be stored on pack 245908, should be blocked with the same factor used for the 90 byte input records, that is, 15 records per block. The drives on which the files are to be mounted should be determined by the programmer.

1	2	3	4	5	6	7	8	9	10	11	12	13	14	15	16	17	18	19	20	21	22	23	24	25	26	27	28	29	30	31	32	33	34	35	36	37	38	39	40	41	42	43	44	45	46	47	48	49	50	51	52	53	54	55

2. Write the job stream to merge two unlabelled tape files which contain 65 byte records blocked 6 records per block. The output file, which is to be stored on a standard labelled tape, should contain records which are in an ascending sequence based upon the customer number field, which is six bytes long beginning in column 5 of the fixed-length records. The devices to be used for the merge operation are to be determined by the programmer.

1	2	3	4	5	6	7	8	9	10	11	12	13	14	15	16	17	18	19	20	21	22	23	24	25	26	27	28	29	30	31	32	33	34	35	36	37	38	39	40	41	42	43	44	45	46	47	48	49	50	51	52	53	54	55

3. Six input files are to be merged into a single output file to be stored on magnetic tape. The characteristics of each file are described below:

a. First file ('CUST MSTR') – Magnetic tape, standard labels, 350 bytes per block.
b. Second file ('CUST TRANS') – Disk, standard labels, Cyl 45 Trk 0 thru Cyl 78 Trk 9, blocked 15 records per block, pack 273480.
c. Third file – Tape, unlabelled, 25 records per block.
d. Fourth file ('CUST TRANS') – Tape, standard labels, 420 bytes per block.
e. Fifth file ('CUST MASTR') – Disk, standard labels, Cyl 125 Trk 0 thru Cyl 130 Trk 9 and Cyl 150 Trk 0 thru Cyl 175 Trk 9, blocked 10 records per block, pack 793921.
f. Sixth file – Tape, unlabelled, one record per block.

All of the input records to the merge program contain 70 bytes per record. They are all a fixed-length. They should be merged on the following fields:

a. Column 35, 6 bytes, binary, ascending;
b. Column 70, 1 byte, character, descending;
c. Column 1, 25 bytes, character, ascending.

The output file should be stored on magnetic tape and contain standard labels.

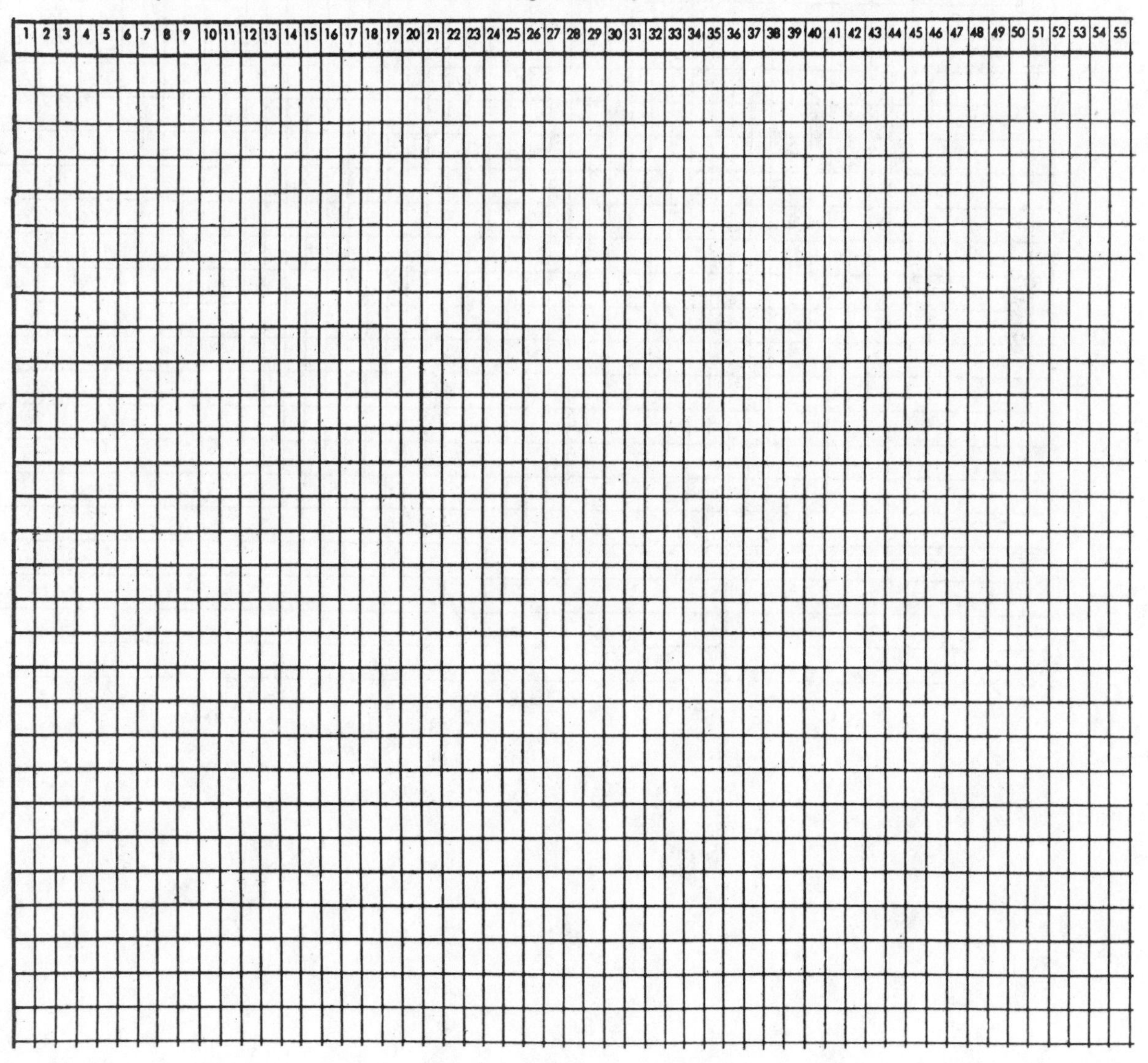

4. Write the job stream to merge three unlabelled tape input files which contain variable-length records into a single file to be stored on disk pack 443298. The maximum length of the input records is 215 bytes with the majority of records containing 140 bytes. The first input file contains a minimum of 4 records per block, the second 7 records per block and the third 5 records per block. The output file should be able to contain a minimum of 8 records per block. The records are to be merged in a descending sequence on the 12 byte field which begins in column 21 of each record. The devices to be used and the extents for the output file are to be determined by the programmer.

1	2	3	4	5	6	7	8	9	10	11	12	13	14	15	16	17	18	19	20	21	22	23	24	25	26	27	28	29	30	31	32	33	34	35	36	37	38	39	40	41	42	43	44	45	46	47	48	49	50	51	52	53	54	55

5. A single, sorted output file is to be created from five input files and the output file is to be sorted in an ascending sequence based upon the character value in bytes 5-24 in each record. The characteristics of each input file are described below.

a. First Input File – Card file, record length is 80 bytes.
b. Second Input File – Tape file, record length is 120 bytes, unlabelled, blocked 12 records per block.
c. Third Input File – Disk file, record length is 120 bytes, blocked 10 records per block, file-ID is 'CUSTOMER ACCOUNT', stored on pack 298374 - Cyl 10 Trk 0 through Cyl 14 Trk 0.
d. Fourth Input File – Tape file, record length is 80 bytes, blocked 6 records per block, standard labels, file-ID is 'CUSTOMER TRANS'.
e. Fifth Input File – Tape file, record length is 120 bytes, blocked 12 records per block, standard labels, file-ID is 'CUST TRANS'.

All of the files contain fixed-length records and only File C (third input file) is sorted in the proper sequence. The following processing rules are to be followed:

a. Card input device is X'00C'.
b. Tape devices are X'180', X'181', and X'182'.
c. Available disk devices are X'191', X'192', and X'193'.
d. The total number of records in all five files does not exceed 8,000.
e. The total number of records in any one file does not exceed 2500.
f. The only exit available for the sort program is used to lengthen records and is named SORT35. It is utilized at Exit 35 of phase 3 of the sort and is link-edited to load into core storage at decimal location 40000.
g. The output file is to be stored on a magnetic tape using standard labels. The record length is 120 bytes and there are 12 records per block. The file-ID should be 'MERGE OUTPUT'.

Write the job stream to accomplish the processing which is to be done in the above problem. Since only Exit 35 is used out of the sort program, it will be necessary to load the card file onto tape or disk using the appropriate utility program.

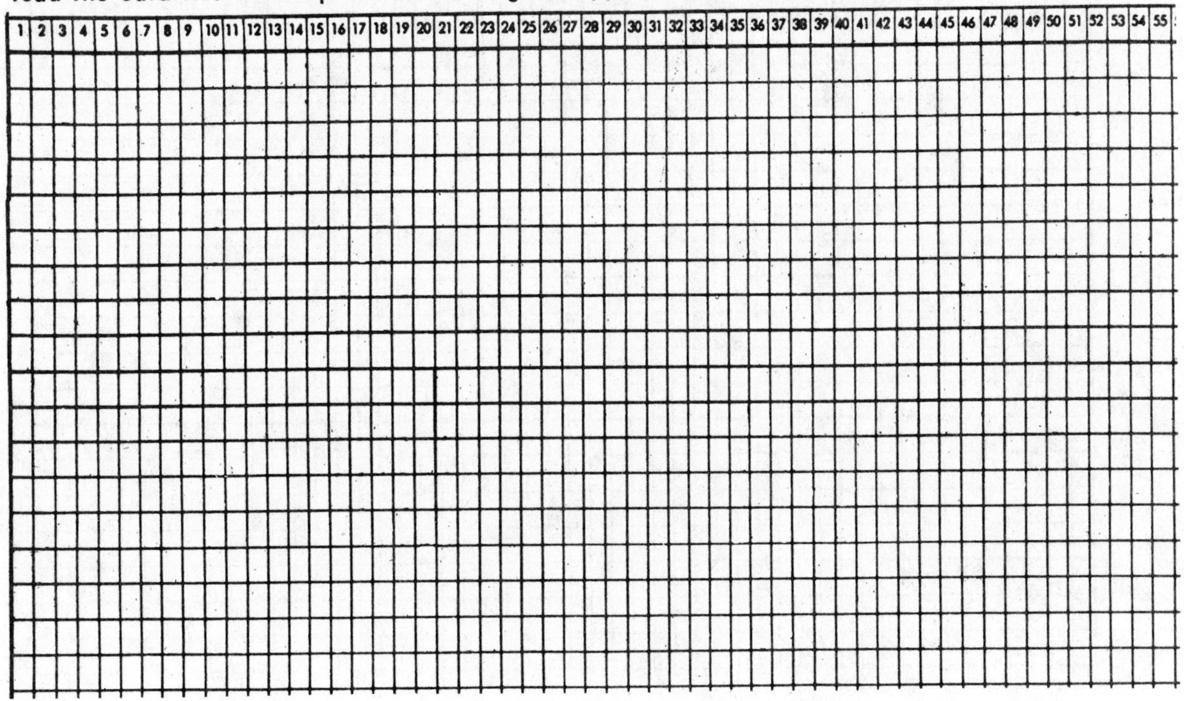

1	2	3	4	5	6	7	8	9	10	11	12	13	14	15	16	17	18	19	20	21	22	23	24	25	26	27	28	29	30	31	32	33	34	35	36	37	38	39	40	41	42	43	44	45	46	47	48	49	50	51	52	53	54	55

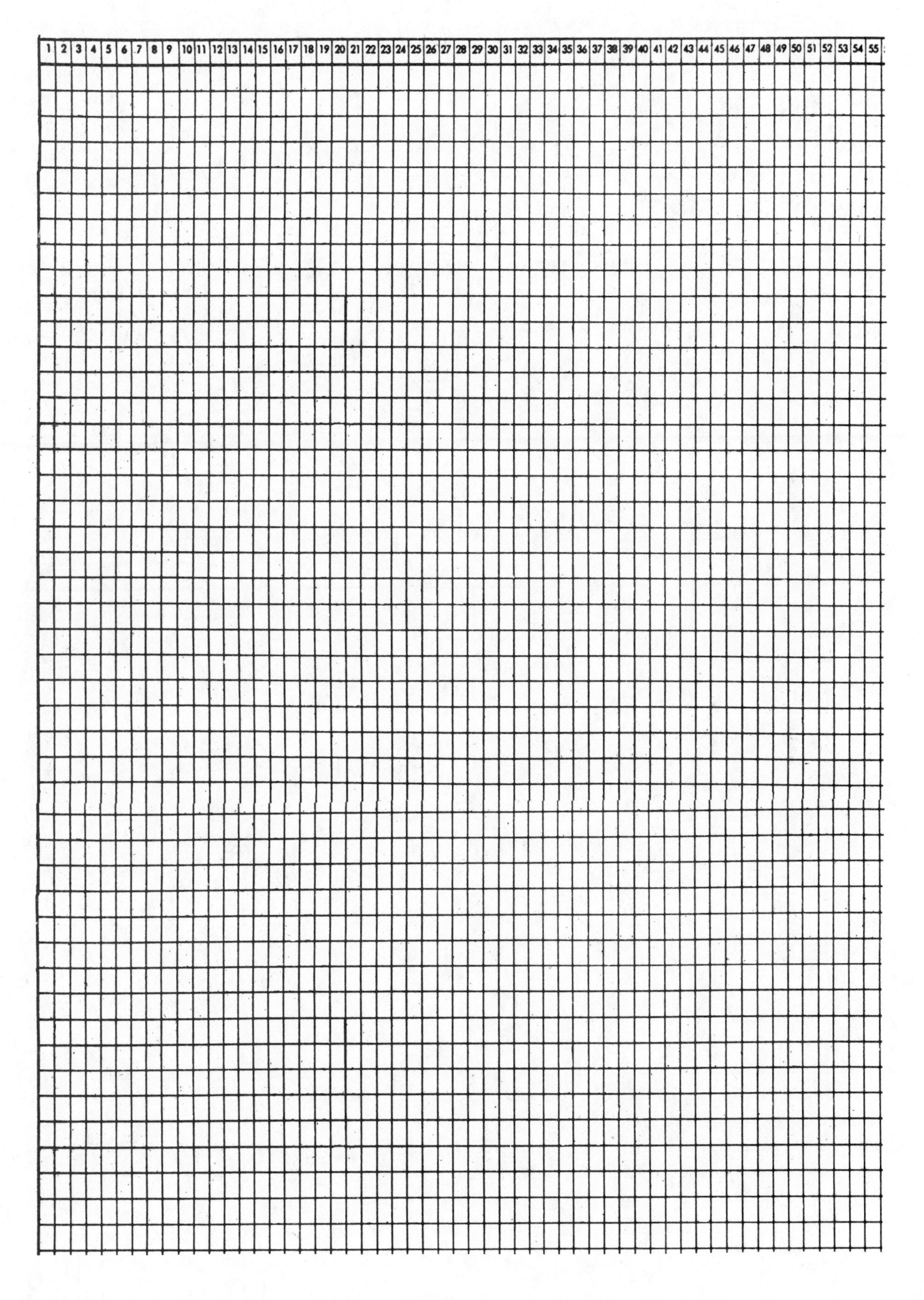
1
2
3
4
5
6
7
8
9
10
11
12
13
14
15
16
17
18
19
20
21
22
23
24
25
26
27
28
29
30
31
32
33
34
35
36
37
38
39
40
41
42
43
44
45
46
47
48
49
50
51
52
53
54
55

CHAPTER 8

MULTIPROGRAMMING

INTRODUCTION

The Disk Operating System provides the capability of executing more than one program in core storage of a System/360 at one time. This process of performing multiple programs at the same time is referred to as MULTIPROGRAMMING. When a System/360 contains 24K or more of core storage, multiprogramming is possible under the control of the core-resident supervisor. The drawing in Figure 8-1 illustrates a "typical" allocation of core storage when multiprogramming processing is to occur on a 64K machine.

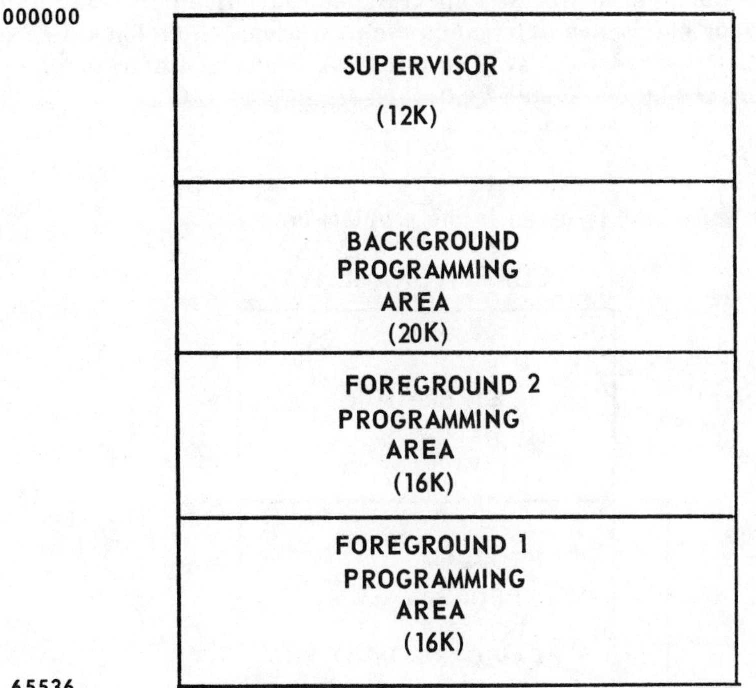

Figure 8-1 Allocation of Core Storage for MULTIPROGRAMMING

Note from the drawing above that core storage is divided into four sections: One section for the core-resident supervisor, one for the "background" programming area, one for the "foreground 2" programming area, and one for the "foreground 1" programming area. Each of these "programming areas" may contain one program which can be operating independent of the other programs within core storage. These programming areas, or PARTITIONS, are named Background, Foreground 2, and Foreground 1 and, when in use, are fixed in length. Thus, from the above example it can be seen that the Background partition is 20K bytes in length, the Foreground 2 partition is 16K bytes in length and the Foreground 1 partition is 16K bytes in length. The core storage illustrated above is 64K bytes in length, which means that it actually contains 65536 bytes which are available for use by the supervisor and user programs.

As noted, each of the programs which are stored in a partition operate under the control of a core-resident supervisor. Each program does not execute actual instructions at the same time. Thus, when a program which is stored in the Foreground 1 partition is executing instructions, the programs in Foreground 2 and Background are not processing at all. It is only when the processing in Foreground 1 must wait for an event to occur that it is possible for a program stored in the Foreground 2 partition or the Background partition to actually execute instructions. The core-resident supervisor controls which program is executing at any particular time. It does this by interpreting and processing interrupts.

INTERRUPTS

An INTERRUPT is an event which occurs on the computer which causes a break in the normal sequence of instruction execution, that is, when an interrupt occurs, the normal execution of machine instructions cease. When an interrupt occurs, the hardware of the System/360 causes a branch of control to a special "Interrupt Handler" routine in the core-resident supervisor which then determines the next steps which should be taken for the allocation of the computer resources. The following example illustrates the step-by-step process which is used by the System/360 when reading a card.

EXAMPLE

Step 1: A "Read" command is given in the problem program.

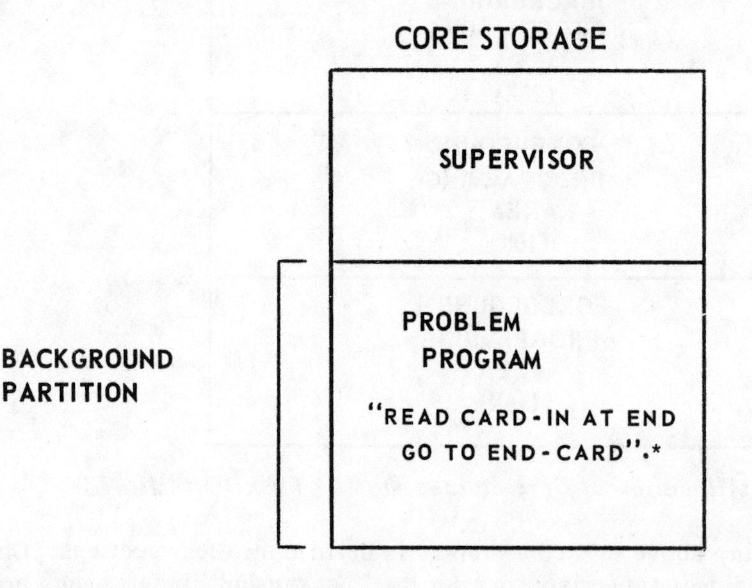

** Although the instruction is written in COBOL source code, it would be stored in core storage as a series of machine-language instructions.*

Figure 8-2 READ Command given to Read a Card

When the Read Command is issued in a problem program, the coding links the user-written statements with the Logical IOCS routines which are included in the program. These routines, in turn, link to the Supervisor which issues the actual command to the card reader.

Step 2: Supervisor issues read command to the channel containing the card reader.

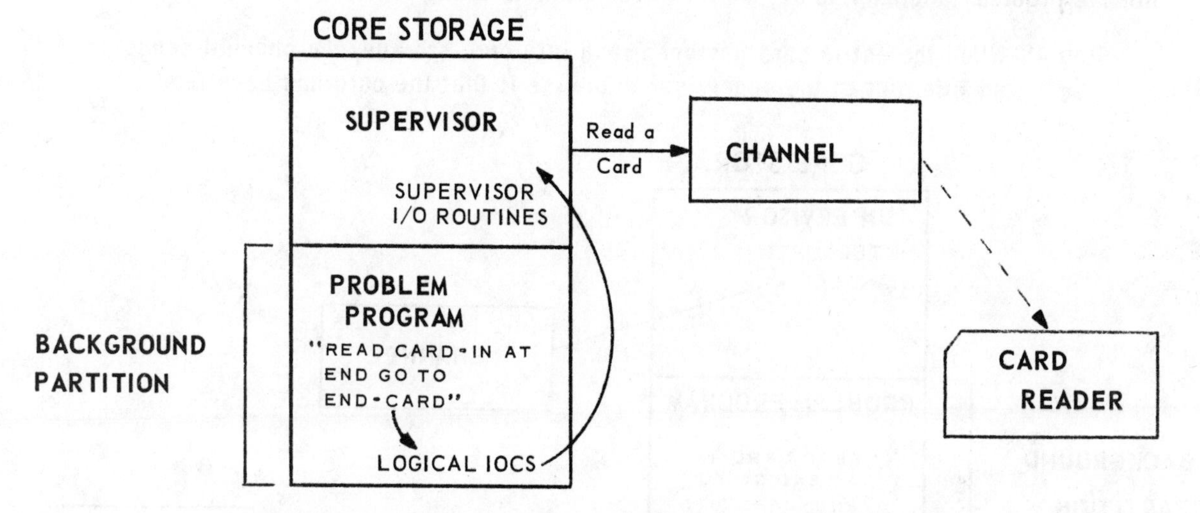

Figure 8-3 Read Card command is issued to channel from Supervisor

When the Logical IOCS routines request the Supervisor to issue the command to the channel to read a card, they also request that the card be read into core storage properly before control is returned to the problem program to process the data read. The channel, which operates like a small independent "computer", reads the data from the card and places it in core storage. The Supervisor, however, since the card is not entirely in core storage, will not give control back to the processing program stored in the background partition. Thus, the computer enters a "Wait" state, that is, it is waiting for the card to be read into core storage.

Step 3: The computer enters the "wait" state while the card is read into core storage.

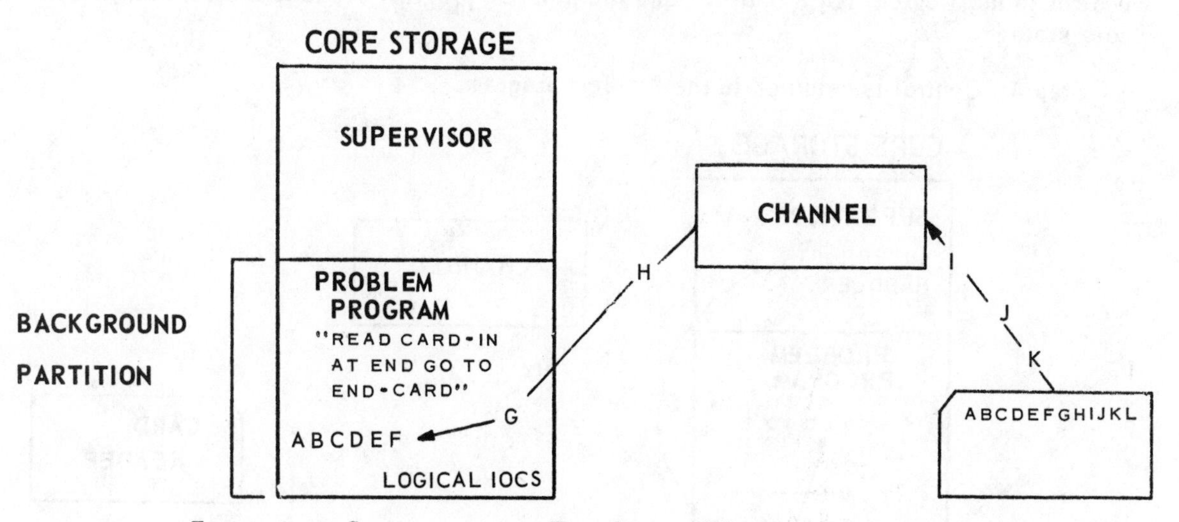

Figure 8-4 Computer enters Wait State while Channel Reads Card

Note from Figure 8-4 that neither the supervisor nor the problem program is executing any instructions while the channel reads the card into core storage. Thus, the computer is said to be in the "wait state", that is, it is waiting for the card to be read. Note also that the channel communicates directly with the I/O area in the problem program area, that is, the data in the card is placed in core storage by the channel and neither the supervisor nor the problem program has any control over the data being read.

Step 4: When the entire card has been read into core storage, the channel sends an interrupt to the supervisor to advise it that the card has been read.

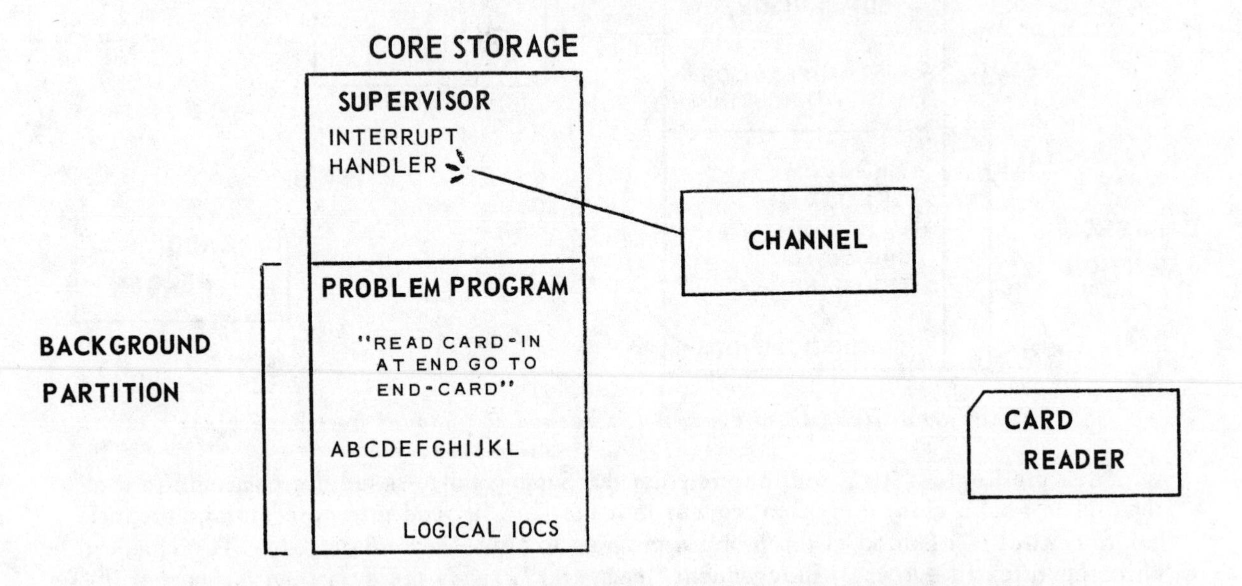

Figure 8-5 Channel Sends INTERRUPT to Signal end of Read Card Operation

When the interrupt is sent from the channel to the supervisor, the supervisor begins executing instructions to determine what the interrupt means and what processing should be performed. Thus, the machine exits from the "wait state" when an interrupt is received. In the example, the supervisor determines that the interrupt was caused by the channel indicating that the card had been read into core storage. Thus, the supervisor will return control to the Logical IOCS routines and the problem program because the card is now in core storage.

Step 4: Control is returned to the problem program.

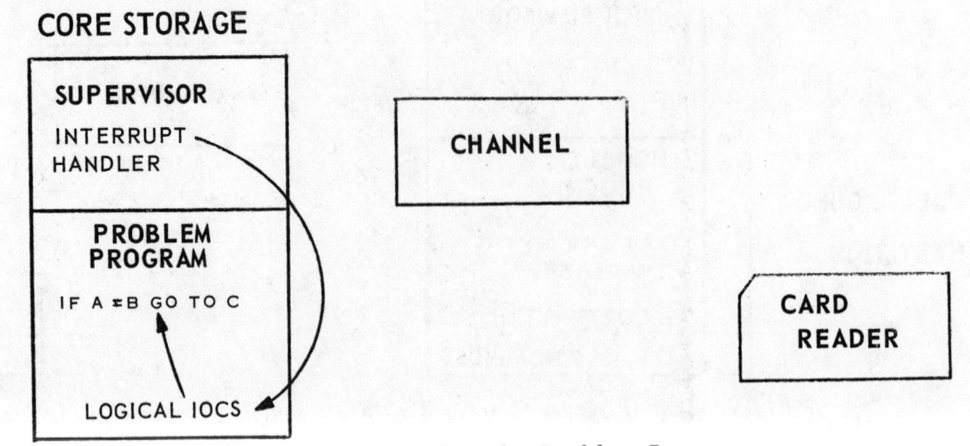

Figure 8-6 Control is returned to the Problem Program

Note from Figure 8-6 that the Interrupt Handler routine in the Supervisor returns control to the Logical IOCS routines which in turn return control to the problem program. The program can then process the data which has been read into core storage from the card in any manner desired. Subsequent reads of more cards will be processed in a similar manner and, if the program issues a WRITE command to a printer, the same basic steps are accomplished except that the data is fetched from core storage by the channel and written on the printer instead of reading the data from the card and placing it in core storage.

Two very important features of the previous example must be noted. First, note that the computer entered the "wait state" after the command to read a card was issued to the channel. Thus, there was no processing being accomplished during the time the card data was being placed in core storage by the channel. This time is essentially lost because nothing productive was being performed by either the supervisor or by a user program. Second, this wait state was ended by the interrupt generated by the channel when the card had been read into core storage. Thus, the interrupt was the means by which processing by the supervisor and the problem program was initiated after the machine had entered the wait state.

MULTIPROGRAMMING OPERATION

The Disk Operating System takes advantage of the fact that any Input/Output causes the computer to wait for completion in order to allow multiprogramming. When more than one program is to be executed in different partitions at the same time, a priority system is established which allows a lower priority program to process data while a higher priority program is waiting for an Input/Output operation to be completed. The following example illustrates the processing which occurs when a program in the Foreground 1 partition and the Background partition are both in core storage to be executed.

EXAMPLE

Step 1: The programs are loaded into core storage and the program in the Foreground 1 partition is given control.

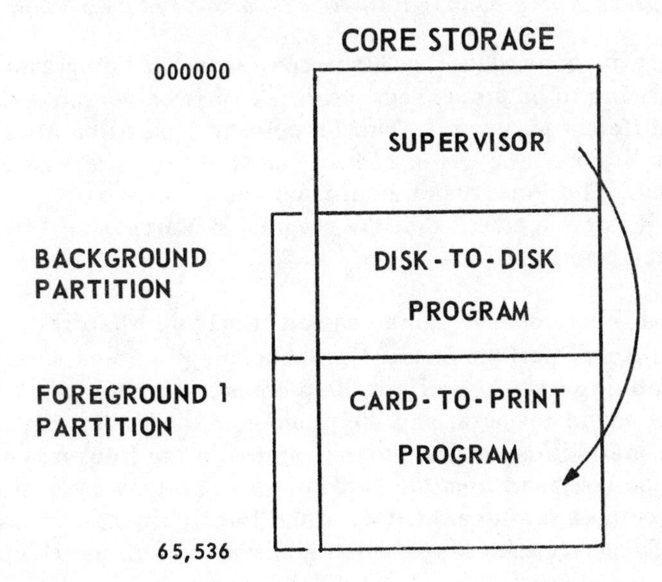

Figure 8-7 Core Storage Allocation for Multiprogramming

Note in Step 1 that the Supervisor is loaded into "low-core", that is, it occupies the lower address in core storage. The Background partition immediately follows the supervisor in core storage. This is always true and if only one partition is to be used, it is always the Background partition. Following the Background partition is the Foreground 1 partition. Note that in this example the Foreground 2 partition is not utilized. It must be determined by the programmer or operator which partitions are to be used for any particular application. Background must always be used but either or both of the Foreground partitions may be used.

Step 2: Control is given to the card-to-print program in the Foreground 1 partition and it issues a Read command to the card reader.

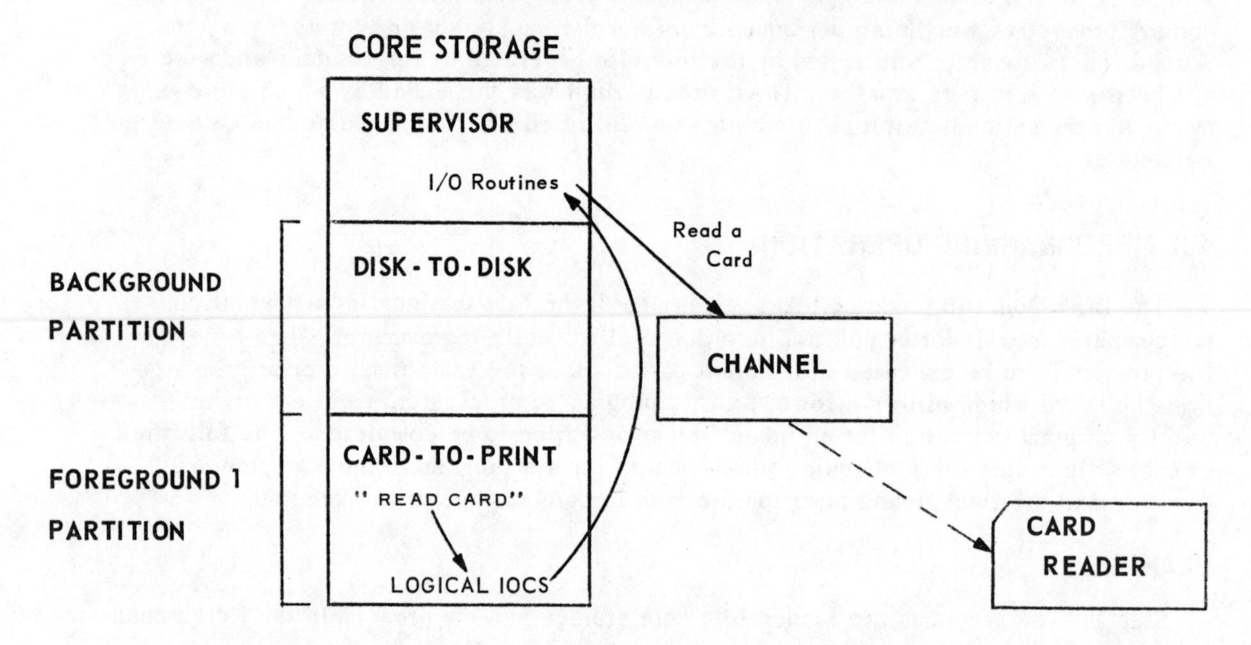

Figure 8-8 Card-To-Print Program in Foreground 1 Partition issues Read Command

In Figure 8-8 it can be seen that the Supervisor has given control to the Foreground 1 partition. In order for multiprogramming to be processed, priorities must be established between the programs stored in the different partitions. The Foreground 1 partition always has the highest priority, that is, it will always be given control ahead of the Foreground 2 partition and the Background partition. The Foreground 2 partition has priority over Background. Thus, in Figure 8-8, it is no accident that the program in Foreground 1 begins processing first. It always has priority.

When the card-to-print program in Foreground 1 gains control, it may do whatever processing it desires. For example, it may perform initial "housekeeping" chores such as setting counters to zero and initializing switches within the program. As long as it is executing instructions, it has control of the computer and the program in the Background partition is idle, that is, none of the instructions in the program stored in the Background partition are being executed. When the command from the card-to-print program to read a card is issued, the Logical IOCS routines turn over control to the Physical IOCS routines which are stored in the Supervisor. These routines issue the actual command to the channel to begin reading a card.

Note that the computer is now in the status as illustrated in the example in Figures 8-2 through 8-6, that is, it now must wait for the card to be read into core storage before the program in the Foreground 1 partition may continue processing. In the previous example, the computer entered the "wait state", waiting for the Input/Output operation to be completed by the channel. In this example, however, it need not enter the wait state because there is another program which must be processed, that is, the program in the Background partition must be processed. Therefore, instead of entering the wait state to await the completion of the card input operation, the Supervisor "Scheduler" routine gives control to the disk-to-disk program which is stored in the Background partition.

Step 3: Control is passed to the Background partition.

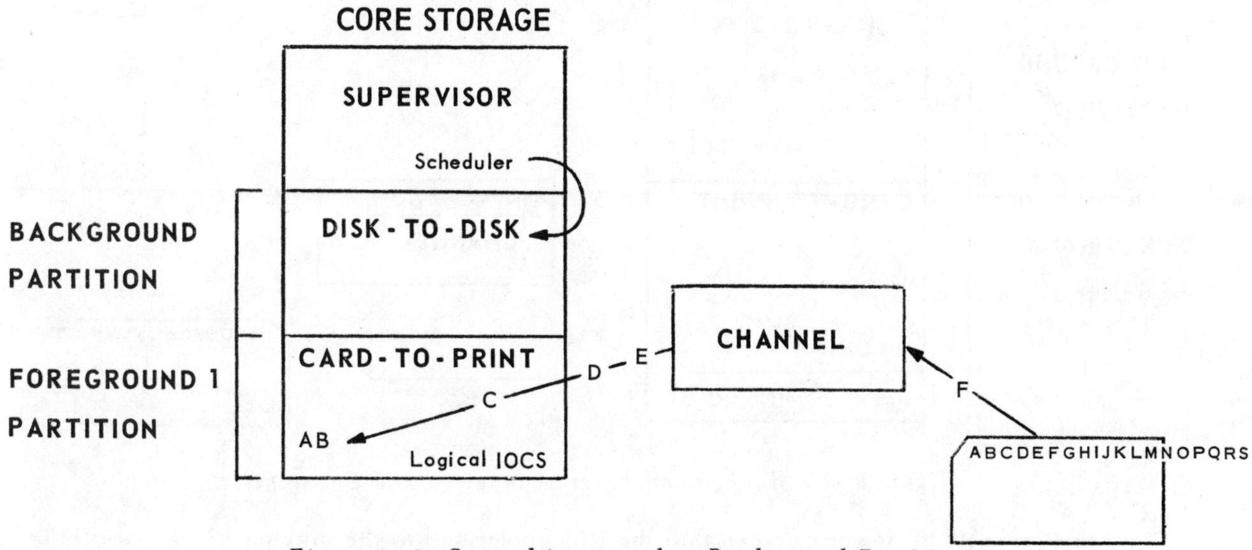

Figure 8-9 Control is passed to Background Partition

Note in Figure 8-9 that the Supervisor Scheduler routine passes control to the program in the Background partition while the channel is reading the contents of the card into core storage in the Foreground 1 partition. This is possible because the channel communicates directly with core storage and need not be concerned with instructions which are being processed on the computer during this data transfer. Thus, while the contents of the card are being placed in the Foreground 1 partition, the disk-to-disk program in the background partition will begin executing instructions. It may, for example, also perform "housekeeping" tasks, initialize values, obtain the date from the Communication Region, etc. It will process any instructions in the program as if it were the only program in core storage.

When the program in the Background partition is ready, it can issue a Read command to the disk input file so that the first input record can be read from the disk.

Step 4: The Background program issues a Read command to a disk file.

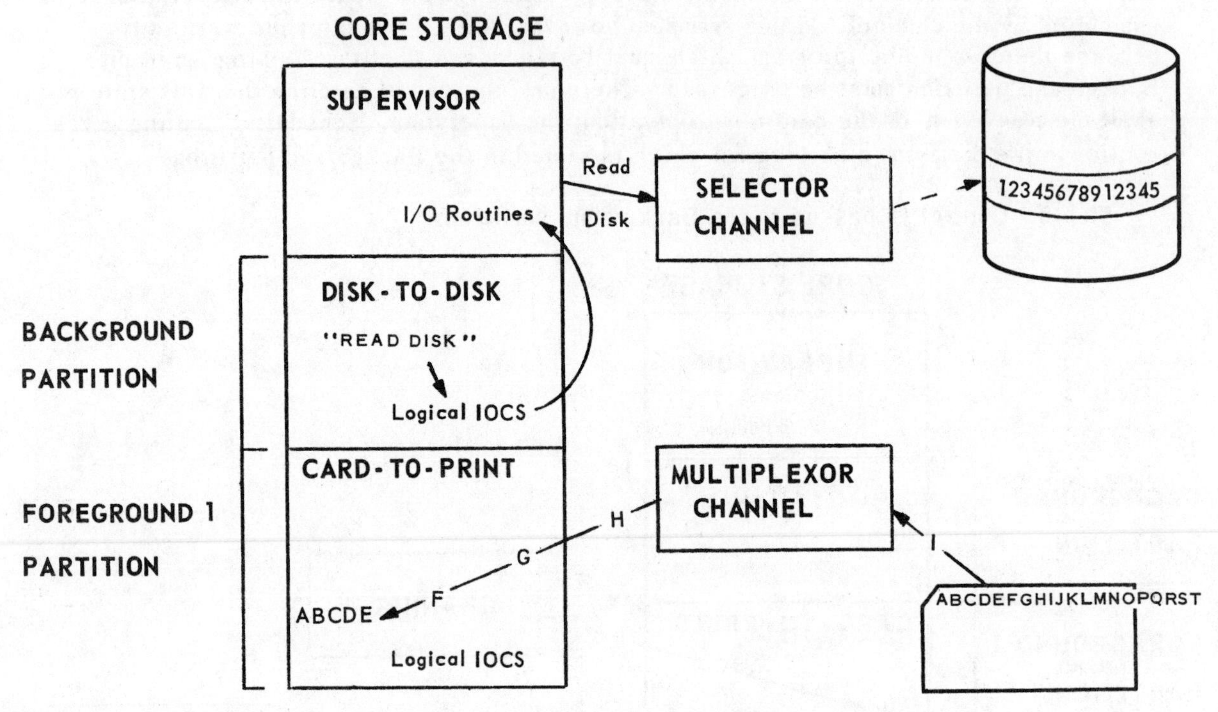

Figure 8-10 Background Program issues Read Command

In Figure 8-10 it can be seen that the Read command to the disk unit is accomplished in the same manner as the read command to the card reader, that is, the Read instruction in the disk-to-disk program links to the logical IOCS routines which in turn link to the physical IOCS routines in the Supervisor which issue the actual command to the channel. Note that now both programs, the card-to-print program in the Foreground 1 partition and the disk-to-disk program in the Background partition, are both waiting for an input/output operation to be completed before continuing with processing. Therefore, the computer must enter the wait state because there is no other processing which can be done.

Step 5: Computer enters wait state.

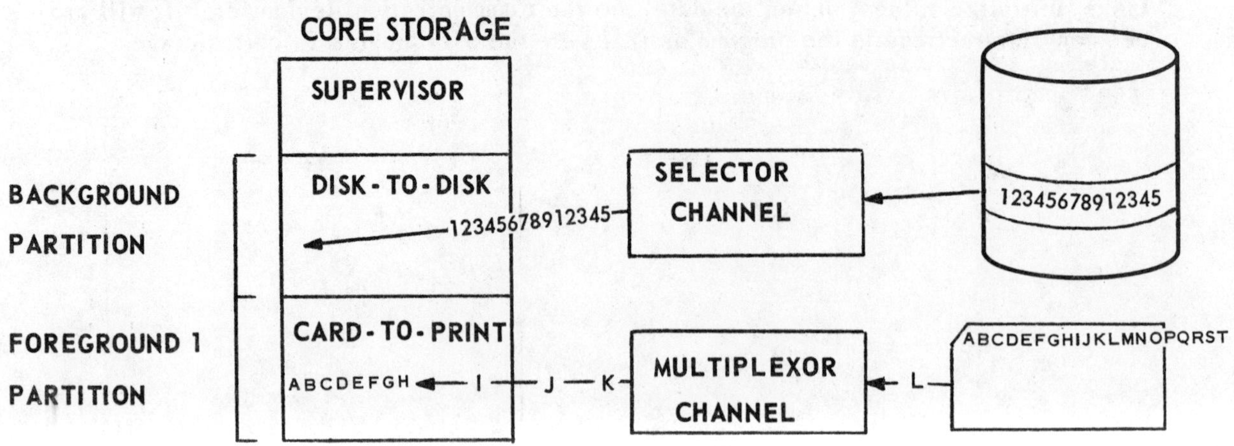

Figure 8-10 Wait State with I/O in both partitions

Note from Figure 8-10 that the entire computer system is in a wait state because both the card-to-print program in the Foreground 1 partition and the disk-to-disk program in the Background partition are waiting for an Input/Output operation to be completed. Note also that the data being read from the card reader is transmitted by the channel a single character at a time but the record being read from the disk is being transferred as a total record. This reflects the difference between the slower multiplexor channel, which is used for card processing, and the faster selector channel, which is used for disk processing.

Since the computer is in a wait state, an interrupt must occur so that processing may begin again. In this example, the disk read will be completed before the slower card read even though the card read was initiated first. Therefore, when the disk record has been read into core storage, the selector channel will send an interrupt to the supervisor.

Step 6: Interrupt occurs from the selector channel.

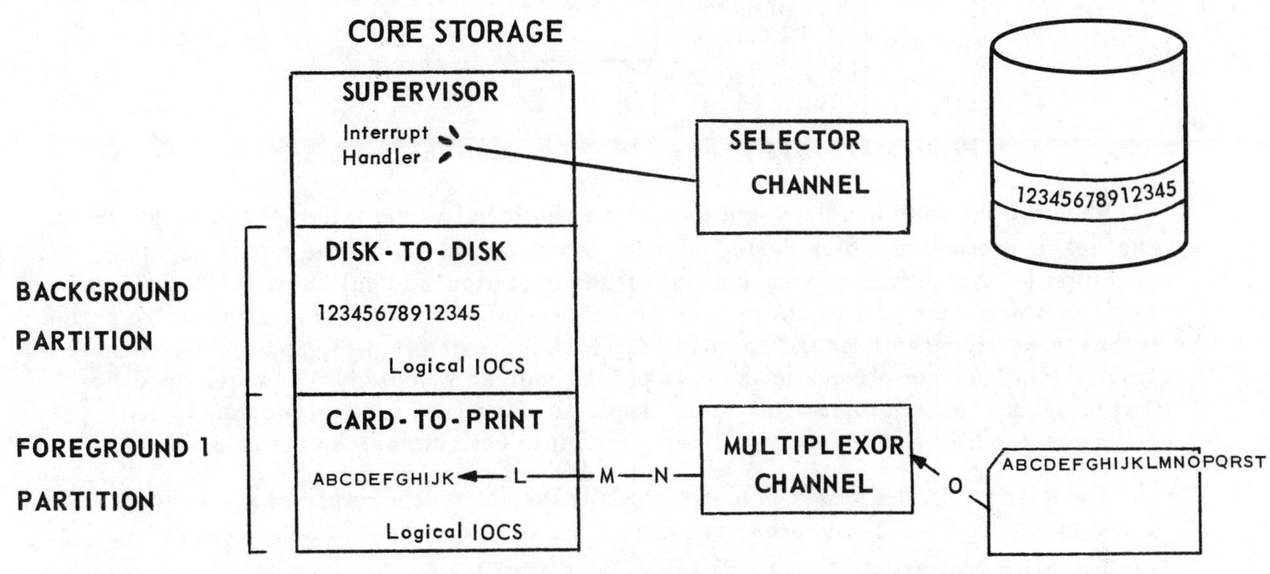

Figure 8-11 Selector Channel sends Interrupt to Supervisor

Note from Figure 8-11 that the selector channel, which was transferring data from the disk to the Background partition in core storage, sends an interrupt to the interrupt handler routine in the Supervisor. At the same time, the multiplexor channel is still involved in the process of transferring data from the punched card to the Foreground 1 partition in core storage. Since this occurs independently of any processing which takes place on the computer, the supervisor interrupt handler routine will begin processing the interrupt from the selector channel.

When the interrupt handler gains control, it must determine the source of the interrupt and the action to be taken. The flowchart in Figure 8-12 illustrates the decisions which must be made by the interrupt handler.

INTERRUPT HANDLER

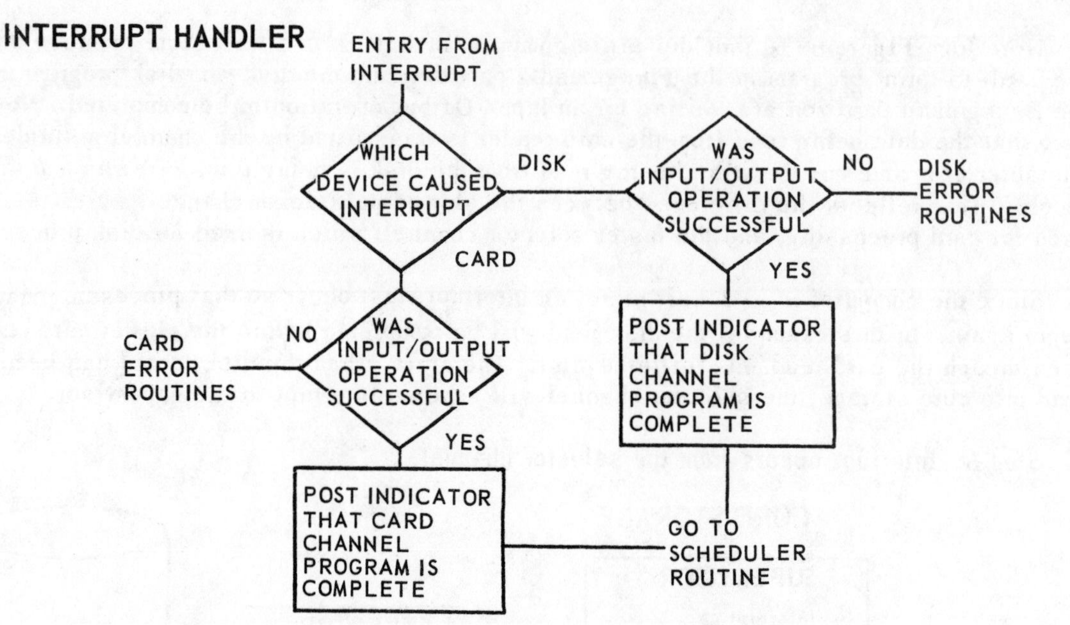

Figure 8-12 Decisions made by Interrupt Handler Routine

When the interrupt handler routine receives control after the interrupt is sent by the channel, it determines which device sent the interrupt and also if the input/output operation for that device was successful. If the operation was not successful, the error routines which are a part of the supervisor are entered to take the proper corrective action. If they were successful, an indicator is set which indicates that the appropriate channel program has been completed. In the example, an indicator would be set which specifies that the disk channel program has been completed. The card channel program is still being executed, ie., the card is still being read into core storage by the channel.

The interrupt handler then goes to the Scheduler Routine which determines which processing should take place next. The flowchart in Figure 8-13 illustrates the Scheduler Routine.

SCHEDULER ROUTINE

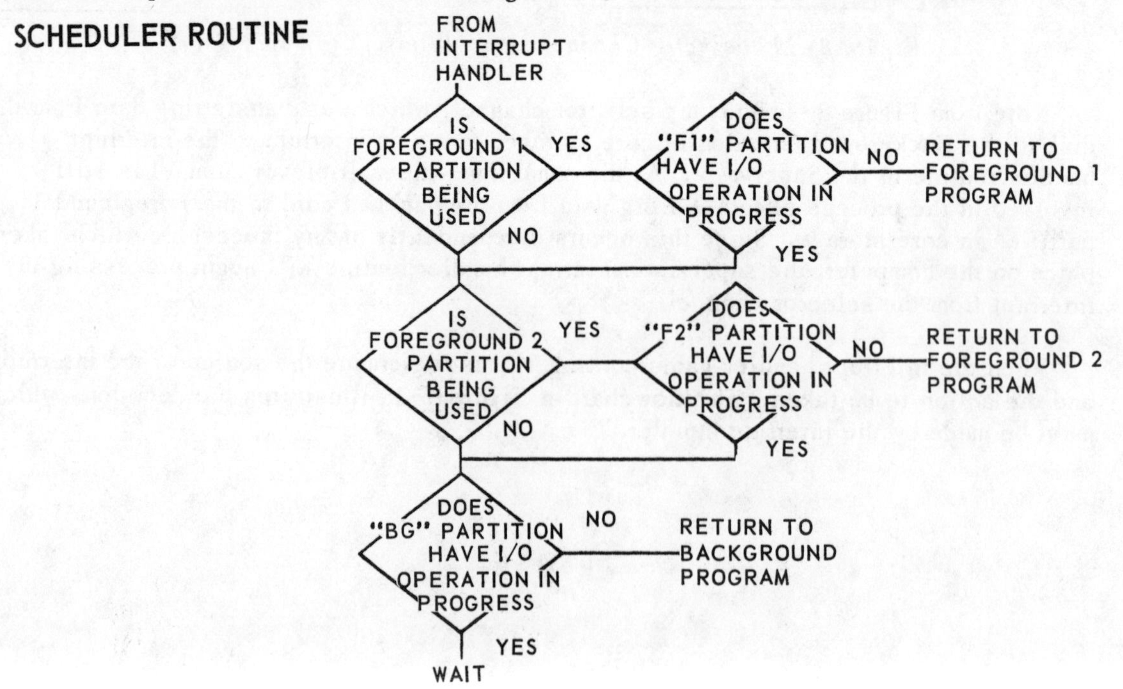

Figure 8-13 Decisions made by Scheduler Routine

Note from the flowchart in Figure 8-13 that the scheduler routine first determines if a particular partition is being used on the machine. Recall that the only required partition is the Background partition, which always must be in use. Note also that the first partition checked is the Foreground 1 partition. This is because the Foreground 1 partition has the highest priority on the machine, that is, if the program being executed in the Foreground 1 partition can be executed, it will always receive control. It is only when the program in Foreground 1 cannot execute because some type of input/output operation is occurring for that partition that the programs stored in other partitions can be executed.

In the example, the program in Foreground 1 cannot be executed because it is still waiting for the card to be read into core storage. Therefore, the scheduler will check if Foreground 2 is being used and it is not. Thus, the Supervisor will return control to the program stored in the Background partition because it does not have any input/output operation in progress, ie. it was the completion of the disk read from the background program which initiated the interrupt processing.

Step 7: Control is returned to the Disk-To-Disk program in the Background partition.

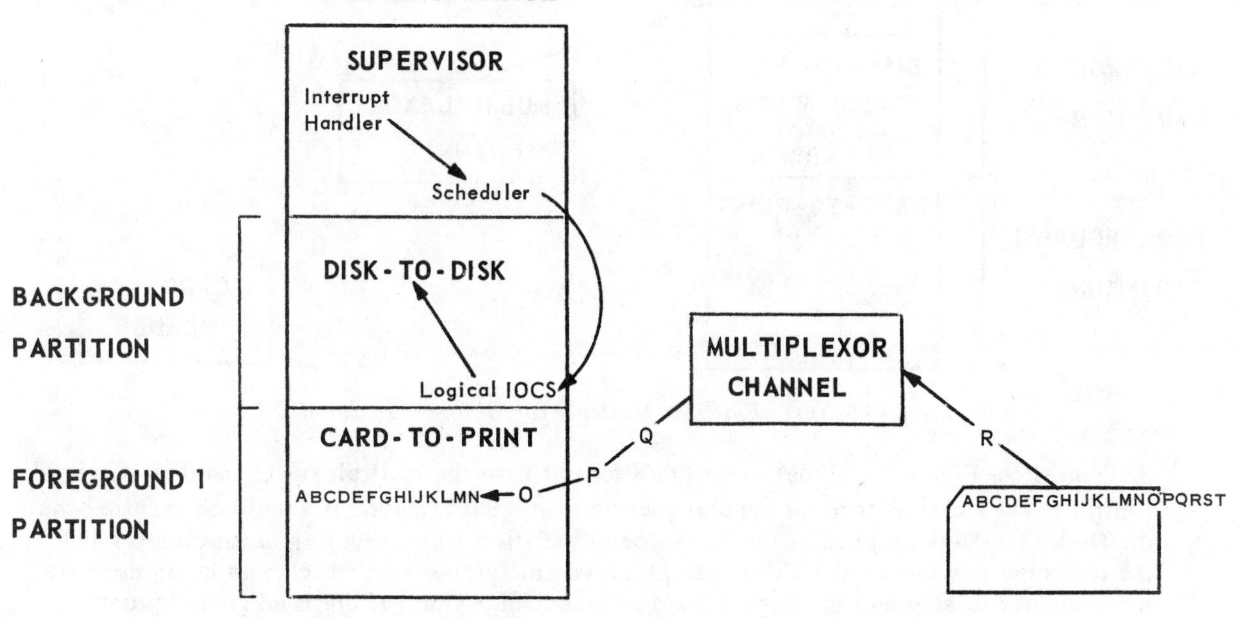

Figure 8-14 Control is returned to Background Program

In Figure 8-14 it can be seen that the scheduler routine has determined that the only program which can be executed in the program stored in the Background partition. The program in the Foreground 1 partition cannot be executed because it is still awaiting the completion of the card read operation. When the Disk-To-Disk program gains control, it will begin processing the data which was read from the disk file as if it were the only program in core storage. Note again that the programs in each of the partitions operate totally independent from one another and neither is aware of the others existence.

The Disk-To-Disk program in this example is to perform a great many calculations on the data each time a new record is read. During its execution, it is not going to issue a large number of read or write commands. The majority of its processing will be involved with instructions executed in core storage in order to manipulate the data which is retrieved from the disk.

The Card-To-Print program in the Foreground 1 partition, on the other hand, will request a large number of input/output operations because its only function is to read a card, format a print line, and print the output on the printer. Therefore, the program in Foreground 1 will be idle a considerable portion of the time it is in core storage because Input/Output operations will be taking place; the background program, however, will be idle only a small portion of the time because its main function is to manipulate the data once it is in core storage.

In the example, the read operation for the input card will be completed when the last column of the card is transferred to core storage. When this occurs, the multiplexor channel will send an interrupt to the interrupt handler routine in the Supervisor.

Step 8: Interrupt is received from the Multiplexor Channel.

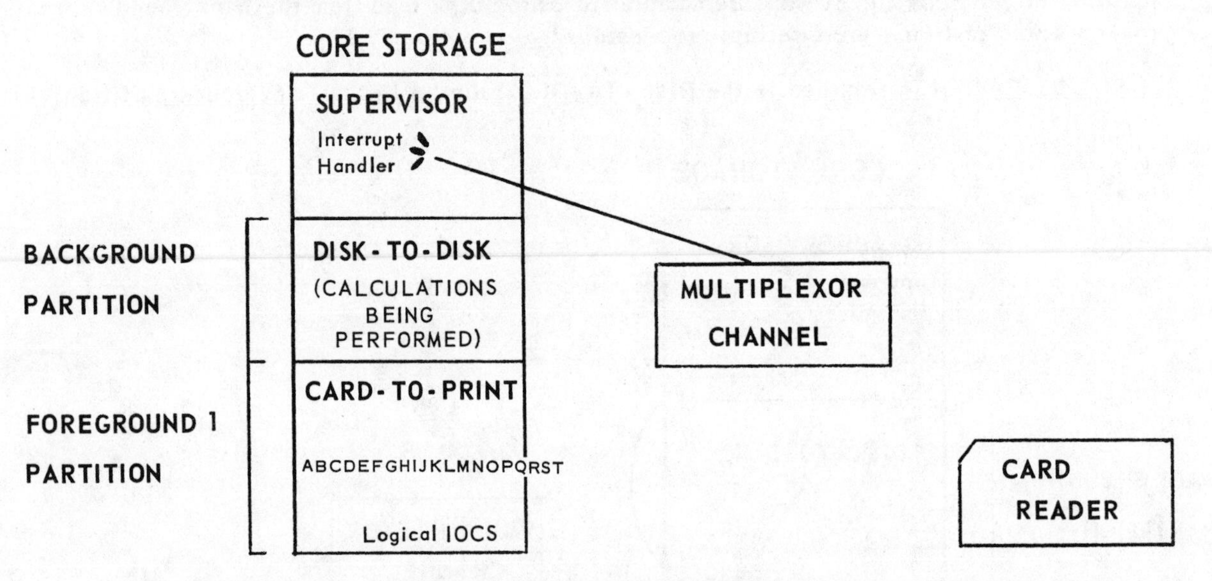

Figure 8-15 Interrupt from Multiplexor Channel is received

Note from Figure 8-15 that when the interrupt from the multiplexor channel is received, control passes to the interrupt handler routine in the Supervisor. It should be recalled that the Disk-To-Disk program in the Background partition was executing instructions when the interrupt was received. The interrupt causes all processing which was being done on the computer to stop and the Supervisor gains control. Thus, if the Background program was in the process of adding two numbers, the addition operation would be completed but no more processing would be done by the program. The Supervisor stores the information which indicates the status of the disk-to-disk program, such as the address of the next instruction to be executed, so that when the program can again begin processing, it will know where to return control.

The interrupt handler in the supervisor processes the interrupt in the manner illustrated in Figure 8-12 and then passes control to the supervisor scheduler which is illustrated in Figure 8-13. The scheduler determines that the Foreground 1 partition is being used and that there is no I/O operation in progress for the Foreground 1 program. This is because the card has been read into core storage and the I/O operation has been posted complete by the interrupt handler. Therefore, the scheduler will pass control to the Card-To-Print program in the Foreground 1 partition.

Step 9: Control is passed to the Foreground 1 partition.

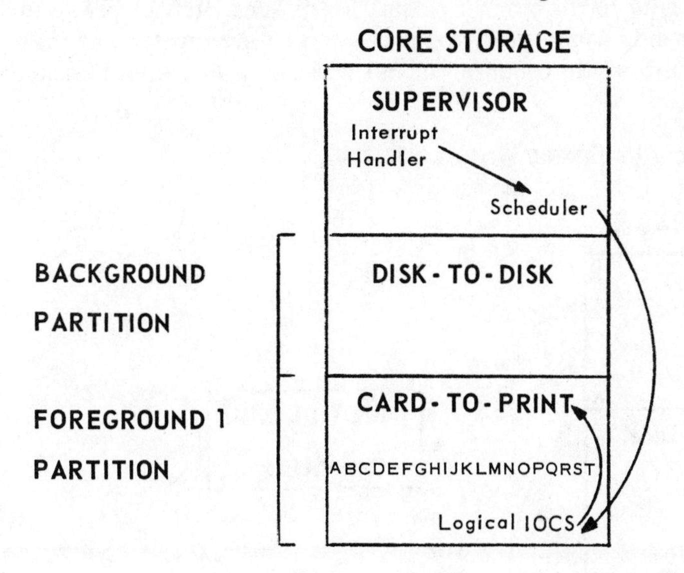

Figure 8-16 Card-To-Print Program regains control

Note in Figure 8-16 that the Supervisor scheduler routine passes control to the Foreground 1 program because it has the highest priority and it is now able to execute instructions because its pending I/O operation has been completed. The program can manipulate the card data which it has read in any manner desired.

Note also that the Disk-To-Disk program which was executing in the Background partition is idle even though it is able to execute, that is, it has no outstanding I/O requests and is not waiting for any conditions to be satisfied before it can begin processing. It does not have control, however, because the Foreground 1 program has priority and whenever it is able to be executed, the Supervisor will again give control to it.

This then, is an extremely important concept concerning multiprogramming, that is, the programs in the lower priority partitions can only be executed when the higher priority programs cannot execute because they are waiting for an Input/Output operation to be completed. Thus it can be seen that if a program which executes very few input/output operations is placed in a high priority partition, the programs in the lower priority partitions will get very little processing time. Therefore, programs such as the card-to-print program, which have a great deal of input/output processing, should be placed in high priority partitions such as Foreground 1 and the programs which do little I/O processing, such as the Disk-To-Disk program, should be placed in a lower priority partition.

In the example, the Card-To-Print program would continue processing the data, perhaps moving the card input fields to the printer output fields, until it was ready to issue another input/output command, such as a Write command to the printer. At that time, the Supervisor Input/Output routine would receive control and issue the actual command to the channel to write on the printer.

Step 10: Card-To-Print Program issues Write command.

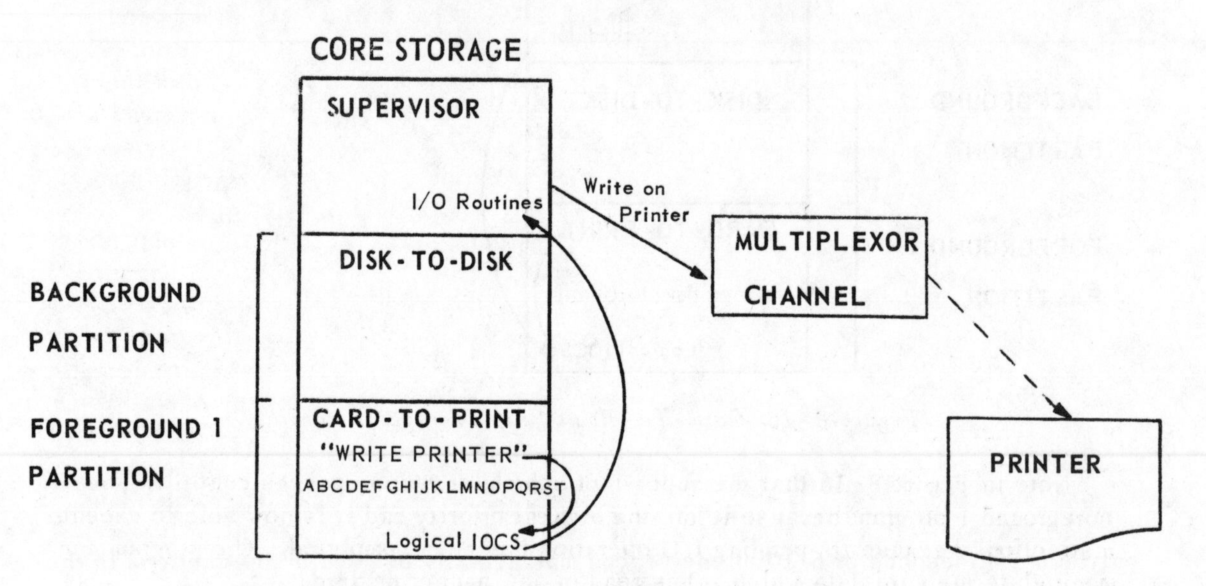

Figure 8-17 Write Command is issued in Card-To-Print Program

When the Write command is issued in the Card-To-Print program which is stored in the Foreground 1 partition, the Supervisor I/O routine is entered and issues the actual command to the channel to begin the writing on the printer. Note that during the time the Card-To-Print program was moving the data from the input area to the output area, the Disk-To-Disk program in the Background Partition was completely idle, that is, no instructions in the Background partition were being executed. When the Write command is issued, however, the Card-To-Print program in the Foreground 1 partition must again wait for an Input/Output operation to be completed before it can process more data. Therefore, when the supervisor scheduler gains control from the I/O routines, it will determine that the Foreground 1 partition cannot be entered because it is waiting for an Input/Output operation (see Figure 8-13). Therefore, it will return control to the Disk-To-Disk program in the background partition because it is not waiting for an input/output operation.

Step 11: Background partition program gains control.

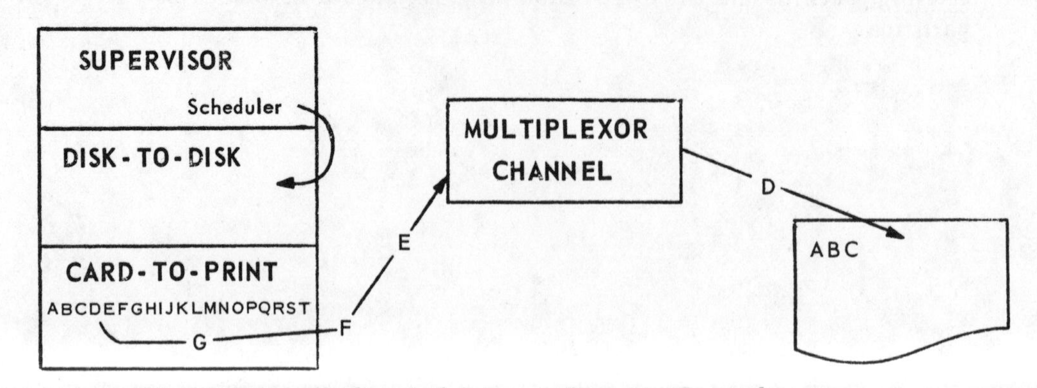

Figure 8-18 Background Program Regains Control

Note from Figure 8-18 that when the channel program begins which controls the printing of the data from the Foreground 1 partition onto the printer that the Card-To-Print program must wait for the input/output operation to be completed. The Disk-To-Disk program, however, was idle not because of an input/output operation but because the Foreground 1 program had priority over it. Therefore, while the Foreground 1 program waits, the Background program can process data as needed. The Disk-To-Disk program will be in control and processing until it issues an input/output command and must wait for I/O completion or until the I/O operation for the Foreground 1 program is complete. If the Foreground 1 I/O is completed, the Card-To-Print program in the Foreground 1 partition will regain control because it has priority. If the Background program issues an I/O command before the printing is complete, the entire system will be in a wait status because neither the Disk-To-Disk nor the Card-To-Print program can process because they are both waiting for an Input/Output operation to be completed.

Again, the important fact to note about the multiprogramming operation is that the Foreground 1 partition has the highest priority, followed by the Foreground 2 partition and then the Background partition. The programs in the lower priority partitions will only gain control for execution when the program in the higher priority partition is waiting for an input/output operation to be completed. It should be noted that this switching from one partition to another partition and the Input/Output operations take place quite rapidly on the computer. For example, the entire process described in Step 1 through 11 in the previous example could take place in less than 1 second on the computer. The supervisor routines and the processing of data in the programs takes very little time. In fact, of the one second used to process Steps 1 through 11, perhaps only 600 milliseconds or 6/10 of one second may actually be used by instructions in the computer. The remaining time would be wait time, that is, time which is spent in the wait state waiting for an input/output operation to be completed.

USES OF MULTIPROGRAMMING

As can be seen from the previous discussion, multiprogramming allows the full resources of the computer to be utilized, that is, the time which is spent in the wait state for Input/Output operations is greatly reduced because when one partition is waiting for I/O, a lower priority partition may be processing. This feature allows a greater "thruput", that is, the amount of work which may be accomplished in a given time period. One of the primary uses of a Foreground partition is to store teleprocessing programs, that is, programs which communicate with terminals which are remotely connected with the computer through the use of telephone lines or other communication hardware. Teleprocessing programs normally require relatively little CPU processing time and a great deal of I/O time in order to communicate with the remote terminals. Thus, it is an ideal type of program to be placed in a high priority Foreground partition because the lower priority partions have a great deal of time for processing.

Other excellent programs for Foreground partitions are programs which are almost totally dedicated to an Input/Output operation, such as utility programs which perform disk initialization and disk copying, programs which utilize relatively slow devices such as the card reader or the card punch, and other programs which do not demand a large amount of processing time from the CPU. Programs such as FORTRAN scientific programs which perform many calculations or programs which utilize a great deal of table-lookup processing are not good programs to be executed in a high priority partition because they do not relinquish control of the CPU very often and the programs in the lower priority partitions, therefore, do not get much processing time.

Multiprogramming may be executed on the System/360 operating under DOS in a number of ways. Chapter 9 contains examples of initiating and using the DOS Multiprogramming capabilities.

CHAPTER 9

MULTIPROGRAMMING OPERATION

INTRODUCTION

As noted in Chapter 8, multiprogramming consists of two or more programs operating independently of one another on the same computer. Under DOS, core storage is divided into three fixed-length "partitions" which are called Foreground 1, Foreground 2, and Background. In the examples of the DOS Utilities and DOS Sort/Merge program, it was assumed that only a single partition was being used. When only one partition is being used, it is always the Background partition. It should be recalled that when programs are executed in the Background partition, the Job Control program is used to link together the jobs and job steps which are processed. Job control must always be used in the Background partition and the use of job control is a BATCHED JOB mode, that is, jobs are batched and run together under the control of the Job Control programs.

The execution of programs in the foreground partitions (Foreground 1 or Foreground 2) may be accomplished in one of two modes: The Batched Job mode, which is the same as in the Background partition, or the Single Program Initiation Mode (SPI). When the Single Program Initiation mode is used, the operator begins the program from the console typewriter (SYSLOG) and when the single program is completed, the operator must start the next program. There is no job continuity such as when job control is used in the Batched Job mode. The choice of Batched Job or SPI depends upon the application to be processed and the parameters and options included in the Supervisor when it is generated. The choices of the installation when the supervisor is generated are that multiprogramming will not be supported at all, that only single program initiation will be supported, or that both SPI and the batched job mode will be supported. Again, it depends upon the nature of the processing in a particular installation as to which of these options is chosen. For the examples presented in this chapter, it will be assumed that both SPI and the batched job mode are supported by the Supervisor.

CORE STORAGE ALLOCATION

In the examples presented in Chapter 8 it can be seen that core storage must be allocated for each of the partitions which are to be used for multiprogramming. In addition, the System/360 on which multiprogramming is to be performed must contain at least 24K if SPI is to be used and at least 32K if both SPI and Batched Job Modes are to be used. The core storage which is to be used for each of the partitions may be determined when the Supervisor is generated and/or by the operator before he initiates the programs in a Foreground partition. The method used in both cases is the use of the ALLOC command The general format of the ALLOC command is illustrated in Figure 9-1.

```
ALLOC F1=nK,F2=nK
```

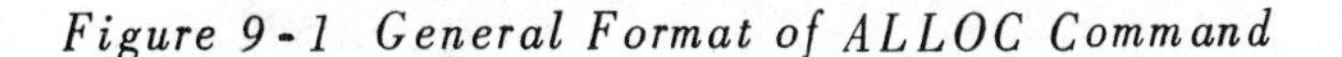

Figure 9-1 General Format of ALLOC Command

In the general format illustrated in Figure 9-1 note that the keyword ALLOC must be specified as shown. The operands are used to indicate the size of the Foreground 1 and Foreground 2 partitions. The following examples illustrate the use of the ALLOC command to allocate core storage for multiprogramming.

EXAMPLE 1

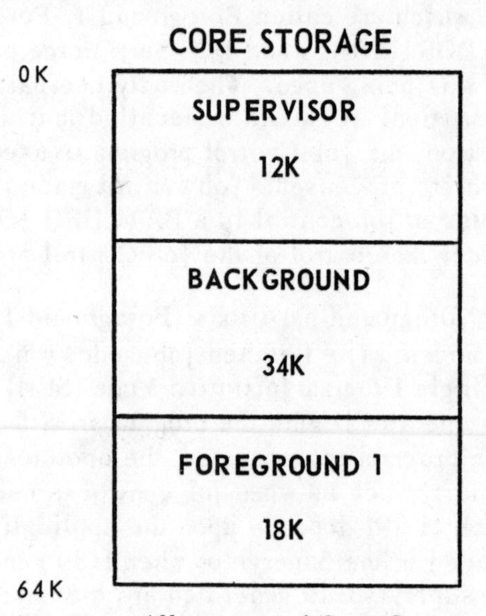

Figure 9-2 Allocation of Core Storage for Multiprogramming

In the example presented above, it is assumed that the System/360 contains 64K of main storage. The Supervisor is allocated 12K because that is the amount of core storage needed to contain all of the routines within the Supervisor. The ALLOC statement allocates 18K of core storage for the Foreground 1 partition. Note that the Foreground 2 partition is allocated zero bytes and the Background partition is not allocated at all. The Background partition, which must always contain a minimum of 10K bytes, always receives the amount of core storage remaining after the Foreground partitions have been allocated.

It should be noted that the Foreground partitions must be allocated in increments of 2K bytes, that is, the value of "n" in the general format in Figure 9-1 must always be an even number.

As noted previously, all three partitions may be allocated for use in a multiprogramming environment. The example in Figure 9-3 illustrates the allocation of all three partitions.

EXAMPLE 2

Command: ALLOC F1=20K,F2=14K

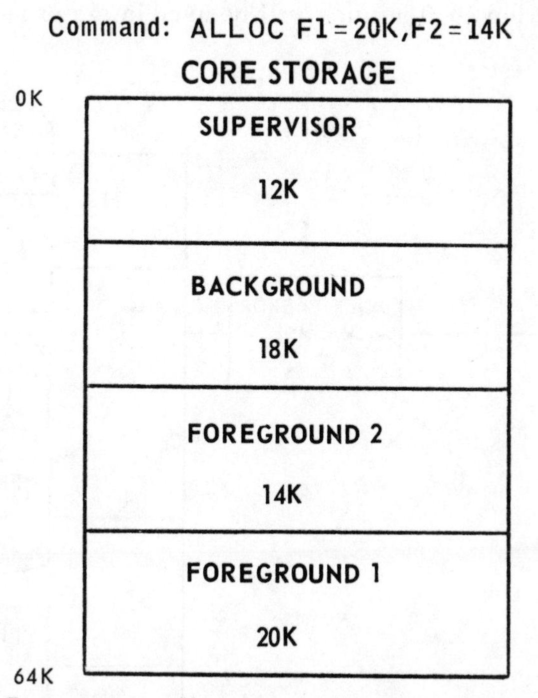

Figure 9-3 Allocation of Foreground 1 and Foreground 2

Note in the ALLOC command used in Figure 9-3 that both Foreground 1 and Foreground 2 are allocated. Note also that both of the allocations specify core in increments of 2K. The Background partition is allocated the remaining core storage after the other partitions are allocated (64 - 12 - 14 - 20 = 18). Again, the Background partition must always be at least 10K bytes in length.

BATCHED JOB MODE

As noted previously, programs which are to be executed in a foreground partition may be executed in either a single program initiation mode or a batched job mode. The programs which are executed in Background are always executed in the batched job mode. When the batched job mode is used, the job control programs operate in the foreground partition in the same manner as the programs which operate in the batched job mode in the background partition, that is, the job control programs are in core storage between the execution of user programs and read the job control statements pertaining to the next program to be executed. Thus, a continual flow of jobs and job steps is facillitated using job control. In order to operate a batched job foreground partition, core storage must first be allocated for the partition as illustrated in the preceding section. The devices must then be made available for the job control program and job control must be loaded into core storage for execution.

Batched job foreground programs may be executed on many different "configurations" of the System/360. The following configuration will be used in the examples to follow.

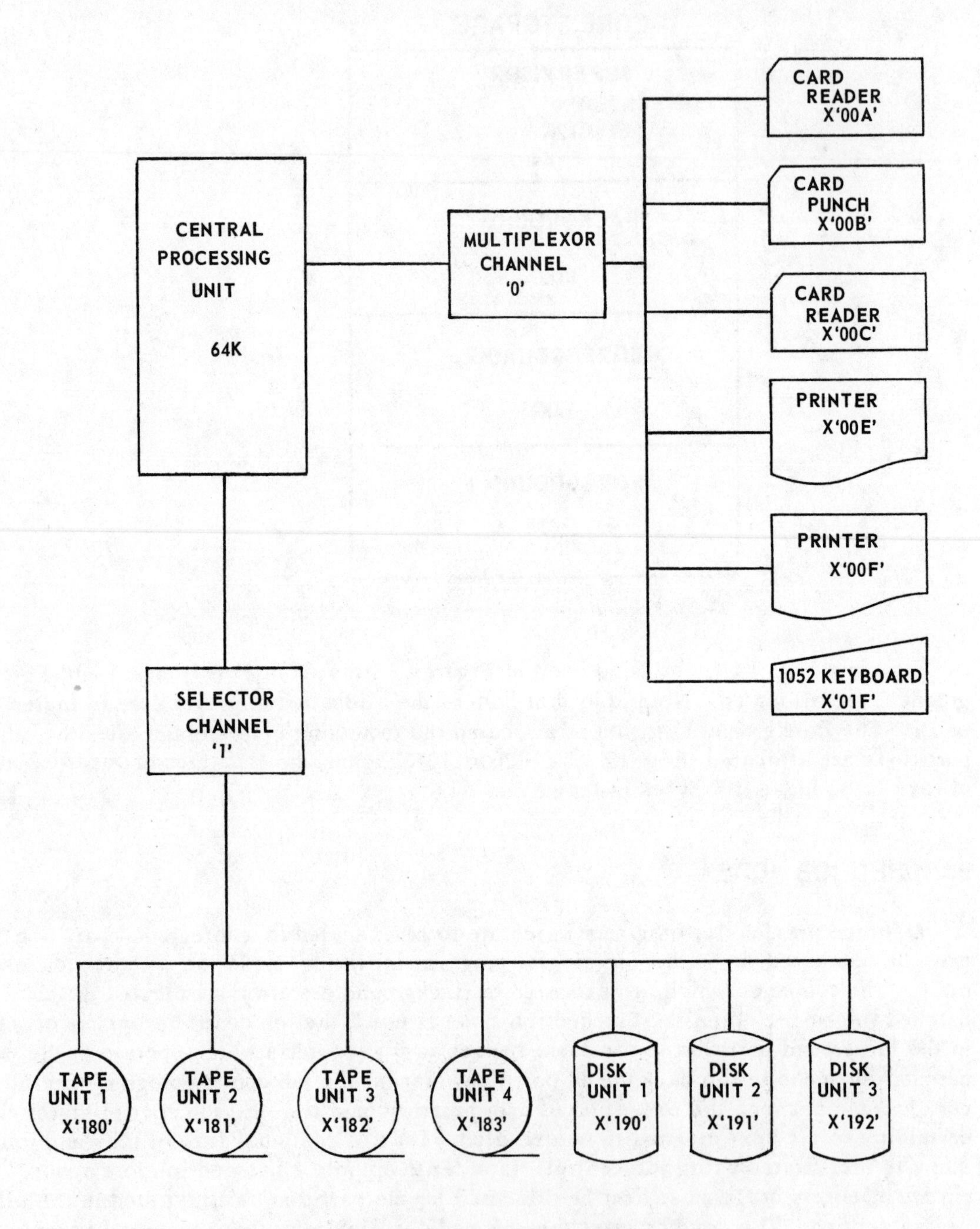

Figure 9-4 System/360 Configuration

Note from the above configuration that there are two card readers (X'00A' and X'00C') and two printers (X'00E' and X'00F'). These devices will be used in order that an input device and output device will be available for the two partitions, Foreground 1 and Background, which will process programs in the batched job mode.

In most installations, the allocation of the foreground partitions is made at "sysgen" time, that is, when the Disk Operating System is "tailored" to the needs of the installation. In addition, the permanent assignments of many of the devices may be made in order that temporary assignments will not need to be made during execution. Included in these assignments are the System Logical Units, that is, the symbolic-unit names which are used by the Operating System programs such as Job Control, compilers, and the Sort/Merge program. The table below illustrates the assignments which have been made for the Supervisor which will be used in the following examples.

PARTITION	SYMBOLIC-UNIT NAME	DEVICE
BG	SYSRDR	X'00A'
BG	SYSIPT	X'00A'
BG	SYSLST	X'00E'
BG, F1	SYSLOG	X'01F'
F1	SYSRDR	X'00C'
F1	SYSIPT	X'00C'
F1	SYSLST	X'00F'
BG	SYS004	X'180'
BG	SYS005	X'181'
F1	SYS004	X'182'
F1	SYS005	X'183'
BG, F1	SYS006	X'191'
BG, F1	SYS007	X'192'

Figure 9-5 Standard Assignments

In the chart of Standard Assignments illustrated above it can be seen that each of the partitions, Background (BG) and Foreground 1 (F1), has its own set of symbolic-unit names, that is, there is a SYSRDR symbolic unit for Background and a SYSRDR symbolic unit name for Foreground 1, etc. This is done so that when the job control is in the Background partition, it will reference the device assigned to SYSRDR and when it is in the Foreground 1 partition, it will reference the device assigned to SYSRDR in F1. A very important rule pertaining to device assignments must always be followed when processing programs in a multiprogramming environment: a Unit-Record device (card reader, card punch, printer) may only be assigned to one partition; a tape device may only be assigned to one partition; a direct-access device may be assigned to more than one partition. Thus, in the chart above, note that the only devices which are assigned to more than one partition are the 1052 typewriter (SYSLOG) and the direct-access disk devices. The card reader X'00A' is assigned to SYSRDR and SYSIPT in the Background partition and cannot be assigned to any symbolic-unit names in any other partition. The reason for this is that if the single card reader were assigned to two partitions, and two different programs were reading cards from the same card reader, there is no control over the sequence of the cards and the program in the Background partition may read a card which is designed for the program in F1 partition, and vice versa. The same is true of the printer (ie. a report would contain printout from both programs) and tape devices. The direct-access devices may be assigned to more than one partition because more than one file may be defined on a direct-access device and each program would reference the proper file.

As noted previously, when programs are processed in a batched-job mode, core storage must first be allocated for the Foreground partition to be used and then the operator must initiate the processing by loading the job control program into the proper partition. In the following example, it is assumed the core storage has been allocated when the Supervisor was generated in a format as illustrated in Figure 9-2, that is, the Background partition contains 34K bytes, and the Foreground 1 partition contains 18K bytes. It should be noted that, in order to process jobs in the batched-job mode, a minimum of 10K bytes must be allocated for the partition to be used and that the Background partition must always be a minimum length of 10K bytes.

When the system is IPLed, that is, when the Supervisor is loaded into core storage and the system is started, job control is automatically loaded into the Background partition. Thus, after IPL, the status of the system would be as illustrated below.

EXAMPLE

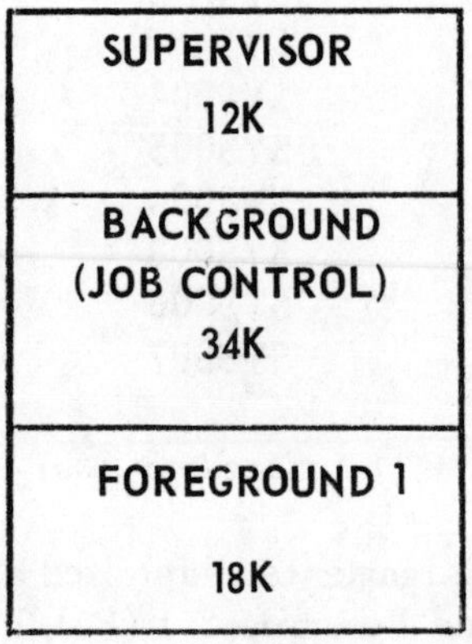

SUPERVISOR 12K
BACKGROUND (JOB CONTROL) 34K
FOREGROUND 1 18K

Figure 9-6 Core Storage Allocation

Note from the illustration above that core storage is allocated for two partitions but only the Background partition is being used, ie. the Job Control program has been loaded and is ready to read job control cards from the device assigned to SYSRDR (X'00A').

When initiating Foreground programs, the operator uses the device assigned to SYSLOG for communicating with the operating system. Therefore, it is important to be aware of messages and techniques used on SYSLOG. The following messages and operator entries would be made on SYSLOG when the system is IPLed.

EXAMPLE

```
0110A GIVE IPL CONTROL STATEMENTS
set date=10/04/71,clock=11/35/00 (B)
0120I DOS IPL COMPLETE
BG 1100A READY FOR COMMUNICATIONS
BG
```

Note: The (B) is shown for illustration purposes only and would not be typed on the SYSLOG sheet.

Figure 9-7 Console (SYSLOG) Messages at IPL

In the example illustrated in Figure 9-7, the IPL routine prints messages 0I10A which instructs the operator to enter the information needed to begin operation. In the example, the SET statement is used to enter the date and the time of day. The operator then returns control to the program by depressing the End-Of-Block key (Ⓑ), and the program prints out message 0I20I indicating that the IPL process is complete and that the Job Control program will be loaded into core storage. The message 1I00A is printed by the job control program to indicate that it is ready to read job control statements from the device assigned to SYSRDR. Note that the prefix BG precedes the message from job control. This is used to indicate which partition is communicating with the operator. BG is used for the Background partition, F2 is used for the Foreground 2 partition and F1 is used for the Foreground 1 partition.

When the operator has completed the IPL procedure, the job stream for the program to be executed in the Background partition may be loaded in the device assigned to SYSRDR to be read by job control. In the example being used, assume that the background job to be processed is the Disk-To-Disk program illustrated in Chapter 8. The jobstream for that program is illustrated below.

EXAMPLE

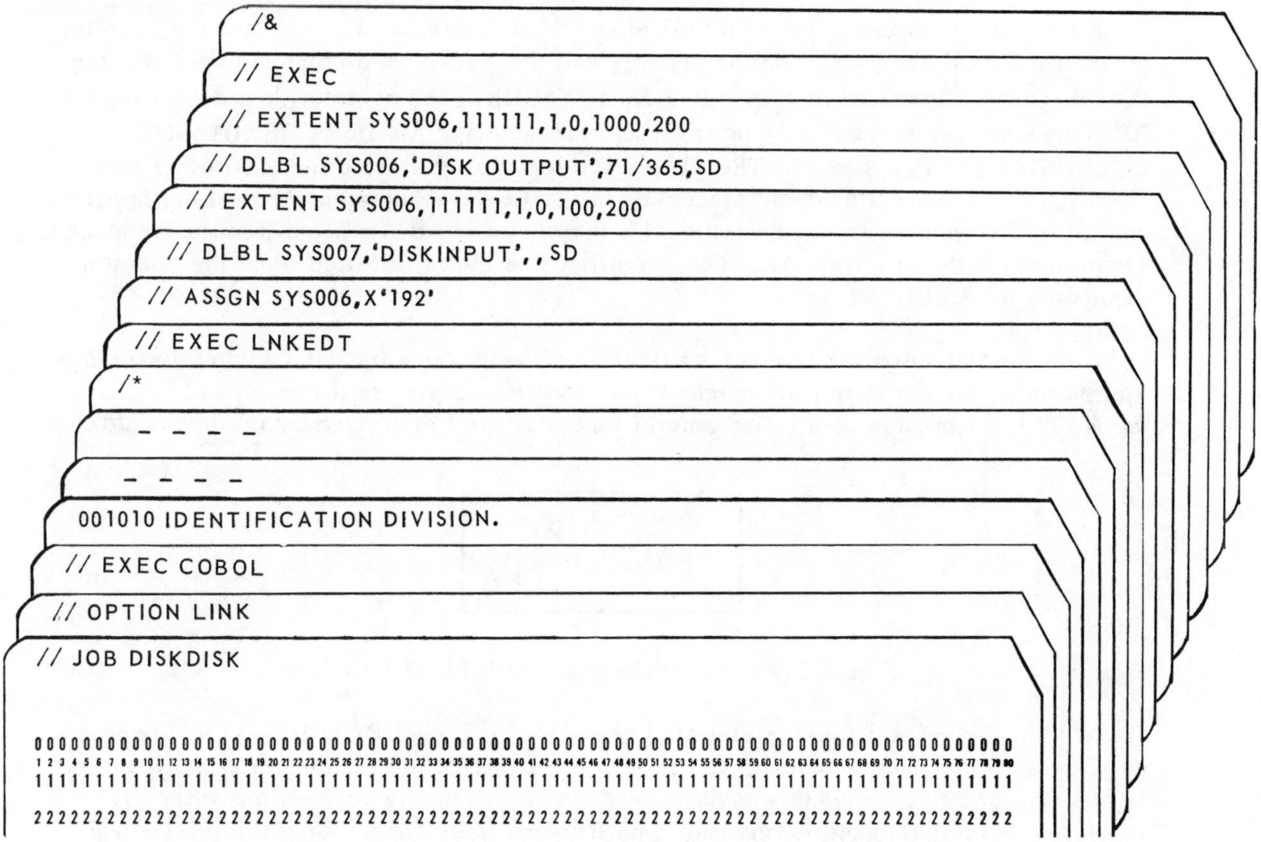

```
/&
// EXEC
// EXTENT SYS006,111111,1,0,1000,200
// DLBL SYS006,'DISK OUTPUT',71/365,SD
// EXTENT SYS006,111111,1,0,100,200
// DLBL SYS007,'DISKINPUT',,SD
// ASSGN SYS006,X'192'
// EXEC LNKEDT
/*
- - - -
- - - -
001010 IDENTIFICATION DIVISION.
// EXEC COBOL
// OPTION LINK
// JOB DISKDISK
```

Figure 9-8 Job Stream to be Executed in Background Partition

Note from the job stream illustrated above that the COBOL program is to be compiled, link-edited, and executed. When the operator hits the EOB (Ⓑ) key on SYSLOG, following the Ready For Communications message, job control will begin to read the job stream from the device assigned to SYSRDR for the Background partition. In the example, this device is X'00A'.

It should be noted at this point that the programs in the Background partition have begun to execute, that is, job control is reading the job stream and the COBOL compiler will begin to compile the program. There are, however, no programs executing in the Foreground partition. This is because the operator must initiate the job control program in the Foreground partition. This initiation is done using the BATCH command and is illustrated in Figure 9-9.

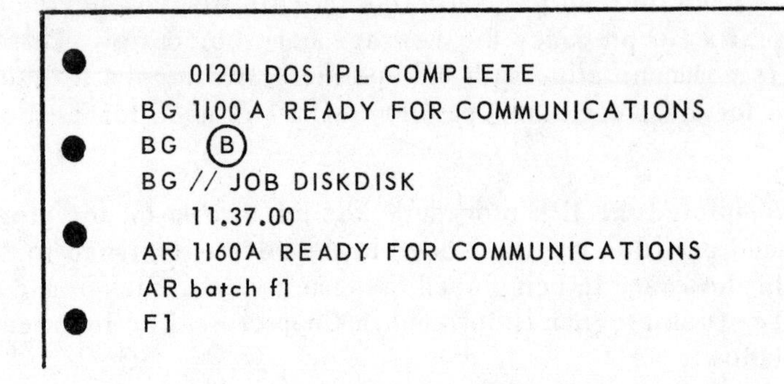

```
   0I20I DOS IPL COMPLETE
BG 1100A READY FOR COMMUNICATIONS
BG (B)
BG // JOB DISKDISK
   11.37.00
AR 1160A READY FOR COMMUNICATIONS
AR batch f1
F1
```

Figure 9-9 Initiation of Job Control in Foreground 1 Partition

Note from the console typewriter (SYSLOG) sheet illustrated above that the operator pressed the EOB key ((B)) on the console and the Background program began reading the job stream illustrated in Figure 9-8 from SYSRDR. The operator then depressed the REQUEST key on the console typewriter and the message AR 1160A READY FOR COMMUNICATIONS appeared. The AR prefix identifies the Attention Routine, a series of routines which are stored and processed in the Logical Transient Area of the Supervisor and allow the operator to communicate with the system. All Attention Routine messages are prefixed with the letters AR. The Attention Routine is initiated when the operator depresses the REQUEST key.

After the Attention Routine has written the message, it waits for the operator to enter a command. In order to initiate batched-job mode processing in a Foreground partition, the BATCH command is used. The general format of the Batch command is illustrated below.

```
BATCH {F1}
      {F2}
```

Figure 9-10 General Format of BATCH Command

Note in the general format illustrated above that the word BATCH must be specified as shown. Either the operand F1 or F2 must be specified to indicate which partition is to be used for batched-job processing. In the example in Figure 9-9, the entry "f1" is used to specify that the Foreground 1 partition is to be used. When the job control program is loaded into core storage, it enables the operator to enter control statements by printing the prefix F1 and then waiting for operator action.

After the Batch command has been used and job control has been loaded into core storage, the status of the system is as illustrated below.

EXAMPLE

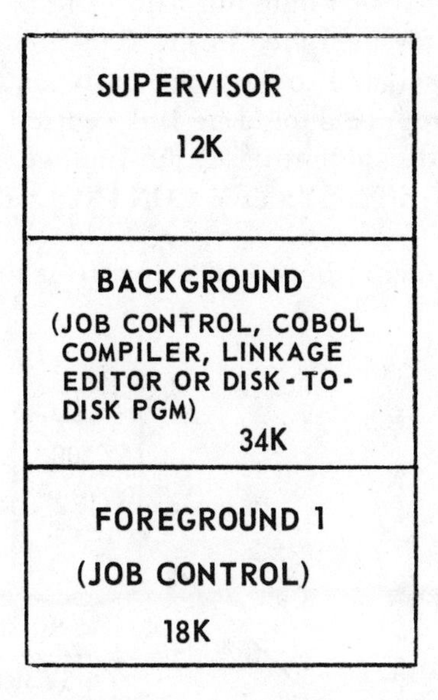

Figure 9-11 Core Storage Allocation After Job Control Loaded in F1

In the illustration above it can be seen that the Job Control program has been loaded into the Foreground 1 partition in core storage. It is now ready to read job control statements from the device assigned to SYSRDR (X'00C'). The job stream which will be read from SYSRDR in the Foreground 1 partition when the operator EOB's is illustrated below.

EXAMPLE

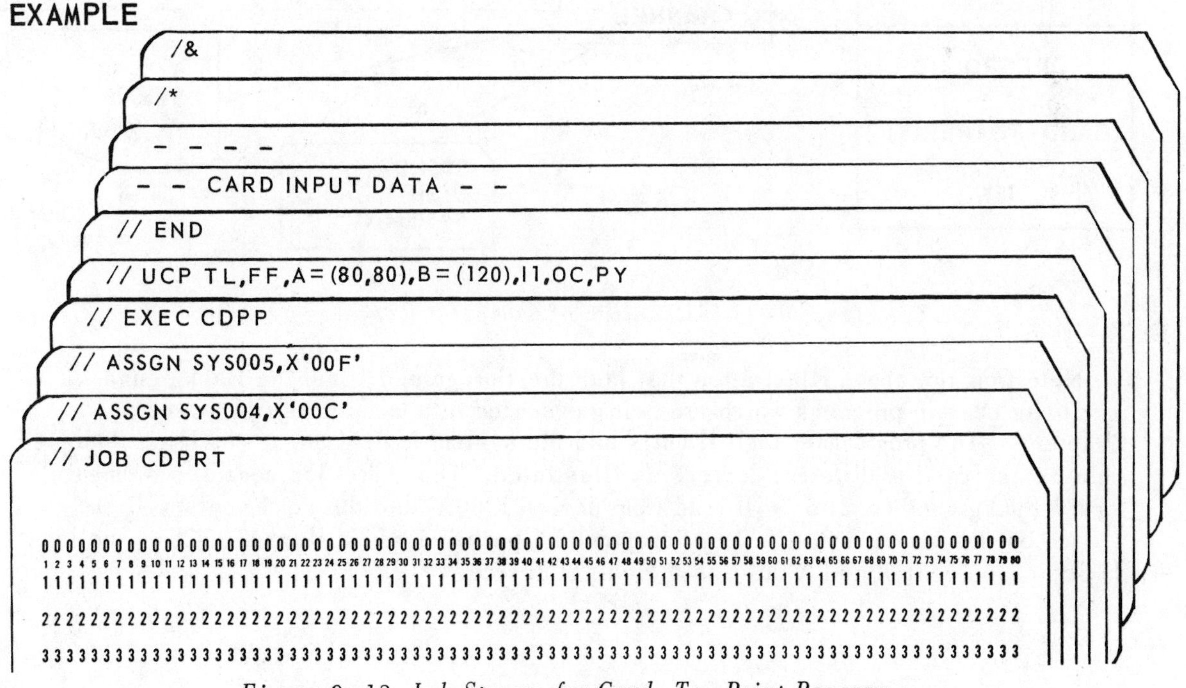

Figure 9-12 Job Stream for Card-To-Print Program

Note from the Job Stream illustrated in Figure 9-12 that the Card-To-Print DOS Utility program is used to read the cards from the card reader (X'00C') and print them on the printer (X'00F'). Note that the devices used are the devices which are available to the Foreground 1 partition. Job control would not allow the assign card to assign SYS005 to X'00E' because X'00E' is already assigned in the Background partition. Again, unit-record and tape devices may be assigned to only one partition. It should be noted also that the Card-To-Print Utility Program should be link-edited to be processed in the Foreground 1 partition. A detailed explanation of the linkage-editor control statements may be found in the IBM SRL C24-5036 SYSTEM CONTROL AND SERVICE PROGRAMS.

When the Card-To-Print program is begun, the computer resources are allocated as illustrated below.

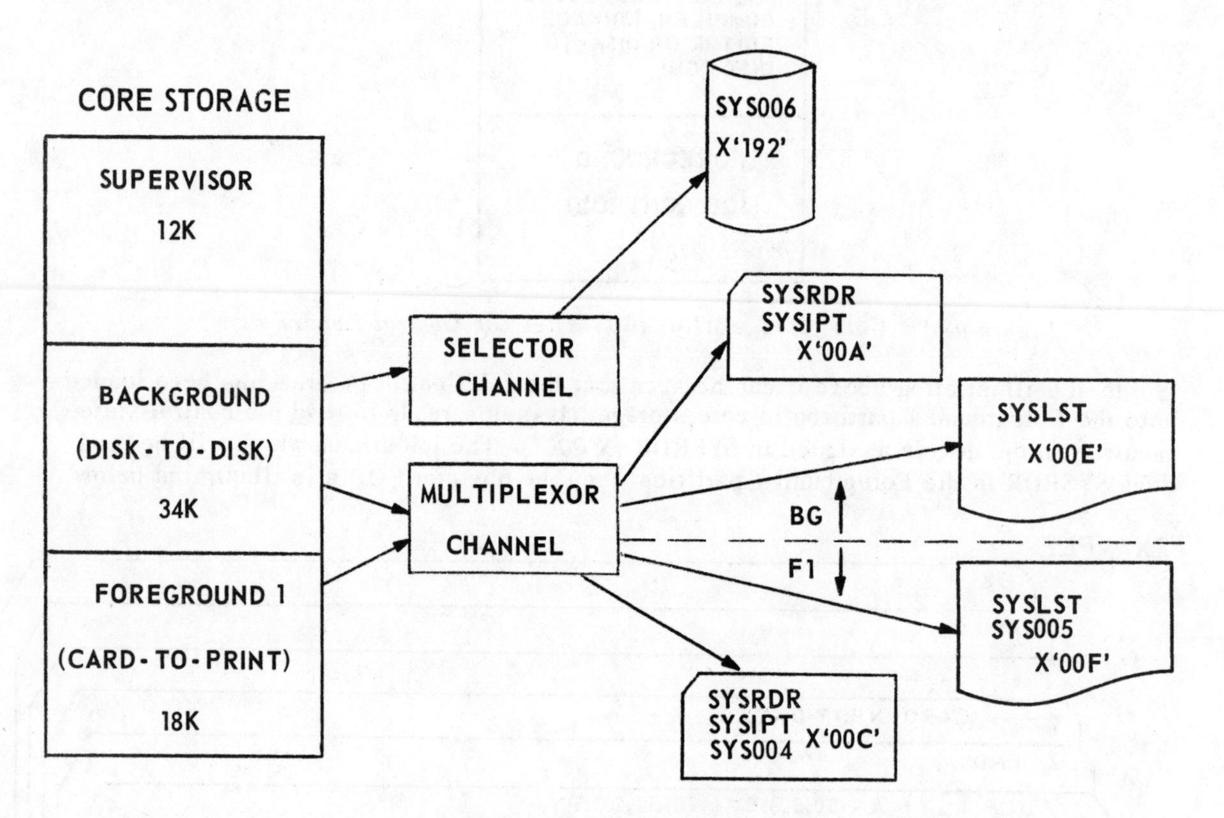

Figure 9-13 Allocation of Computer Resources

Note from the above illustration that both the Foreground 1 and the Background partitions contain programs which are being executed in a manner as illustrated in Chapter 8. The programmer logical units and the system logical units which are assigned must be assigned to different devices as illustrated. Thus, the job control statements for the background partition will read from device X'00A' and the statements will be written on the device X'00E'. The Foreground 1 partition will utilize X'00C' as the input device to job control and X'00F' as the output device.

When either of the job streams being processed has been completed, a message will appear on SYSLOG. In this example, assume that the Card-To-Print program being executed in Foreground 1 is the first to be completed. The sheet on SYSLOG would appear as illustrated in Figure 9-14.

EXAMPLE

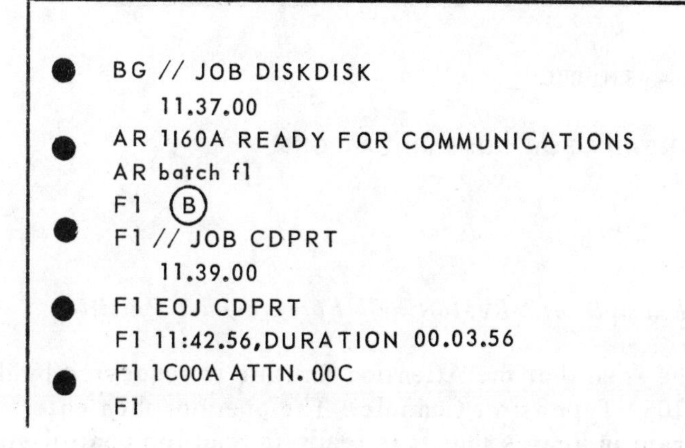

Figure 9-14 SYSLOG Output

Note in the above example that job control has written message 1C00A ATTN. 00C on SYSLOG because there are no more job control cards in SYSRDR. Therefore, it is informing the operator that it is ready to read more control cards on device X'00C', which is assigned to SYSRDR. If the operator has more jobs to submit, he would place the job streams on SYSRDR and then EOB to continue processing.

If no more job streams are to be executed in the Foreground 1 partition, however, the operator must decide on one of several courses of action. Normally, one of the following situations will occur: 1) More jobs are to be executed in the Foreground 1 partition but the job streams are not ready for processing; 2) Another job is to be processed in the Foreground 1 partition, but it is to use the Single Program Initiation mode, not the batched-job mode; 3) No more programs are to be processed in the Foreground 1 partition.

Each one of the above situations calls for different operator action. If more jobs are to be executed in the Foreground 1 partition but at a later time, the operator should enter the STOP command. This is illustrated in Figure 9-15.

EXAMPLE

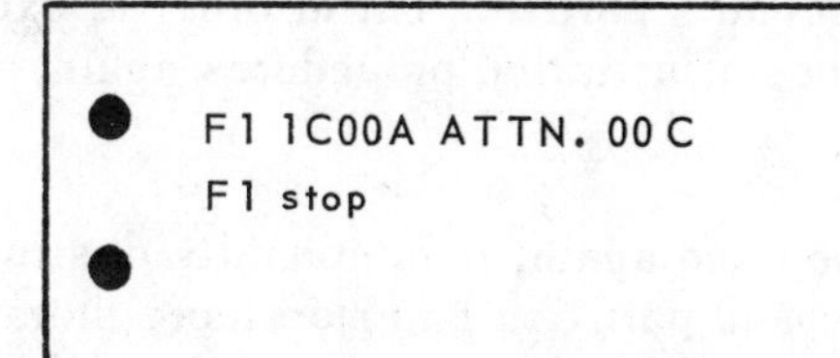

Figure 9-15 Example of STOP Command

When the STOP command is used, all batch processing in the partition is suspended, that is, it is put in a status of not being able to execute a job stream. Note that the Stop command is issued to the Job Control program so Job Control is still in the partition. It is, however, inactive and cannot read any job control cards from SYSRDR.

As noted, the Stop command is normally used when a partition is to be left allocated and job control is to be kept in core storage, but no processing is to occur at the moment. In order to "restart" job control, the Batch command is used. This is illustrated in Figure 9-16.

EXAMPLE

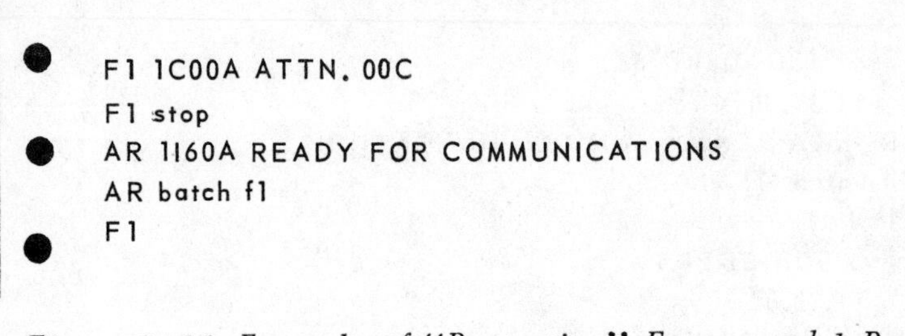

Figure 9-16 Example of "Restarting" Foreground 1 Partition

In Figure 9-16 it can be seen that the Attention Routine is requested by depressing the REQUEST key on the 1052 Typewriter Console. The operator then enters the command Batch f1 and job control again indicates that it is ready to read job control statements. Processing would then begin as illustrated in Figure 9-9.

If the Foreground 1 partition is to be used for a program operating in the Single Program Initiation mode or if the Foreground 1 partition is no longer needed, the first command to be issued by the operator must be the UNBATCH command. This is illustrated in Figure 9-17.

EXAMPLE

Figure 9-17 Example of Unbatch Command

When the UNBATCH command is used, the entire partition is released, that is, the job control program is no longer considered active in the partition and any single program initiation commands or partition allocation commands may be used. The partition is returned to the same status as when the system is initially IPL'ed (see Figure 9-6). The core storage is still allocated to the Foreground 1 partition but in order to execute any program in the partition, the operator must begin initiation procedures again, such as is illustrated in Figure 9-9.

If the Foreground 1 partition is not to be used again, it is normally desirable to allocate zero bytes for it so that the background partition has more core storage in which to store programs. The ALLOC command is used for this purpose and is illustrated in Figure 9-18.

EXAMPLE

```
F1 1C00A ATTN. 00C
F1 unbatch
AR 1160A READY FOR COMMUNICATIONS
AR alloc f1=0k
AR (B)
```

Figure 9-18 Example of ALLOC Command

Note in the example above that the ALLOC statement is used to allocate zero bytes of core storage for the Foreground 1 partition. Therefore, the Background partition will be allocated 52K bytes of core storage because it uses all of the core not used by Foreground partitions. Note that since no core storage is allocated to the Foreground partitions, the only processing which may occur is in the Background partition. In order for Foreground processing to be initiated, core storage must again be allocated for the desired Foreground partition. The detailed rules for allocating foreground partitions when a background partition is active may be found in IBM SRL GC24-5022 DOS OPERATING GUIDE.

SINGLE PROGRAM INITIATION MODE

As noted previously, the foreground partitions may be used for batched-job programs or programs executed in the single program initiation mode (SPI). Program which are executed in the SPI mode are normally programs which will be processing for a long duration of time, such as a teleprocessing program which will be executed during a continuous 10 hour period. It is normally not used for programs which could also be processed in a batched job mode because the initiation of SPI is more difficult for the operator and the feature of processing one program after another is not available.

In order to process programs in the SPI mode, core storage must be allocated for the partition the same as when the batched-job mode is used. It should be noted, however, that the minimum core storage requirement is 2K, not 10K. This is because the job control program, which requires 10K, is not used. The following example of SPI usage will assume that core storage is allocated in the manner illustrated in Figure 9-2, that is, the Background partition has 34K and the Foreground 1 partition has 18K.

As with the batched-job mode, the first function which must be performed is the initiation of the SPI processing routine. This routine is stored in the Logical Transient area of the Supervisor. The START command is used to load it into core storage. This is illustrated in the example in Figure 9-19.

EXAMPLE

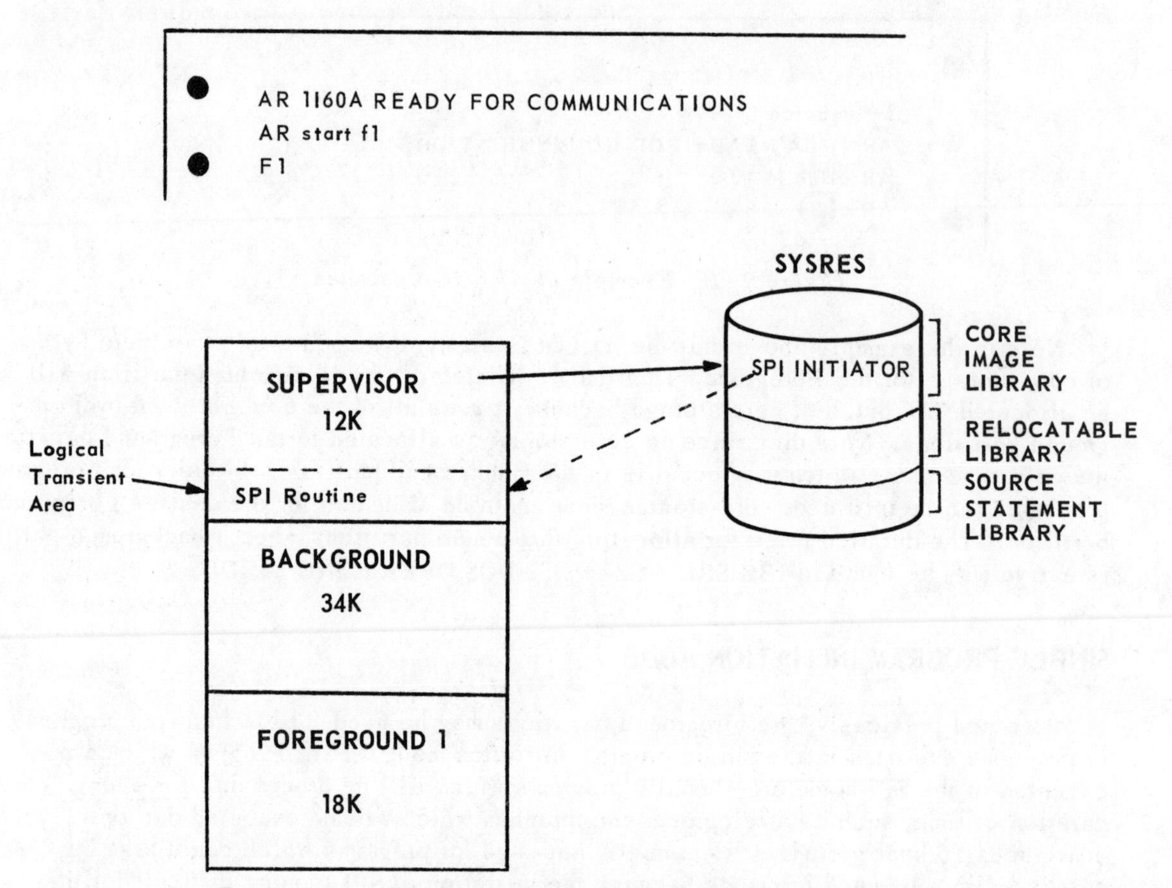

Figure 9-19 Single Program Initiation Routine is Loaded

Note from the above example that the START f1 command is given after the Attention routine has been loaded using the REQUEST key on the console. The Start command instructs the Attention Routine to load the SPI routine into the Logical Transient Area of the Supervisor and to give control to it. The SPI routine then writes the prefix F1 on SYSLOG and waits for the operator to enter commands.

The control statements which are read by the SPI routine may be entered either through the console typewriter or through a card reader. In this example, the control statements will be read on the card reader X'00C' since it is available to the Foreground 1 program. When a card reader is to be used, the READ command must be given to the SPI routine. This is illustrated in Figure 9-20.

EXAMPLE

```
AR 1160A READY FOR COMMUNICATIONS
AR start f1
F1 read x'00c'
```

Figure 9-20 Example of READ Command

When the Read command is issued, the SPI routine reads the statements from the card reader X'00C'. Note that the only operand of the Read command is the hardware device address of the card reader to be used. As with the batched job mode of execution, this card reader may not be assigned to another partition. In this example, X'00C' is assigned to the Foreground 1 partition.

The following example illustrates the control statements which are to be read from X'00C'.

EXAMPLE

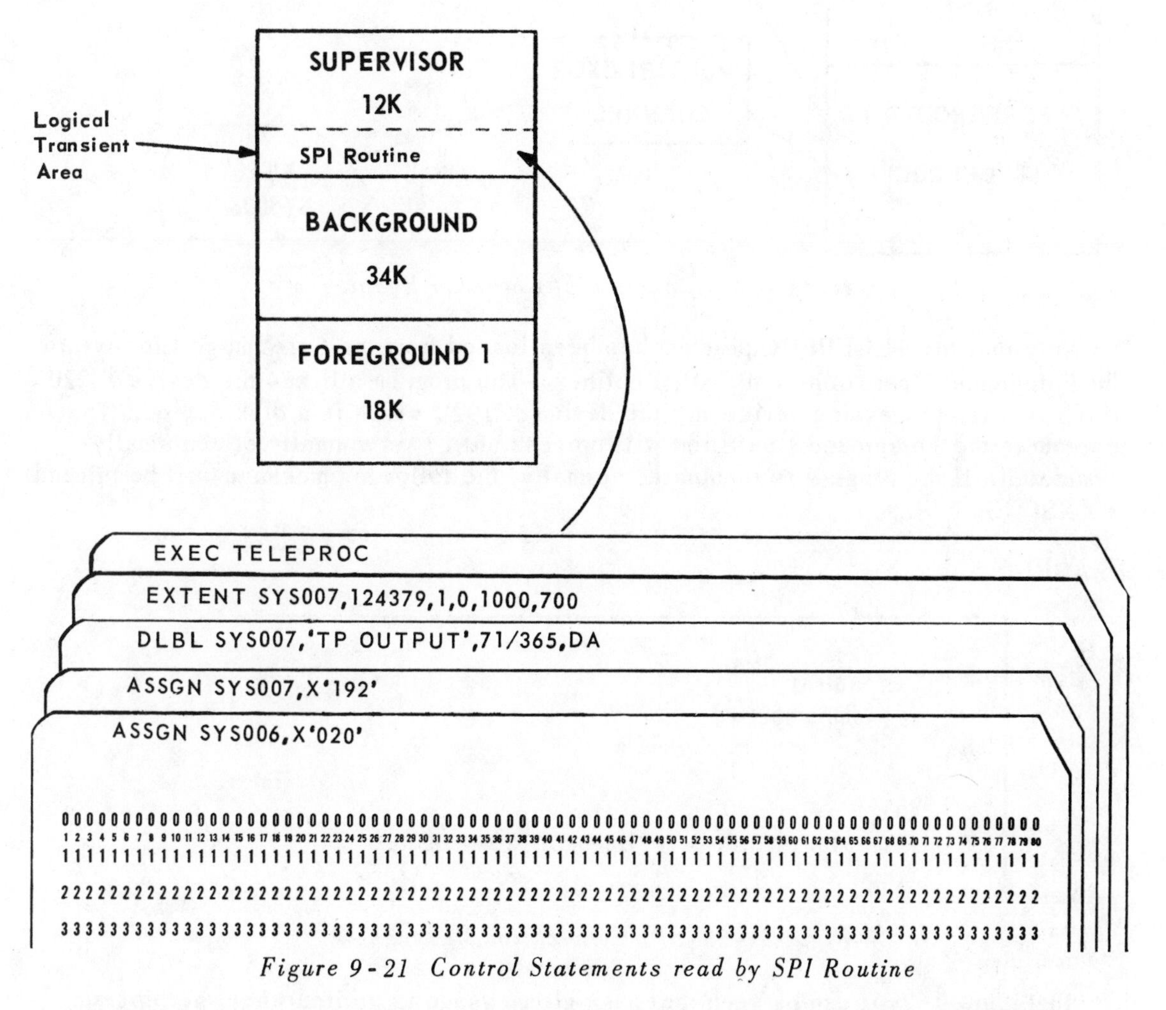

Figure 9-21 Control Statements read by SPI Routine

Note from the example above that the control statements are read from X'00C' into the Logical Transient area to be processed by the SPI Routines. Note also that the normal // characters in column 1 and column 2 of the control cards are not included on the cards read by the SPI Routines. The keywords which identify the statement must begin in column 2 and the remaining format of the control statement is the same as when they are processed by Job Control. In the example, the ASSGN cards are used to assign a teleprocessing line (X'020') and a disk device (X'192'). The DLBL and EXTENT cards are submitted to define the file on the disk to be used by the program. The EXEC card causes the TELEPROC program to be loaded into core storage from the Core Image Library and execution of the program begins.

The core storage allocation after the program begins is illustrated in Figure 9-22.

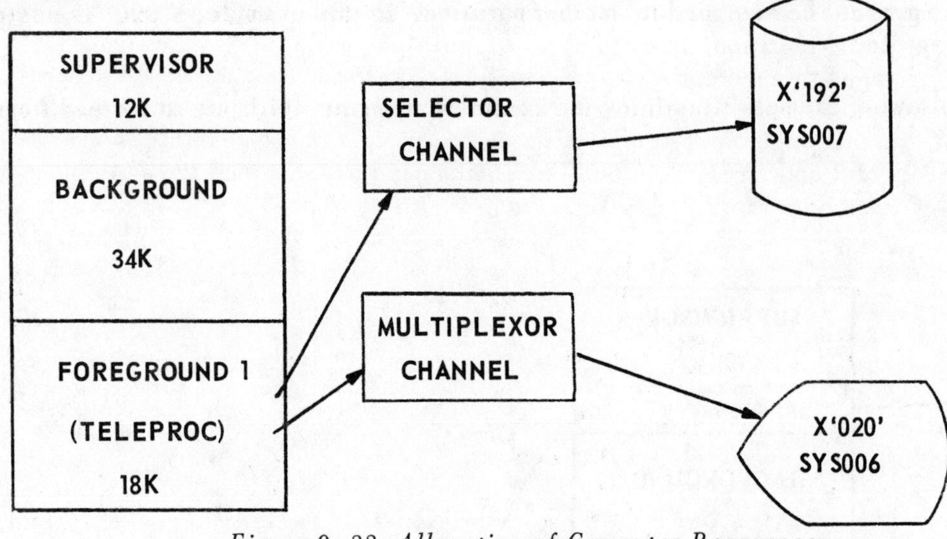

Figure 9-22 Allocation of Computer Resources

Note that the TELEPROC program has been loaded from the Core Image Library into the Foreground 1 partition by the SPI Routines. The program utilizes the device X'020' which is a teleprocessing device and the device X'192', which is a disk device. The program in the Foreground 1 partition will process until it is normally or abnormally terminated. If the program is terminated normally, the following message will be printed on SYSLOG.

EXAMPLE

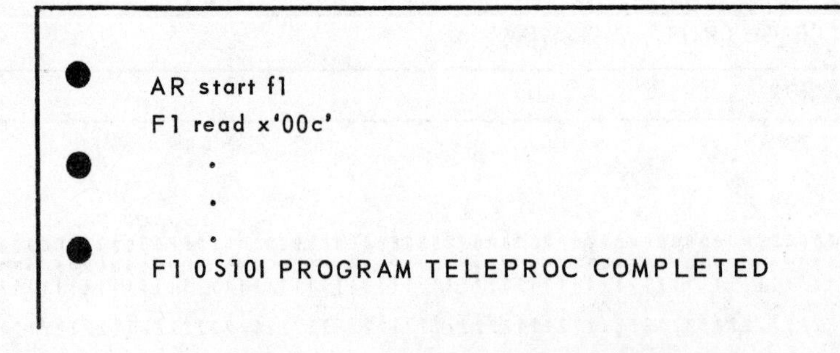

Figure 9-23 SPI Termination Message

In Figure 9-23 it can be seen that a single message is printed when the program operating in the SPI mode is complete. When the program is complete, the partition is still allocated in core storage but there is no linkage to another program, such as when the batched-job mode is used. Thus, in order for another program to be started in the Foreground 1 partition, the operator must enter either the BATCH command to begin batched-job processing or the START command to begin processing another program in the SPI mode. Note again that when a program operating in the SPI mode is complete, the operator must take specific steps in order to begin using the partition again. There is no continuity between programs in the SPI mode.

SYSTEM LOGICAL UNITS ON TAPE

As has been noted, unit-record and magnetic tape devices may be assigned to only one partition, that is, a single card reader may not be assigned to both the Background partition and the Foreground 1 partition. In the previous example of the batched-job mode of processing in a Foreground partition, it was assumed that the computer contained two card readers. In many installations, this is not true, that is, many computers only have one card reader and one printer. When this occurs, it is not possible to use the card reader for SYSRDR and SYSIPT in both the Background partition and the Foreground 1 partition. Since Job Control reads its control statements from SYSRDR, it must be possible to assign SYSRDR to a device other than a card reader. The Disk Operating System allows SYSRDR and SYSIPT to be assigned to magnetic tape and, if desired, to direct-access device. A special entry must be specified when the Supervisor is generated in order for these system logical units to be assigned to disk. In the example which follows, it will be assumed that only a single card reader, X'00C', is contained on the computer and only a single printer, X'00E'. It will be required, therefore, to assign SYSRDR, SYSIPT, and SYSLST to tape devices in order that job control may process programs in the Foreground 1 partition. In addition, it will be assumed that core storage is allocated as illustrated in Figure 9-2.

When SYSRDR and SYSIPT are to be assigned to a magnetic tape drive, the first step must be to load the tape with the job stream which will be read by job control when it is in the Foreground 1 partition. The DOS Utility Card-To-Tape program may be used for this function. The job stream in Figure 9-24 illustrates the statements necessary to load the tape.

EXAMPLE

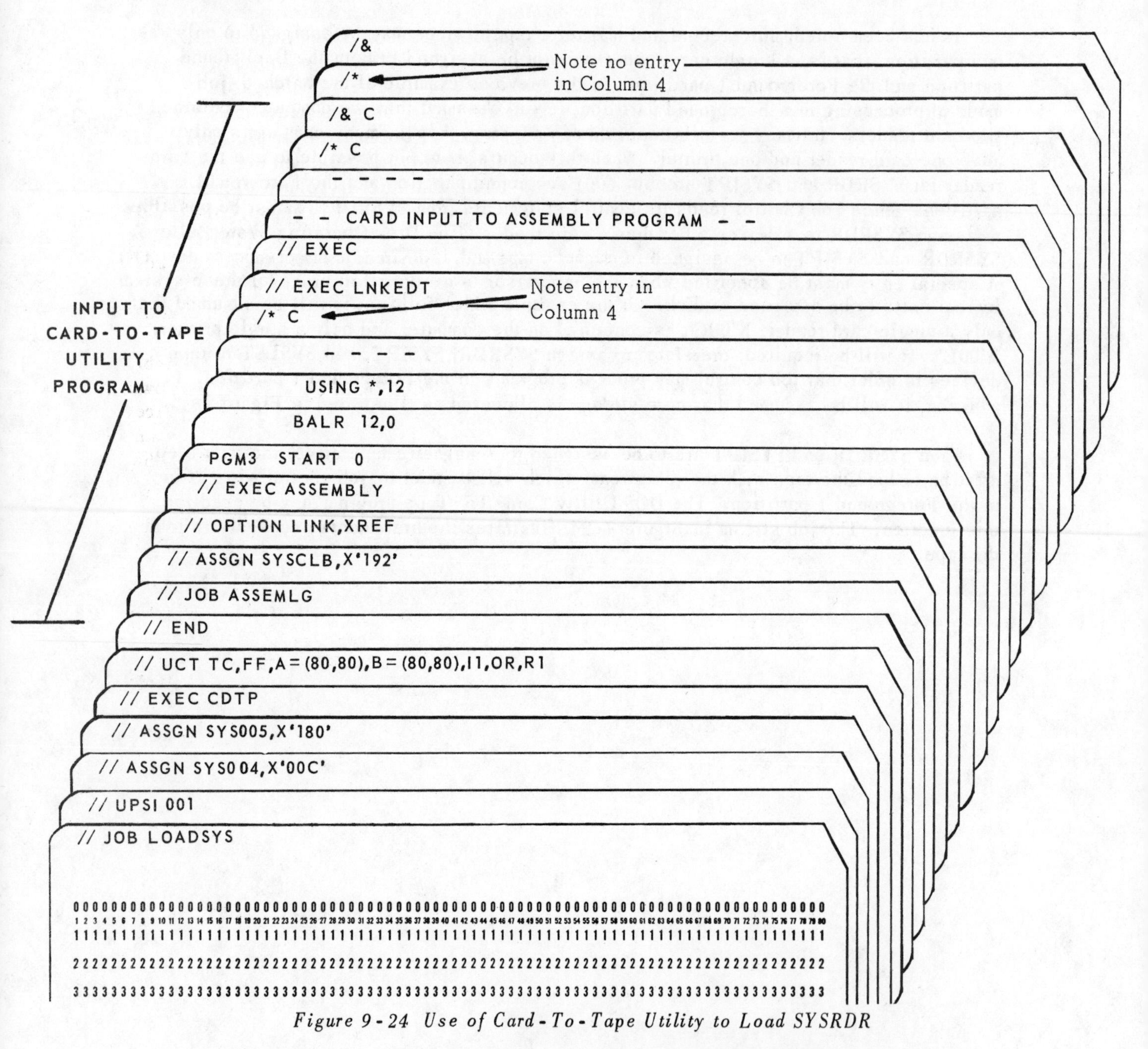

Figure 9-24 Use of Card-To-Tape Utility to Load SYSRDR

In the example above, the Card-To-Tape Utility program is used to load the job stream which will be read by Job Control in the Foreground 1 partition onto a tape. Note that the card data which will be placed on the tape follows the //END Utility Control card. The input to the Utility program is the job stream containing the job control cards, Assembler source deck, and input data to the Assembler program which will be read when the job stream is processed in the Foreground 1 partition.

It should be recalled that the /* card is used to indicate to the Utility program that the end of the card data has been reached. When loading a job stream, however, the /* card may often be included within the job stream which is to be loaded. The Utility program can determine the difference between the /* card which indicates end-of-file and the /* which is to be included in the job stream by the value contained in Column 4 of the card. If a non-blank character is in column 4 of the /* card, the Utility program treats the card as part of the input data and not as the end-of-file indicator. Note in the example in Figure 9-24 that the /* cards following the source deck and following input data to the assembler program contain a "C" in column 4. With this non-blank character in column 4, these /* cards are treated as data by the Utility program, not as the end-of-file card. The /* card to indicate end-of-data to the Card-To-Tape Utility program contains a blank in column 4.

After the job stream has been loaded onto magnetic tape, the batched-job mode of processing must be initiated in the Foreground 1 partition. This is accomplished through the 1052 Console Typewriter (SYSLOG) using the Batch command.

EXAMPLE

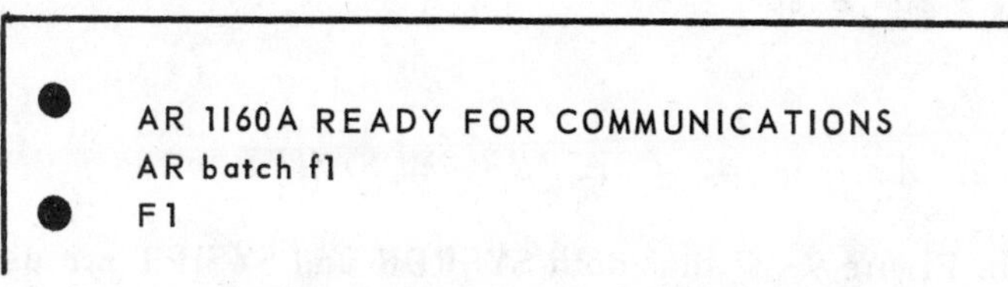

Figure 9-25 Use of Batch Command

When the operator depresses the REQUEST Key on the console typewriter, the Attention Routine is activated and writes the message 1I60A on SYSLOG. The operator then responds with the batch command which loads the job control program into the Foreground 1 partition and job control then readies SYSLOG for the operator to enter instructions. The core storage allocation after the batch command is illustrated in Figure 9-26.

EXAMPLE

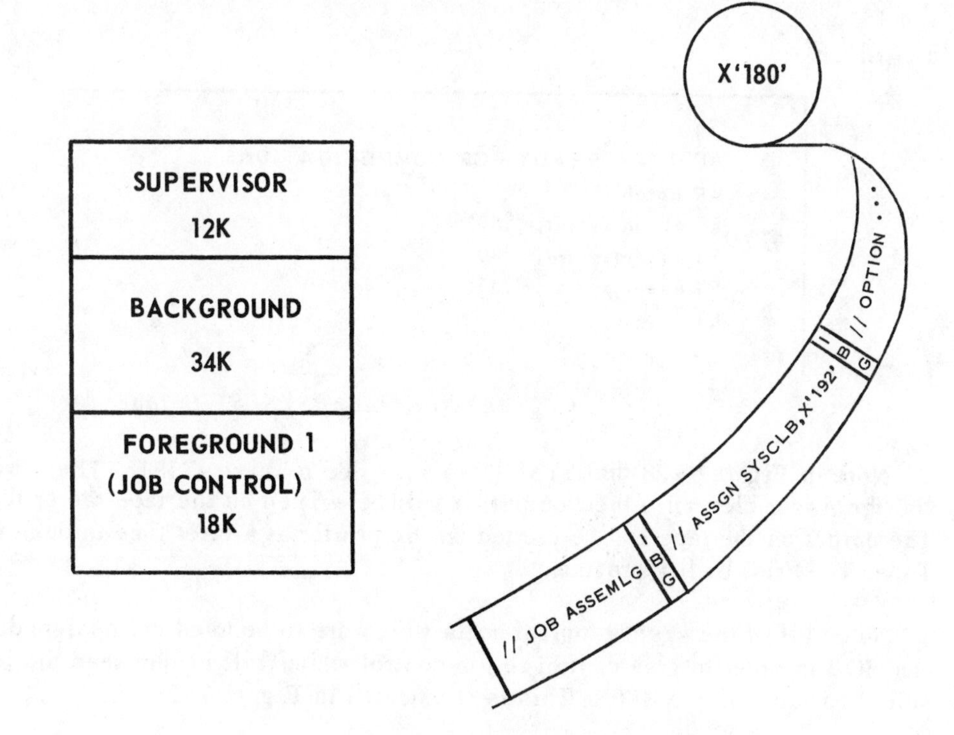

Figure 9-26 Core Storage Allocation and Loaded SYSRDR Tape

Note from the illustration in Figure 9-26 that the core storage allocation has the Supervisor in 12K, the Background partition is 34K bytes in length, and the Foreground 1 partition is 18K bytes in length. It should be recalled that the minimum size of a partition which is to use the batched-job mode is 10K bytes. Note also from the illustration that the tape on drive X'180' contains the job stream which is to be read by job control in the Foreground 1 partition. Therefore, the operator must make the assignment of SYSRDR and SYSIPT to drive X'180' before job control begins reading the control statements. The ASSGN statements which are used by the operator are illustrated below.

EXAMPLE

```
AR 1160A READY FOR COMMUNICATIONS
AR batch f1
F1 assgn sysrdr,x'180'
F1 assgn sysipt,x'180'
F1
```

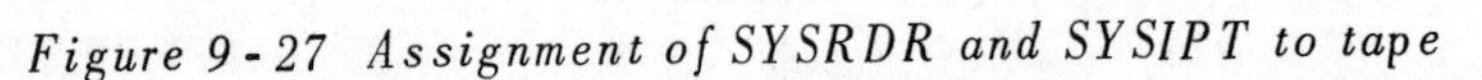

Figure 9-27 Assignment of SYSRDR and SYSIPT to tape

Note in the example in Figure 9-27 that both SYSRDR and SYSIPT are assigned to tape drive X'180'. The symbolic-unit SYSRDR is used by job control to read the job control statements and the symbolic-unit SYSIPT is used by the Assembler to read the source statements. Thus, both symbolic names must be assigned to drive X'180'. It should be recalled that job control writes all of its printer output on the device assigned to SYSLST. Since only one printer is on the computer in this example, it is necessary to also assign SYSLST to a tape drive so that job control can write the statements on tape. The assignment of SYSLST is illustrate below.

EXAMPLE

```
AR 1160A READY FOR COMMUNICATIONS
AR batch f1
F1 assgn sysrdr,x'180'
F1 assgn sysipt,x'180'
F1 assgn syslst,x'181'
F1
```

Figure 9-28 Assignment of SYSLST to tape

Note in Figure 9-28 that SYSLST is assigned to drive X'181'. Thus, when job control and the Assembler write their output, it will be written on the tape rather than the printer. The output on the tape can be printed on the printer at a later time through the use of the Tape-To-Print Utility program.

Since all of the system logical units which are to be used are assigned, the operator can EOB in order to give control to job control which will in turn read the job stream stored on tape drive X'180'. This is illustrated in Figure 9-29.

EXAMPLE

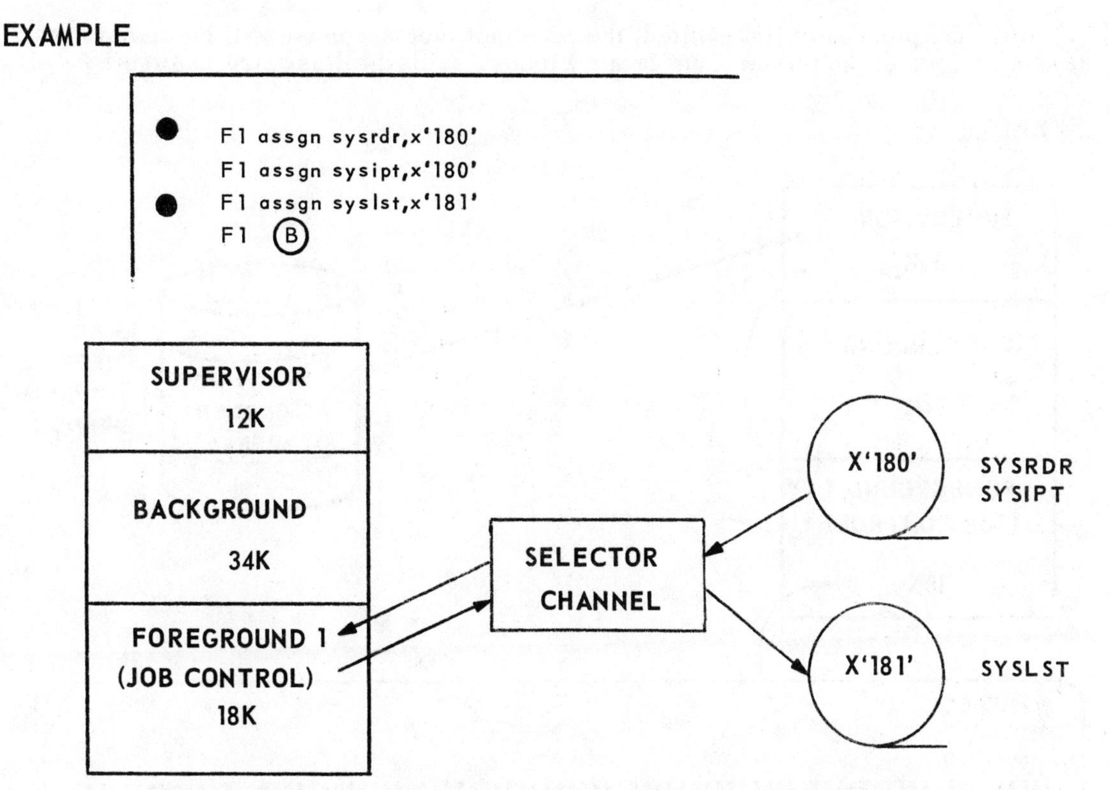

Figure 9-29 Job Control Execution in Foreground 1 Partition

Note from Figure 9-29 that when job control begins processing, it reads its input data from X'180', which is assigned to SYSRDR and it writes its output data on X'181', which is assigned to SYSLST. When the // EXEC ASSEMBLY control card is read, the Assembler will be loaded into core storage and the source program will be assembled. Since the option LINK is requested, the output of the Assembly will be written on the SYSLNK file. This assumes that the SYSLNK file has been defined in the Supervisor and that labels are contained on the standard label cylinder for the SYSLNK file. Care must be taken that if SYSLNK extents for the Background partition are the same as the SYSLNK extents for the Foreground 1 partition, there is not another linkage editor run in the Background partition. If this occurs, the results are unpredictable. Note also in the job stream illustrated in Figure 9-24 that the symbolic-unit name SYSCLB is assigned to drive X'192'. This is the assignment for a private Core Image Library and must be made if linkage editing is to be performed in a Foreground partition. The system Core Image Library may not be used by the Linkage Editor in the Foreground partitions. This assignment again assumes that labels have been stored on the label cylinder for the private Core Image Library on drive X'192'.

After the program is link-edited, the resultant program phase will be stored in the temporary area of the private Core Image Library. This is illustrated in Figure 9-30.

EXAMPLE

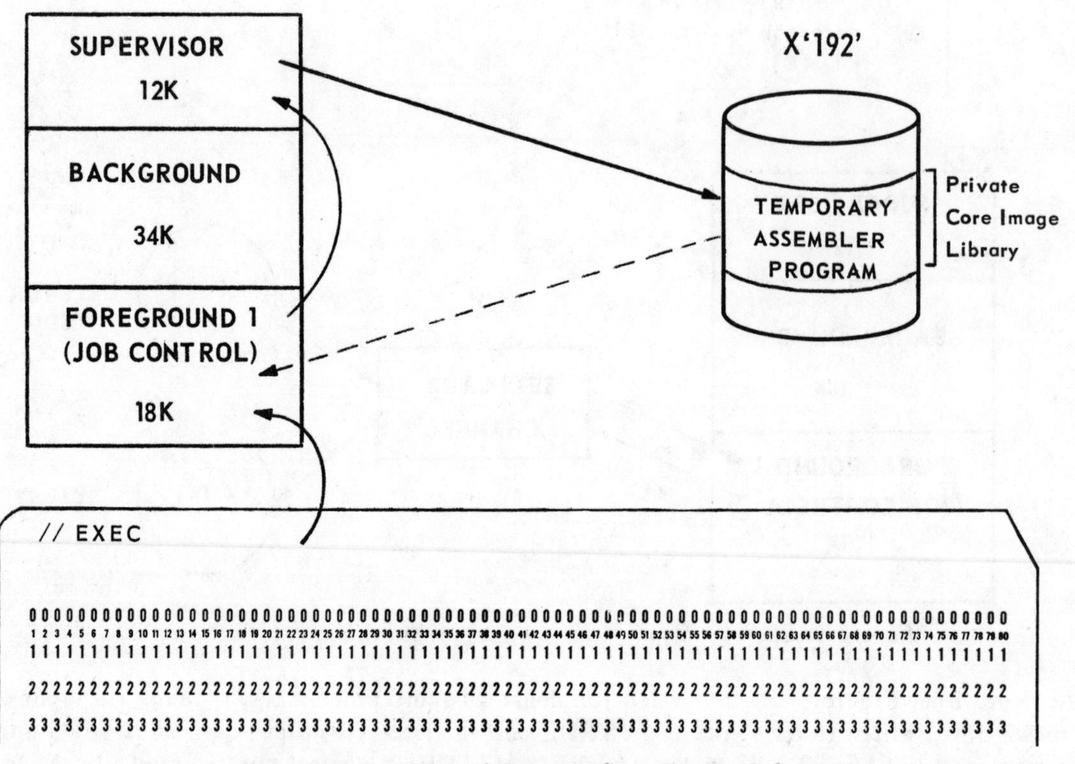

Figure 9-30 Use of Private Core Image Library

Note in the example above that the program which was stored in the temporary area of the Private Core Image Library by the Linkage Editor is loaded when the // EXEC card is read by job control. When the program is loaded into core storage, it will begin reading data which is stored on the tape immediately following the Execute card. It will read from the tape because the Assembler program specified that its input would be stored on SYSIPT (DEVADDR=SYSIPT). The printed output from the program will be written on SYSLST (DEVADDR=SYSLST) immediately following the output from the job control program. Thus, the tape on drive X'181' which is assigned to SYSLST, will contain both the job control and linkage editor statements and the output from the problem program.

When the program is completed, that is, when it reads the /* C card which indicates end-of-file, it will return control to the Supervisor which in turn loads the Job Control program. Job control will read the /& card which indicates end of job and then will read the tapemark which was placed on the tape by the Card-To-Tape Utility program. This is illustrated in Figure 9-31.

EXAMPLE

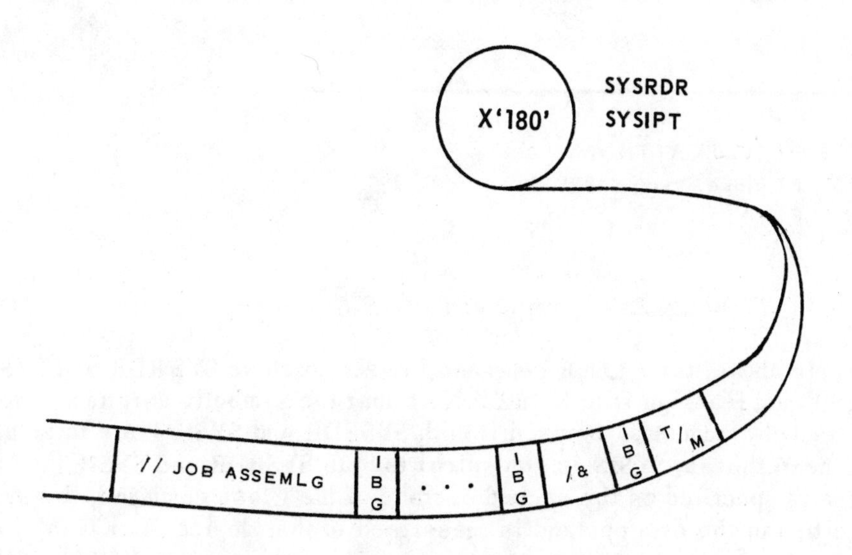

Figure 9-31 SYSRDR, SYSIPT Input Tape

Note from the illustration in Figure 9-31 that following the record containing the /& is a tape mark which was written when the file was closed in the Card-To-Tape program. This tape mark indicates to job control that there are no more job control statements on the tape assigned to SYSRDR. When this occurs, the following message is printed on SYSLOG.

EXAMPLE

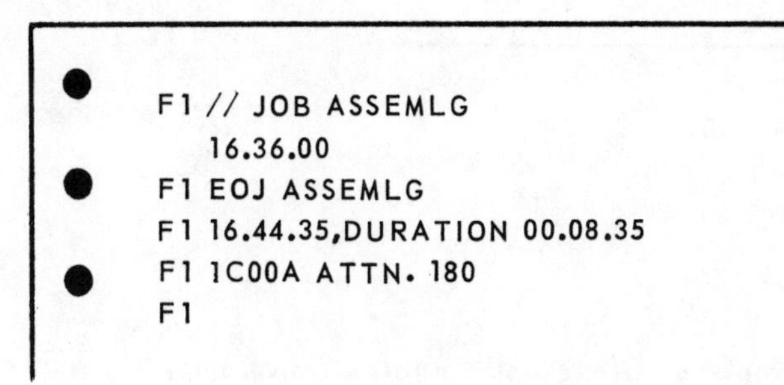

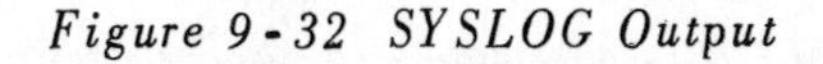

Figure 9-32 SYSLOG Output

When the message 1C00A is received on SYSLOG, the operator has several options. If another tape containing another job stream is to be mounted on X'180' for processing in the Foreground 1 partition, the operator should mount the tape and then EOB the keyboard. This will cause job control to begin reading the job stream on the new tape.

If a tape containing a job stream is to be mounted on another tape drive, the operator should use the CLOSE statement which is illustrated in Figure 9-33.

EXAMPLE

```
F1 1C00A ATTN.180
F1 close sysin,x'182'
F1
```

Figure 9-33 Example of CLOSE Command

In the example above, the CLOSE command is used to close SYSRDR and SYSIPT on tape drive X'180' and reassign it to X'182'. Note that the symbolic unit name used is SYSIN. This symbolic name may be used if both SYSRDR and SYSIPT are to be used in the same statement, that is, SYSIN is equivalent to both SYSRDR and SYSIPT. When a hardware device is specified as the second operand of the Close command, the symbolic-unit name specified in the first operand is reassigned to that device after it is closed on the device currently assigned to the symbolic unit. Thus, the tape on X'180' would be closed and rewound and the symbolic unit name SYSIN (SYSRDR and SYSIPT) would be assigned to X'182'. The operator should have mounted the tape which is to be input to Job Control on X'182' prior to issuing the Close command. It should be noted also that SYSLST will continue to be on drive X'181' because it has not been closed and reassigned.

If the Foreground 1 partition is not to be used anymore, or if an SPI mode is to be used for the partition, the operator should enter the following statements.

EXAMPLE

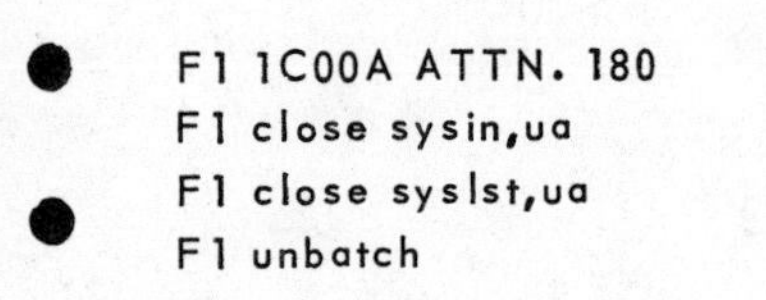

```
F1 1C00A ATTN. 180
F1 close sysin,ua
F1 close syslst,ua
F1 unbatch
```

Figure 9-34 Example of Close and Unbatch Commands

In the example above, the Close command is used to both close SYSIN and SYSLST from their current assignments and to UnAssign them, that is, to cause SYSIN and SYSLST to not be assigned to any device. When the second operand UA is utilized with the Close Command, it causes the symbolic-unit specified in the first operand to be unassigned. The Unbatch command is then issued to free the partition for other processing such as SPI mode processing, or for reallocation to a different size.

It should be noted that whenever tape and/or disk files have been opened and used in a Foreground partition, they must be closed before the unbatch command is issued. Thus, if the Unbatch command were issued prior to the close commands, the system would have indicated an error. When SYSIN is closed, the tape is rewound on the drive. When SYSLST is closed, a tape mark is written on the tape to indicate end-of-file and then the tape is rewound. This tape may then be used as input to the Tape-To-Print Utility program to print the contents on the printer.

The third alternative which the operator has when the end of the job stream is found on the tape assigned to SYSRDR is to leave the job control program in core storage but not to initiate any new processing immediately. When this is to occur, the STOP command should be entered as illustrated below.

EXAMPLE

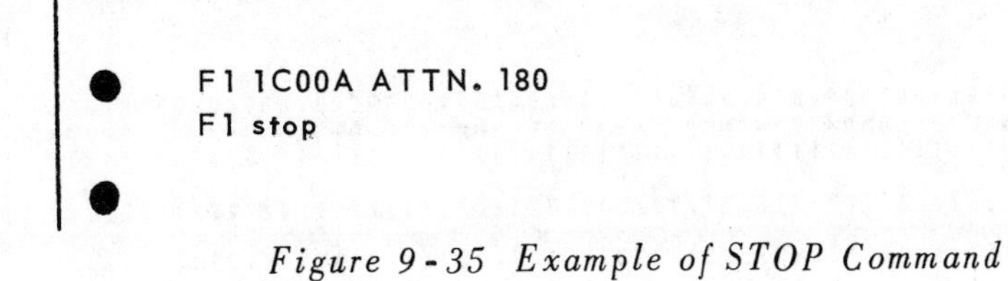

Figure 9-35 Example of STOP Command

When the Stop command is used, the Job Control program remains in core storage and the assignments remain the same. The only effect of the stop command is to free the console typewriter so that other partitions may communicate with the operator. Note that following the 1C00A message, the typewriter waits for an operator response. Until the operator responds, the keyboard is locked and no other partition may use it to type a message. Thus, if the operator did not respond, the other partitions may not be able to continue processing because they are waiting for SYSLOG. Therefore, the Stop command is used to free the keyboard when no immediate processing is to take place in the Foreground partition. In order to begin processing again in Foreground 1, the operator would get the Attention Routine by depressing REQUEST key and enter the Batch command. It should be noted that when the Stop command is issued, job control remains in the partition. Thus, a SPI mode program may not be processed following a stop command. The Unbatch command must be used to free the partition for SPI processing.

PRINTING SYSLST

When the tape assigned to SYSLST is closed, a tape mark is written on the tape and the tape can then be used as input to the Tape-To-Print Utility program to print the records which were written on SYSLST by job control, the Assembler, the Linkage Editor, and the problem program. The job stream which could be used to print the SYSLST tape created in the previous example is illustrated in Figure 9-36.

EXAMPLE

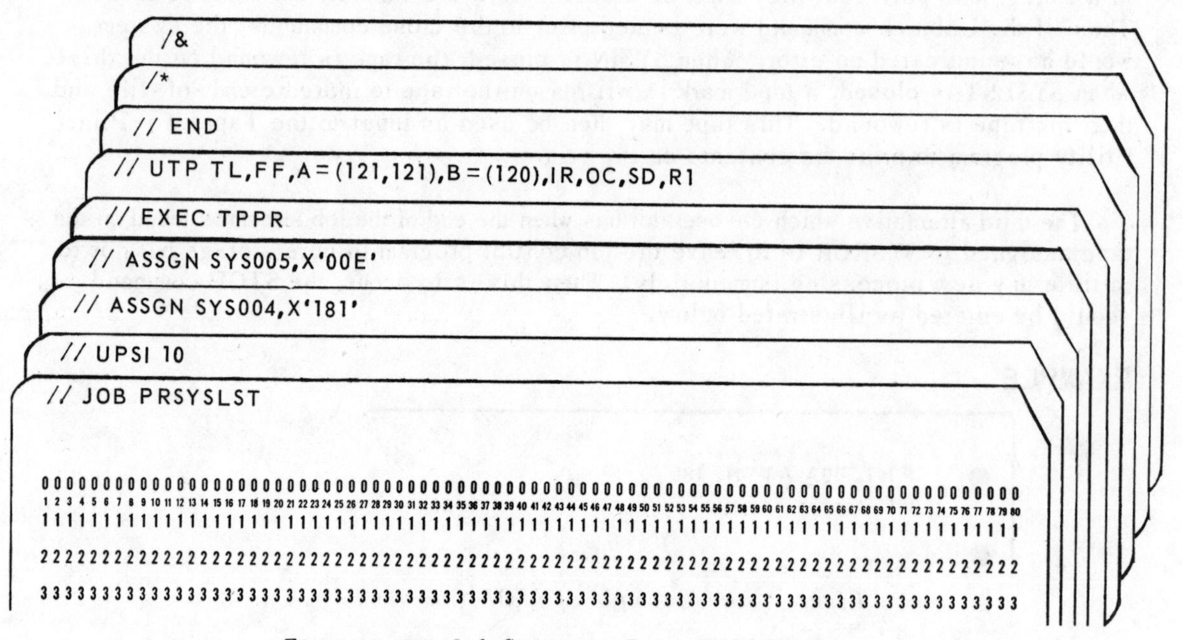

Figure 9-36 Job Stream to Print SYSLST Tape

Note in the job stream illustrated in Figure 9-36 that the Tape-To-Print Utility program is used. The input record length is 121 bytes because the SYSLST tape always contains records with one byte used for carriage control and 120 bytes used for data. They are always blocked 1 record per block. The carriage control operand (SD) specifies that the "D-Type" carriage control characters should be used. These control characters correspond to the ASA carriage control characters which are written on the SYSLST tape. When the program is executed, the spacing on the printer will be the same as if SYSLST were assigned to a printer during the execution of the job streams which created the output.

It should be noted also that the utility program illustrated above may be executed in either the Background partition or one of the Foreground partitions. It is normally executed in the Background partition because the single printer on the system is normally assigned to the Background partition.

SUMMARY

The use of two or three partitions on System/360 computers operating under the Disk Operating System significantly increases the amount of processing which may be accomplished on the machine in a given time period. The use of multiprogramming depends upon the applications to be processed and the operational capabilities of the computer installation. It can be said, however, that in general the use of multiprogramming will significantly increase the efficiency and processing capabilities of a System/360 installation.

APPENDICES

A. Utility Modifier Statement

B. Sort/Merge Control Statements

C. File-To-File Utility Messages

D. Sort/Merge Messages

E. Test Data

APPENDIX A

UTILITY MODIFIER STATEMENT

The following tables contain the formats and entries for the Utility Modifier Statements utilized with the DOS Utility Programs. A detailed explanation of the entries may be found in Chapters 2, 3, and 4.

CARD-TO-DISK

PARAMETER	POSSIBLE FORMS	ENTRIES	EXPLANATION
Function Tt	TC TF TR TRF	T	The initial T identifies this as the type of function parameter.
		C	Copy
		F	Field Select
		R	Reblock
		RF	Reblock and Field Select
Format Ff	FF	F	The initial F of this form identifies this as the format parameter.
		F	The second F of the form must be indicated for fixed-length records.
Input Description	A=(n,m)	A=	The letter and symbol indicate this is the input-description parameter.
		(n,m)	For fixed-length input records, the input record length (the letter n) and the input block length (the letter m) must be enclosed in parentheses and separated by a comma.
Output Description	B=(n,m)	B=	This letter and symbol indicate this is the output-description parameter.
		(n,m)	For fixed-length output records, the output record length (the letter n) and the output block length (the letter m) must be enclosed in parentheses and separated by a comma.
	B=(K=l,D=l)	B=	This letter and symbol indicate this is the output-description parameter.
		(K=l,D=l)	For fixed-length disk output records with keys, the letter K and symbol = must precede the length of the key field. The letter D and symbol = must precede the length of tne data field. These two fields must be separated by a comma and enclosed in parentheses.
Card Input Ix	I1 I2	I	The first letter in these forms identifies this parameter.
		1	EBCDIC Input
		2	Binary Input
Disk Check Ox	OY ON	O	The first letter in these forms identifies this parameter.
		Y	Write-disk check.
		N	Do not write-disk check.
Sequence-Numbering Q=(x,y)	Q=(x,y)	Q=	This letter and symbol identify this parameter.
		x	This represents the first position of a field in a card (relative to one) for sequence-numbering (1 or 2 digits).
		,	Separator
		y	This represents the length of the field (maximum 10). The (x,y) portion of this parameter must be enclosed in parentheses.
First Record Rx	Rx	R	The first letter in this form identifies this parameter.
		x	This represents the position of the first logical input record to be output (x-1 records will be by-passed). If the file is to be copied, the function parameter must be indicated to be reblocked and the input and output file description parameters must contain identical values.
Device Type Description	E=(e)	E=	This letter and symbol indicate this is the device type description parameter.
		(e)	For output devices the valid entries are 2311and 2314. This entry must be enclosed in parentheses.

CARD-TO-PRINT/CARD

PARAMETER	POSSIBLE FORMS	ENTRIES	EXPLANATION
Function Tt	TB TBF TC TD TF TL TLF	T	The initial T identifies this as the type of function parameter.
		B	Both Print and Punch
		BF	Both Print and Punch with Field Select
		C	Copy (punch output only)
		D	Display
		F	Field Select (punch output only)
		L	List
		LF	List and Field Select
Format Ff	FF	F	The initial F of this form identifies this as the format parameter.
		F	The second F of the form must be indicated for fixed-length records.
Input Description	A=(n, m)	A=	This letter and symbol indicate this is the input-description parameter.
		(n, m)	For fixed-length input records, the input record length (the letter n) and the input block length (the letter m) must be enclosed in parentheses and separated by a comma.
Output Description	B=(n, m)	B=	This letter and symbol indicate this is the output-description parameter.
		(n, m)	For fixed-length output records, the output record length (the letter n) and the output block length (the letter m) must be enclosed in parentheses and separated by a comma.
	B=(p)	B=	This letter and symbol indicate this as the output-description parameter.
		(p)	For printer output, the size of the print line (120, 132, or 144) must be entered. A print line size of 120 is forced if TD is specified.
Card Input Ix	I1 I2	I	The first letter in these forms identifies this parameter.
		1	EBCDIC Input
		2	Binary Input
Printer or Punch Output Ox	O1 O2 OX OC	O	The first letter in these forms identifies this parameter. For printer output, the type of output indicated by the field-select parameter (hexadecimal or character) overrides this parameter.
		1	EBCDIC Output (punch only)
		2	Binary Output (punch only)
		X	Hexadecimal Output (printer only)
		C	Character Output (printer only)
Page Numbering Px	PY PN	P	The first letter in these forms identifies this parameter.
		Y	Number pages.
		N	Do not number pages.
Sequence-Numbering Q=(x,y)	Q=(x,y)	Q=	This letter and symbol identify this parameter.
		x	This represents the first position of a field in a card (relative to one) for sequence-numbering (1 or 2 digits).
		,	Separator
		y	This represents the length of the field (maximum 10). The (x,y) portion of this parameter must be enclosed in parentheses.
First Record Rx	Rx	R	The first letter in this form identifies this parameter.
		x	This represents the position of the first logical input record to be output (x-1 records will be by-passed). If the file is to be copied, the function parameter must be indicated to be reblocked and the input and output file description parameters must contain identical values.
Spacing and Stacker Control Sx	S1 S2 S3	S	The first letter in these forms identifies this parameter.
		1	Printer Output: Single Spacing Punch Output: Select Stacker 1 Printer and Punch: Printer Control Only
		2	Printer Output: Double Spacing Punch Output: Select Stacker 2 Printer and Punch: Printer Control Only
		3	Printer Output: Triple Spacing Punch Output: Invalid Printer and Punch: Printer Control Only

CARD -TO-TAPE

PARAMETER	POSSIBLE FORMS	ENTRIES	EXPLANATION
Function Tt	TC TF TR TRF	T	The initial T identifies this as the type of function parameter.
		C	Copy
		F	Field Select
		R	Reblock
		RF	Reblock and Field Select
Format Ff	FF	F	The leading F of this form identifies this as the format parameter.
		F	The second F of the form must be indicated for fixed-length records.
Input Description	A=(n, m)	A=	This letter and symbol indicate this is the input-description parameter.
		(n, m)	For fixed-length input records, the input record length (the letter n) and the input block length (the letter m) must be enclosed in parentheses and separated by a comma.
Output Description	B=(n, m)	B=	This letter and symbol indicate this is the output-description parameter.
		(n, m)	For fixed-length output records, the output record length (the letter n) and the output block length (the letter m) must be enclosed in parentheses and separated by a comma.
Card Input Ix	I1 I2	I	The first letter in these forms identifies this parameter.
		1	EBCDIC Input
		2	Binary Input
Rewind Output Ox	OR ON OU	O	The first letter in these forms identifies this parameter. The rewind option for the output tape is active both before and after data transfer.
		R	Rewind both before and after data transfer.
		N	Do not rewind either before or after data transfer.
		U	Rewind before and rewind and unload after data transfer.
Sequence- Numbering Q=(x,y)	Q=(x,y)	Q=	This letter and symbol identify this parameter.
		x	This represents the first position of a field in a card (relative to one) for sequence-numbering (1 or 2 digits).
		,	Separator.
		y	This represents the length of the field (maximum 10). The (x,y) portion of this parameter must be included in parentheses.
First Record Rx	Rx	R	The first letter in this form identifies this parameter.
		x	This represents the position of the first logical input record to be output (x-1 records will be by-passed). If the file is to be copied, the function parameter must be indicated to be reblocked and the input and output file description parameters must contain identical values.

DATA CELL-TO-DATA CELL

PARAMETER	POSSIBLE FORMS	ENTRIES	EXPLANATION
Function Tt	TC TF TR TRF	T	The initial T identifies this as the type of function parameter.
		C	Copy
		F	Field Select
		R	Reblock
		RF	Reblock and Field Select
Format Ff	FF FV FU	F	The leading F of these three possible forms identifies this as the format parameter.
		F	The second F must be indicated for fixed-length records.
		V	The letter V must be indicated for variable-length records.
		U	The letter U must be indicated for undefined records.
Input Description	A=(n,m)	A=	This letter and symbol indicate this is the input-description parameter.
		(n,m)	For fixed-length input records, the input record length (the letter n) and the input block length (the letter m) must be enclosed in parentheses and separated by a comma. For field select with variable-length records the letter n indicates the size of the fixed portion of each record, and the letter m indicates the maximum block size.
	A=(g)	A=	This letter and symbol indicate this is the input-description parameter.
		(g)	For undefined input records or variable input records without field select, the maximum block length must be enclosed in parentheses.
	A=(K=l,D=l)	A=	This letter and symbol indicate this is the input-description parameter.
		(K=l,D=l)	For fixed-length data cell input records with keys, the letter K and symbol = must precede the length of the key field. The letter D and symbol = must precede the length of the data field. These two fields must be separated by a comma and enclosed in parentheses.
Output Description	B=(n,m)	B=	This letter and symbol indicate this is the output-description parameter.
		(n,m)	For fixed-length output records, the output record length (the letter n) and the output block length (the letter m) must be enclosed in parentheses and separated by a comma. For field select with variable-length records the letter n indicates the size of the fixed portion of each record, and the letter m indicates the maximum block size.
	B=(K=l,D=l)	B=	This letter and symbol indicate this is the output description parameter.
		(K=l,D=l)	For fixed-length data cell output records with keys, the letter K and symbol = must precede the length of the key field. The letter D and symbol = must precede the length of the data field. These two fields must be separated by a comma and enclosed in parentheses.
	B=(g)	B=	This letter and symbol indicate this is the output-description parameter.
		(g)	For undefined output records or variable output records without field select, the maximum block length must be enclosed in parentheses.
Data Cell Check Ox	OY ON	O	The first letter in these forms identifies this parameter.
		Y	Write-data cell check (forced for this program).
		N	Do not write-data cell check (ignored for this program).
First Record Rx	Rx	R	The first letter in this form identifies this parameter.
		x	This represents the position of the first logical input record to be output (x-l records will be by-passed). If the file is to be copied, the function parameter must be indicated to be reblocked and the input and output file description parameters must contain identical values.
Device Type Description	E=(c, d)	E=	This letter and symbol indicate this is the device type description parameter.
		(c, d)	For input devices (the letter c) and output devices (the letter d) the valid entry is 2321. These entries must be enclosed in parentheses and separated by a comma.
	E=(e)	E=	This letter and symbol indicate this is the device type description parameter.
		(e)	For input and output devices the valid entry is 2321. This entry must be enclosed in parentheses.

DATA CELL-TO-DISK

PARAMETER	POSSIBLE FORMS	ENTRIES	EXPLANATION
Function Tt	TC TF TR TRF	T	The initial T identifies this as the type of function parameter.
		C	Copy
		F	Field Select
		R	Reblock
		RF	Reblock and Field Select
Format Ff	FF FV FU	F	The leading F of these three possible forms identifies this as the format parameter.
		F	The second F of the first possible form must be indicated for fixed-length records.
		V	The letter V must be indicated for variable-length records.
		U	The letter U must be indicated for undefined records.
Input Description	A=(n,m)	A=	This letter and symbol indicate this is the input-description parameter.
		(n,m)	For fixed-length input records, the input record length (the letter n) and the input block length (the letter m) must be enclosed in parentheses and separated by a comma. For field select with variable length records the letter n indicates the size of the fixed portion of each record, and the letter m indicates the maximum block size.
	A=(g)	A=	This letter and symbol indicate this is the input-description parameter.
		(g)	For undefined input records or variable input records without field select, the maximum block length must be enclosed in parentheses.
	A=(K=l,D=l)	A=	This letter and symbol indicate this is the input-description parameter.
		(K=l,D=l)	For fixed-length data cell input records with keys, the letter K and symbol = must precede the length of the key field. The letter D and symbol = must precede the length of the data field. These two fields must be separated by a comma and enclosed in parentheses.
Output Description	B=(n,m)	B=	This letter and symbol indicate this is the output-description parameter.
		(n,m)	For fixed-length output records, the output record length (the letter n) and the output block length (the letter m) must be enclosed in parentheses and separated by a comma. For field select with variable length records the letter n indicates the size of the fixed portion of each record, and the letter m indicates the maximum block size.
	B=(K=l,D=l)	B=	This letter and symbol indicate this is the output description parameter.
		(K=l,D=l)	For fixed-length disk output records with keys, the letter K and symbol = must precede the length of the key field. The letter D and symbol = must precede the data field. These two fields must be separated by a comma and enclosed in parentheses.
	B=(g)	B=	This letter and symbol indicate this is the output-description parameter.
		(g)	For undefined output records or variable output records without field select, the maximum block length must be enclosed in parentheses.
Disk Check Ox	OY ON	O	The first letter in these forms identifies this parameter.
		Y	Write-disk check.
		N	Do not write-disk check.
First Record Rx	Rx	R	The first letter in this form identifies this parameter.
		x	This represents the position of the first logical input record to be output (x-1 records will be bypassed). If the file is to be copied, the function parameter must be indicated to be reblocked and the input and output file description parameters must contain identical values.
Device Type Description	E=(c,d)	E=	This letter and symbol indicate this is the device type description parameter.
		(c,d)	For input devices (the letter c) the valid entry is 2321. For output devices (the letter d) the valid entries are 2311 and 2314. These entries must be enclosed in parentheses and separated by a comma.

DATA CELL-TO-PRINTER

PARAMETER	POSSIBLE FORMS	ENTRIES	EXPLANATION
Function Tt	TD TL TLF	T	The initial T identifies this as the type of function parameter.
		D	Display
		L	List
		LF	List and Field Select
Format Ff	FF FV FU	F	The leading F of these three possible forms identifies this as the format parameter.
		F	The second F must be indicated for fixed-length records.
		V	The letter V must be indicated for variable-length records.
		U	The letter U must be indicated for undefined records. With undefined records, the only valid entry above is data display.
Input Description	A=(n,m)	A=	This letter and symbol indicate this is the input-description parameter.
		(n,m)	For fixed-length input records, the input record length (the letter n) and the input block length (the letter m) must be enclosed in parentheses and separated by a comma. For field select with variable length records the letter n indicates the size of the fixed portion of each record, and the letter m indicates the maximum block size.
	A=(K=l,D=l)	A=	This letter and symbol indicate this is the input-description parameter.
		(K=l,D=l)	For fixed-length data cell input records with keys, the letter K and symbol = must precede the length of the key field. The letter D and symbol = must precede the length of the data field. These two fields must be separated by a comma and enclosed in parentheses.
	A=(g)	A=	This letter and symbol indicate this is the input-description parameter.
		(g)	For undefined input records or variable input records without field select, the maximum block length must be enclosed in parentheses.
Output Description	B=(p)	B=	This letter and symbol indicate this is the output-description parameter.
		(p)	For printer output, the size of the print line (120, 132, or 144) must be entered. A print line size of 120 is forced, if TD is specified.
	B=(n,p)	B=	This letter and symbol indicate this is the output description parameter.
		(n,p)	For field select of variable length records with printer output records the fixed portion of each output record (the letter n) and the size of the print line (the letter p) must be enclosed in parentheses and separated by a comma.
Printer Output Ox	OX OC	O	The first letter in these forms identifies this parameter. The character printout is forced for data list. The type of output indicated by the field-select parameter (hexadecimal or character) overrides this parameter.
		X	Hexadecimal Printout
		C	Character Printout
Page-Numbering Px	PY PN	P	The first letter in these forms identifies this parameter.
		Y	Number pages (forced for data display).
		N	Do not number pages (forced for first character forms control).
First Record Printed Rx	Rx	R	The first letter in these forms identifies this parameter.
		x	This represents the position of the first logical record to be printed; x-1 records will be bypassed.

DATA CELL-TO-TAPE

PARAMETER	POSSIBLE FORMS	ENTRIES	EXPLANATION
Function Tt	TC TF TR TRF	T	The initial T identifies this as the type of function parameter.
		C	Copy
		F	Field Select
		R	Reblock
		RF	Reblock and Field Select
Format Ff	FF FV FU	F	The leading F of these three possible forms identifies this as the format parameter.
		F	The second F of the first possible form must be indicated for fixed-length records.
		V	The letter V must be indicated for variable-length records.
		U	The letter U must be indicated for undefined records.
Input Description	A=(n,m)	A=	This letter and symbol indicate this is the input description parameter.
		(n,m)	For fixed-length input records, the input record length (the letter n) and the input block length (the letter m) must be enclosed in parentheses and separated by a comma. For field select with variable length records the letter n indicates the size of the fixed portion of each record, and the letter m indicates the maximum block size.
	A=(K=l,D=l)	A=	This letter and symbol indicate this is the input-description parameter.
		(K=l,D=l)	For fixed-length data cell input records with keys, the letter K and symbol = must precede the length of the key field. The letter D and symbol = must precede the length of the data field. These two fields must be separated by a comma and enclosed in parentheses.
	A=(g)	A=	This letter and symbol indicate this is the input-description parameter.
		(g)	For undefined input records or variable input records without field select, the maximum block length must be enclosed in parentheses.
Output Description	B=(n,m)	B=	This letter and symbol indicate this is the output-description parameter.
		(n,m)	For fixed-length output records, the output record length (the letter n) and the output block length (the letter m) must be enclosed in parentheses and separated by a comma. For field select with variable length records the letter n indicates the size of the fixed portion of each record, and the letter m indicates the maximum block size.
	B=(g)	B=	This letter and symbol indicate this is the output-description parameter.
		(g)	For undefined output records or variable output records without field select, the maximum block length must be enclosed in parentheses.
Rewind Output	OR ON OU	O	The first letter in these forms identifies this parameter. The rewind option for the output tape is active both before and after data transfer.
		R	Rewind both before and after data transfer.
		N	Do not rewind either before or after data transfer.
		U	Rewind before and rewind and unload after data transfer.
First Record Rx	Rx	R	The first letter in this form identifies this parameter.
		x	This represents the position of the first logical input record to be output (x-1 records will be bypassed). If the file is to be copied, the function parameter must be indicated to be reblocked and the input and output file description parameters must contain identical values.
Device Type Description	E=(e)	E=	This letter and symbol indicate this is the device type description parameter.
		(e)	For input devices the valid entry is 2321. This entry must be enclosed in parentheses.

DISK-TO-CARD

PARAMETER	POSSIBLE FORMS	E TRIES	EXPLANATION
Function Tt	TC TF TR TRF	T	The initial T identifies this as the type of function parameter.
		C	Copy
		F	Field Select
		R	Reblock
		RF	Reblock and Field Select
Format Ff	FF	F	The leading F of this form identifies this as the format parameter.
		F	The second F of the form must be indicated for fixed-length records.
Input Description	A=(n,m)	A=	This letter and symbol indicate this is the input-description parameter.
		(n,m)	For fixed-length input records, the input record length (the letter n) and the input block length (the letter m) must be enclosed in parentheses and separated by a comma.
	A=(K=l,D=l)	A=	This letter and symbol indicate this is the input-description parameter.
		(K=l,D=l)	For fixed-length disk input records with keys, the letter K and symbol = must precede the length of the key field. The letter D and symbol must precede the length of the data field. These two fields must be separated by a comma and enclosed in a parentheses.
Output Description	B=(n,m)	B=	This letter and symbol indicate this is the output-description parameter.
		(n,m)	For fixed-length output records, the output record length (the letter n) and the output block length (the letter m) must be enclosed in parentheses and separated by a comma.
Output Mode Ox	O1 O2	O	The first letter in these forms identifies this parameter.
		1	EBCDIC Punching
		2	Binary Punching
Sequence-Numbering Q=(x,y)	Q=(x,y)	Q=	This letter and symbol identify this parameter.
		x	This represents the first position of a field in a card (relative to one) for sequence-numbering (1 or 2 digits).
		,	Separator
		y	This represents the length of the field (maximum 10). The (x,y) portion of this parameter must be enclosed in parentheses.
First Record Rx	Rx	R	The first letter in this form identifies this parameter.
		x	This represents the position of the first logical input record to be output (x-1 records will be by-passed). If the file is to be copied, the function parameter must be indicated to be reblocked and the input and output file description parameters must contain identical values.
Stacker Select Sx	S1	S	The first letter in these forms identifies this parameter.
	S2	1	Select Pocket 1
	S3	2	Select Pocket 2
		3	First Character Stacker Select
Device Type Description	E=(e)	E=	This letter and symbol indicate this is the device type description parameter.
		(e)	For input devices the valid entries are 2311 and 2314. This entry must be enclosed in parentheses.

DISK-TO-DATA CELL

PARAMETER	POSSIBLE FORMS	ENTRIES	EXPLANATION
Function Tt	TC TF TR TRF	T	The initial T identifies this as the type of function parameter.
		C	Copy
		F	Field Select
		R	Reblock
		RF	Reblock and Field Select
Format Ff	FF FV FU	F	The leading F of these three possible forms identifies this as the format parameter.
		F	The second F of the first possible form must be indicated for fixed-length records.
		V	The letter V must be indicated for variable-length records.
		U	The letter U must be indicated for undefined records.
Input Description	A=(n,m)	A=	This letter and symbol indicate this is the input-description parameter.
		(n,m)	For fixed-length input records, the input record length (the letter n) and the input block length (the letter m) must be enclosed in parentheses and separated by a comma. For field select with variable length records the size of the fixed portion of each record, and the letter m indicates the maximum block size.
	A=(g)	A=	This letter and symbol indicate this is the input-description parameter.
		(g)	For undefined input records or variable input records without field select, the maximum block length must be enclosed in parentheses.
	A=(K=I,D=I)	A=	This letter and symbol indicate this is the input-description parameter.
		(K=I,D=I)	For fixed-length disk input records with keys, the letter K and symbol = must precede the length of the key field. The letter D and symbol = must precede the length of the data field. These two fields must be separated by a comma and enclosed in parentheses.
Output Description	B=(n,m)	B=	This letter and symbol indicate this is the output-description parameter.
		(n,m)	For fixed-length output records, the output record length (the letter n) and the output block length (the letter m) must be enclosed in parentheses and separated by a comma. For field select with variable length records the letter n indicates the size of the fixed portion of each record, and the letter m indicates the maximum block size.
	B=(K=I,D=I)	B=	This letter and symbol indicate this is the output-description parameter.
		(K=I,D=I)	For fixed-length data cell output records with keys, the letter K and symbol must precede the length of the key field. The letter D and symbol must precede the length of the data field. These two fields must be separated by a comma and enclosed in parentheses.
	B=(g)	B=	This letter and symbol indicate this is the output-description parameter.
		(g)	For undefined output records or variable output records without field select, the maximum block length must be enclosed in parentheses.
Data Cell Check Ox	OY ON	O	The first letter in these forms identifies this parameter.
		Y	Write-data cell check (forced for this program).
		N	Do not write-data cell check (ignored for this program).
First Record Rx	Rx	R	The first letter in this form identifies this parameter.
		x	This represents the position of the first logical input record to be output (x-1 records will be by-passed). If the file is to be copied, the function parameter must be indicated to be reblocked and the input and output file description parameters must contain identical values.
Device Type Description	E=(c, d)	E=	This letter and symbol indicate this is the device type description parameter.
		(c, d)	For input devices (the letter c) the valid entries are 2311 and 2314. For output devices (the letter d) the valid entry is 2321. These entries must be enclosed in parentheses and separated by a comma.

DISK - TO - DISK

PARAMETER	POSSIBLE FORMS	ENTRIES	EXPLANATION
Function Tt	TC TF TR TRF	T	The initial T identifies this as the type of function parameter.
		C	Copy
		F	Field Select
		R	Reblock
		RF	Reblock and Field Select
Format Ff	FF FV FU	F	The leading F of these three possible forms identifies this as the format parameter.
		F	The second F of the first possible form must be indicated for fixed-length records.
		V	The letter V must be indicated for variable-length records.
		U	The letter U must be indicated for undefined records.
Input Description	A=(n,m)	A=	This letter and symbol indicate this is the input-description parameter.
		(n,m)	For fixed-length input record length (the letter n) and the input block length (the letter m) must be enclosed in parentheses and separated by a comma. For field select with variable length records the letter n indicates the size of the fixed portion of each record, and the letter m indicates the maximum block size.
	A=(g)	A=	This letter and symbol indicate this is the input-description parameter.
		(g)	For undefined input records or variable input records without field select, the maximum block length must be enclosed in parentheses.
	A=(K=l,D=l)	A=	This letter and symbol indicate this is the input-description parameter.
		(K=l,D=l)	For fixed-length disk input records with keys, the letter K and symbol = must precede the length of the key field. The letter D and symbol = must precede the length of the data field. These two fields must be separated by a comma and enclosed in parentheses.
Output Description	B=(n,m)	B=	This letter and symbol indicate this is the output-description parameter.
		(n,m)	For fixed-length output records, the output record length (the letter n) and the output block length (the letter m) must be enclosed in parentheses and separated by a comma. For field select with variable length records the letter n indicates the size of the fixed portion of each record, and the letter m indicates the maximum block size.
	B=(K=l,D=l)	B=	This letter and symbol indicate this is the output-description parameter.
		(K=l,D=l)	For fixed-length disk output records with keys, the letter K and symbol = must precede the length of the key field. The letter D and symbol = must precede the length of the data field. These two fields must be separated by a comma and enclosed in parentheses.
	B=(g)	B=	This letter and symbol indicate this is the output-description parameter.
		(g)	For undefined output records or variable output records without field select, the maximum block length must be enclosed in parentheses.
Disk Check Ox	OY ON	O	The first letter in these forms identifies this parameter.
		Y	Write-disk check.
		N	Do not write-disk check.
First Record Rx	Rx	R	The first letter in this form identifies this parameter.
		x	This represents the position of the first logical input record to be output (x-1 records will be bypassed). If the file is to be copied, the function parameter must be indicated to be reblocked and the input and output file description parameters must contain identical values.
Device Type Description	E=(c, d)	E=	This letter and symbol indicate this is the device type description parameter.
		(c, d)	For input devices (the letter c) and output devices (the letter d) the valid entries are 2311 and 2314. These entries must be enclosed in parentheses and separated by a comma.
	E=(e)	E=	This letter and symbol indicate this is the device type description parameter.
		(e)	For input and output devices the valid entries are 2311 or 2314. This entry must be enclosed in parentheses.

DISK-TO-PRINT

PARAMETER	POSSIBLE FORMS	ENTRIES	EXPLANATION
Function Tt	TD TL TLF	T	The initial T identifies this as the type of function parameter.
		D	Display
		L	List
		LF	List and Field Select
Format Ff	FF FV FU	F	The leading F of these three possible forms identifies this as the format parameter.
		F	The second F must be indicated for fixed-length records.
		V	The letter V must be indicated for variable-length records.
		U	The letter U must be indicated for undefined records. With undefined records, the only valid entry above is data display.
Input Description	A=(n, m)	A=	This letter and symbol indicate this is the input-description parameter.
		(n, m)	For fixed-length input records, the input record length (the letter n) and the input block length (the letter m) must be enclosed in parentheses and separated by a comma. For field select with variable length records the letter n indicates the size of the fixed portion of each record, and the letter m indicates the maximum block size.
	A=(K=l, D=l)	A=	This letter and symbol indicate this is the input-description parameter.
		(K=1,D=1)	For fixed-length disk input records with keys, the letter K and symbol = must precede the length of the key field. The letter D and symbol = must precede the length of the data field. These two fields must be separated by a comma and enclosed in parentheses.
	A=(g)	A=	This letter and symbol indicate this is the input-description parameter.
		(g)	For undefined input records or variable input records without field select length must be enclosed in parentheses.
Output Description	B=(p)	B=	This letter and symbol indicate this is the output-description parameter.
		(p)	For printer output, the size of the print line (120, 132, or 144) must be entered. A print line size of 120 is forced if TD is specified.
	B=(n, p)	B=	This letter and symbol indicate this is the output-description parameter.
		(n, p)	For field select of variable length records with printer output records, the fixed portion of each output record (the letter n) and the size of the print line (the letter p) must be enclosed in parentheses and separated by a comma.
Printer Output Ox	OX OC	O	The first letter in these forms identifies this parameter. The type of output indicated by the field-select parameter (hexadecimal or character) overrides this parameter.
		X	Hexadecimal Printout (for data display only)
		C	Alphameric Printout (forced for data list mode)
Page-Numbering Px	PY PN	P	The first letter in these forms identifies this parameter.
		Y	Number pages (forced for data display)
		N	Do not number pages (forced for first character forms control)
First Record Rx	Rx	R	The first letter in these forms identifies this parameter.
		x	This represents the position of the first logical record to be printed; x - 1 records will be bypassed.
Spacing Sx	S1 S2 S3 SA SB SC SD	S	The first letter in these forms identifies this parameter.
		1	Single Spacing (forced for data display)
		2	Double Spacing
		3	Triple Spacing
		A	Type A First Character Forms Control
		B	Type B First Character Forms Control
		C	Type C First Character Forms Control
		D	Type D First Character Forms Control
Device Type Description	E=(e)	E=	This letter and symbol indicate this is the device type description parameter.
		(e)	For input devices the valid entries are 2311 and 2314. This entry must be enclosed in parentheses.

DISK-TO-TAPE

PARAMETER	POSSIBLE FORMS	ENTRIES	EXPLANATION
Function Tt	TC TF TR TRF	T	The initial T identifies this as the type of function parameter.
		C	Copy
		F	Field Select
		R	Reblock
		RF	Reblock and Field Select
Format Ff	FF FV FU	F	The leading F of these three possible forms identifies this as the format parameter.
		F	The second F of the first possible form must be indicated for fixed-length records.
		V	The letter V must be indicated for variable-length records.
		U	The letter U must be indicated for undefined records.
Input Description	A=(n,m)	A=	This letter and symbol indicate this is the input-description parameter.
		(n,m)	For fixed-length input records, the input record length (the letter n) and the input block length (the letter m) must be enclosed in parentheses and separated by a comma. For field select with variable length records the letter n indicates the size of the fixed portion of each record, and the letter m indicates the maximum block size.
	A=(K=l, D=l)	A=	This letter and symbol indicate this is the input-description parameter.
		(K=l, D=l)	For fixed-length disk input records with keys, the letter K and symbol = must precede the length of the key field. The letter D and symbol = must precede the length of the data field. These two fields must be separated by a comma and enclosed in parentheses.
	A=(g)	A=	This letter and symbol indicate this is the input-description parameter.
		(g)	For undefined input records or variable input records without field select, the maximum block length must be enclosed in parentheses.
Output Description	B=(n,m)	B=	This letter and symbol indicate this is the output-description parameter.
		(n,m)	For fixed-length output records, the output record length (the letter n) and the output block length (the letter m) must be enclosed in parentheses and separated by a comma. For field select with variable length records the letter n indicates the size of the fixed portion of each record, and the letter m indicates the maximum block size.
	B=(g)	B=	This letter and symbol indicate this is the output-description parameter.
		(g)	For undefined output records or variable output records without field select, the maximum block length must be enclosed in parentheses.
Rewind Output Ox	OR ON OU	O	The first letter in these forms identifies this parameter. The rewind option for the output tape is active both before and after data transfer.
		R	Rewind both before and after data transfer.
		N	Do not rewind either before or after data transfer.
		U	Rewind before and rewind and unload after data transfer.
First Record Rx	Rx	R	The first letter in this form identifies this parameter.
		x	This represents the position of the first logical input record to be output (x-1 records will be bypassed). If the file is to be copied, the function parameter must be indicated to be reblocked and the input and output file description parameters must contain identical values.
Device Type Description	E=(e)	E=	This letter and symbol indicate this is the device type description parameter.
		(e)	For input devices the valid entries are 2311 and 2314. This entry must be enclosed in parentheses.

TAPE-TO-CARD

PARAMETER	POSSIBLE FORMS	ENTRIES	EXPLANATION
Function Tt	TC TF TR TRF	T	The initial T identifies this as the type of function parameter.
		C	Copy
		F	Field Select
		R	Reblock
		RF	Reblock and Field Select
Format Ff	FF	F	The leading F of this form identifies this as the format parameter.
		F	The second F of the form must be indicated for fixed-length records.
Input Description	A=(n,m)	A=	This letter and symbol indicate this is the input-description parameter.
		(n,m)	For fixed-length input records, the input record length (the letter n) and the input block length (the letter m) must be enclosed in parentheses and separated by a comma.
Output Description	B=(n,m)	B=	This letter and symbol indicate this is the output-description parameter.
		(n,m)	For fixed-length output records, the output record length (the letter n) and the output block length (the letter m) must be enclosed in parentheses and separated by a comma.
Rewind Input Ix	IR IN IU	I	The first letter in these forms identifies this parameter.
		R	Rewind both before and after data transfer.
		N	Do not rewind either before or after data transfer.
		U	Rewind before and rewind and unload after data transfer.
Sequence-Numbering Q=(x,y)	Q=(x,y)	Q=	This letter and symbol identify this parameter.
		x	This represents the first position of a field in a card (relative to one) for sequence-numbering (1 or 2 digits).
		,	Separator
		y	This represents the length of the field (maximum 10). The (x,y) parts of this parameter must be enclosed in parentheses. Absence of this parameter indicates no sequence numbers.
First Record Rx	Rx	R	The first letter in this form identifies this parameter.
		x	This represents the position of the first logical input record to be output (x-1 records will be bypassed). If the file is to be copied, the function parameter must be indicated to be reblocked and the input and output file description parameters must contain identical values.
Stacker Control Sx	S1 S2 S3	S	The first letter in these forms identifies this parameter.
		1	Select Pocket 1
		2	Select Pocket 2
		3	First Character Stacker Control
Output Mode Ox	O1 O2	O	The first letter in these forms identifies this parameter.
		1	EBCDIC Punching
		2	Binary Punching

TAPE-TO-DATA CELL

PARAMETER	POSSIBLE FORMS	ENTRIES	EXPLANATION
Function Tt	TC TF TR TRF	T	The initial T identifies this as the type of function parameter.
		C	Copy
		F	Field Select
		R	Reblock
		RF	Reblock and Field Select
Format Ff	FF FV FU	F	The leading F of these three possible forms identifies this as the format parameter.
		F	The second F must be indicated for fixed-length records.
		V	The letter V must be indicated for variable-length records.
		U	The letter U must be indicated for undefined records.
Input Description	A=(n,m)	A=	This letter and symbol indicate this is the input-description parameter.
		(n,m)	For fixed-length input records, the input record length (the letter n) and the input block length (the letter m) must be enclosed in parentheses and separated by a comma.
			For field select with variable length records, the letter n indicates the size of the fixed portion of each record, and the letter m indicates the maximum block size.
	A=(g)	A=	This letter and symbol indicate this is the input-description parameter.
		(g)	For undefined input records or variable input records without field select, the maximum block length must be enclosed in parentheses.
Output Description	B=(n,m)	B=	This letter and symbol indicate this is the output-description parameter.
		(n,m)	For fixed-length output records, the output record length (the letter n) and the output block length (the letter m) must be enclosed in parentheses and separated by a comma.
			For field select with variable length records, the letter n indicates the size of the fixed portion of each record, and the letter m indicates the maximum block size.
	B=(K=l,D=l)	B=	This letter and symbol indicate this is the output-description parameter.
		(K=l,D=l)	For fixed-length data cell output records with keys, the letter K and symbol = must precede the length of the key field. The letter D and symbol = must precede the length of the data field. These two fields must be separated by a comma and enclosed in parentheses.
	B=(g)	B=	This letter and symbol indicate this is the output-description parameter.
		(g)	For undefined output records or variable output records without field select, the maximum block length must be enclosed in parentheses.
Rewind Input Ix	IR IN IU	I	The first letter in these forms identifies this parameter. The rewind option for the input tape is active both before and after data transfer.
		R	Rewind both before and after data transfer.
		N	Do not rewind either before or after data transfer.
		U	Rewind before and rewind and unload after data transfer.
Data Cell Check Ox	OY ON	O	The first letter in these forms identifies this parameter.
		Y	Write-data cell check (forced for this program).
		N	Do not write-data cell check (ignored for this program).

TAPE-TO-DISK

PARAMETER	POSSIBLE FORMS	ENTRIES	EXPLANATION
Function Tt	TC TF TR TRF	T	The initial T identifies this as the type of function parameter.
		C	Copy
		F	Field Select
		R	Reblock
		RF	Reblock and Field Select
Format Ff	FF FV FU	F	The leading F of these three possible forms identifies this as the format parameter.
		F	The second F must be indicated for fixed-length records.
		V	The letter V must be indicated for variable-length records.
		U	The letter U must be indicated for undefined records.
Input Description	A=(n,m)	A=	This letter and symbol indicate this is the input-description parameter.
		(n,m)	For fixed-length input records, the input record length (the letter n) and the input block length (the letter m) must be enclosed in parentheses and separated by a comma. For field select with variable length records the letter n indicates the size of the fixed portion of each record, and the letter m indicates the maximum block size.
	A=(g)	A=	This letter and symbol indicate this is the input-description parameter.
		(g)	For undefined input records or variable input records without field select, the maximum block length must be enclosed in parentheses.
Output Description	B=(n,m)	B=	This letter and symbol indicate this is the output-description parameter.
		(n,m)	For fixed-length output records, the output record length (the letter n) and the output block length (the letter m) must be enclosed in parentheses and separated by a comma. For field select with variable length records, the letter n indicates the size of the fixed portion of each record, and the letter m indicates the maximum block size.
	B=(K=l,D=l)	B=	This letter and symbol indicate this is the output-description parameter.
		(K=l,D=l)	For fixed-length disk output records with keys, the letter K and symbol = must precede the length of the key field. The letter D and symbol = must precede the length of the data field. These two fields must be separated by a comma and enclosed in parentheses.
	B=(g)	B=	This letter and symbol indicate this is the output-description parameter.
		(g)	For undefined output records or variable output records without field select, the maximum block length must be enclosed in parentheses.
Rewind Input Ix	IR IN IU	I	The first letter in these forms identifies this parameter. The rewind option for the input tape is active both before and after data transfer.
		R	Rewind both before and after data transfer.
		N	Do not rewind either before or after data transfer.
		U	Rewind before and rewind and unload after data transfer.
Disk Check Ox	OY ON	O	The first letter in these forms identifies this parameter.
		Y	Write-disk check.
		N	Do not write-disk check.
First Record Rx	Rx	R	The first letter in this form identifies this parameter.
		x	This represents the position of the first logical input record to be output (x-1 records will be bypassed). If the file is to be copied, the function parameter must be indicated to be reblocked and the input and output file description parameters must contain identical values.
Device Type Description	E=(e)	E=	This letter and symbol indicate this is the device type description parameter.
		(e)	For output devices the valid entries are 2311 and 2314. This entry must be enclosed in parentheses.

TAPE-TO-PRINTER

PARAMETER	POSSIBLE FORMS	ENTRIES	EXPLANATION
Function Tt	TD TL TLF	T	The initial T identifies this as the type of function parameter.
		D	Display
		L	List
		LF	List and Field Select
Format Ff	FF FV FU	F	The leading F of these three possible forms identifies this as the format parameter.
		F	The second F must be indicated for fixed-length records.
		V	The letter V must be indicated for variable-length records.
		U	The letter U must be indicated for undefined records. With undefined records, the only valid entry above is data display.
Input Description	A=(n,m)	A=	This letter and symbol indicate this is the input-description parameter.
		(n,m)	For fixed-length input records, the input record length (the letter n) and the input block length (the letter m) must be enclosed in parentheses and separated by a comma. For field select with variable length records the letter n indicates the size of the fixed portion of each record, and the letter m indicates the maximum block size.
	A=(g)	A=	This letter and symbol indicate this is the input-description parameter.
		(g)	For undefined input records or variable input records without field select, the maximum block length must be enclosed in parentheses.
Output Description	B=(p)	B=	This letter and symbol indicate this is the output-description parameter.
		(p)	For printer output, the size of the print line (120, 132, or 144) must be entered. A print line size of 120 is forced if TD is specified.
	B=(n,p)	B=	This letter and symbol indicate this is the output-description parameter.
		(n,p)	For field select of variable length records with printer output records, the fixed portion of each output record (the letter n) and the size of the print line (the letter p) must be enclosed in parentheses and separated by a comma.
Rewind Input	IR IN IU	I	The first letter in these forms identifies this parameter. The rewind option for the input tape is active both before and after data transfer.
		R	Rewind both before and after data transfer.
		N	Do not rewind either before or after data transfer.
		U	Rewind before and rewind and unload after data transfer.
Print Output Ox	OX OC	O	The first letter in these forms identifies this parameter.
		X	Hexadecimal Printout (for data display only)
		C	Character Printout (forced for data list)
			The type of output indicated by the field-select parameter (hexadecimal or character) overrides this parameter.

TAPE-TO-TAPE

PARAMETER	POSSIBLE FORMS	ENTRIES	EXPLANATION
Function Tt	TC TF TR TRF	T	The first letter in these forms identifies this parameter.
		C	Copy
		F	Field Select
		R	Reblock
		RF	Reblock and Field Select
Format Ff	FF FV FU	F	The leading F of these three possible forms identifies this as the format parameter.
		F	The second F must be indicated for fixed-length records.
		V	The letter V must be indicated for variable-length records.
		U	The letter U must be indicated for undefined records.
Input Description	A=(n, m)	A=	This letter and symbol indicate this is the input-description parameter.
		(n, m)	For fixed-length input records, the input record length (the letter n) and the input block length (the letter m) must be enclosed in parentheses and separated by a comma. For field select with variable length records the letter n indicates the size of the fixed portion of each record, and the letter m indicates the maximum block size.
	A=(g)	A=	This letter and symbol indicate this is the input-description parameter.
		(g)	For undefined input records or variable input records without field select, the maximum block length must be enclosed in parentheses.
Output Description	B=(n, m)	B=	This letter and symbol indicate this is the output-description parameter.
		(n, m)	For fixed-length output records, the output record length (the letter n) and the output block length (the letter m) must be enclosed in parentheses and separated by a comma. For field select with variable length records, the letter n indicates the size of the fixed portion of each record, and the letter m indicates the maximum block size.
	B=(g)	B=	This letter and symbol indicate this is the output-description parameter.
		(g)	For undefined output records or variable output records without field select, the maximum block length must be enclosed in parentheses.
Rewind Option for Input Ix	IR IN IU	I	The first letter in these forms identifies this parameter. The rewind option for the input tape is active both before and after data transfer.
		R	Rewind both before and after data transfer.
		N	Do not rewind either before or after data transfer.
		U	Rewind before and rewind and unload after data transfer.
First Record Rx	Rx	R	The first letter in this form identifies this parameter.
		x	This represents the position of the first logical input record to be output (x-1 records will be bypassed). If the file is to be copied, the function parameter must be indicated to be reblocked and the input and output file description parameters must contain identical values.
Rewind Output Ox	OR ON OU	O	The first letter in these forms identifies this parameter. The rewind option for the output tape is active both before and after data transfer.
		R	Rewind both before and after data transfer.
		N	Do not rewind either before or after data transfer.
		U	Rewind before and rewind and unload after data transfer.

APPENDIX B

SORT/MERGE CONTROL STATEMENTS

The following is a listing of the general formats of the sort/merge control statements. For a detailed explanation of the entries, see Chapter 5 and Chapter 6.

<table>
<tr><th>Statement</th><th>Operands</th><th>Comments</th></tr>
<tr><td>SORT</td><td>FIELDS=(p_n,m_n,f_n,s_n)
or
FIELDS=(p_n,m_n,s_n),FORMAT={CH|BI|ZD|PD|FI|FL}</td><td>Required parameter for a sort</td></tr>
<tr><td></td><td>WORK=n</td><td>Required parameter for a sort</td></tr>
<tr><td></td><td>SIZE=n</td><td>Default of sort capacity</td></tr>
<tr><td></td><td>FILES=n</td><td>Default=1</td></tr>
<tr><td></td><td>CKPT or CHKPT</td><td></td></tr>
<tr><td>MERGE</td><td>FIELDS=(p_n,m_n,f_n,s_n)
or
FIELDS=(p_n,m_n,s_n),FORMAT={CH|BI|ZD|PD|FI|FL}</td><td>Required parameter for a merge</td></tr>
<tr><td></td><td>FILES=n
or
ORDER=n</td><td>Required parameter for a merge</td></tr>
<tr><td>RECORD</td><td>TYPE={F,V}</td><td>Required parameter</td></tr>
<tr><td></td><td>LENGTH=(l_1,l_2,l_3,l ,l) or LENGTH=(l_1)</td><td>Required parameter</td></tr>
<tr><td>MODS</td><td>PHn=(name,loading information,exit,
exit,...),...</td><td></td></tr>
<tr><td>INPFIL</td><td>BLKSIZE=n</td><td>Default=l_1 value in RECORD statement</td></tr>
<tr><td></td><td>BYPASS</td><td></td></tr>
<tr><td></td><td>EXIT</td><td></td></tr>
<tr><td></td><td>VOLUME=(n,...) or VOLUME=n</td><td>Default=1</td></tr>
<tr><td></td><td>OPEN={RWD|NORWD}</td><td>Default=RWD</td></tr>
<tr><td></td><td>CLOSE={RWD|UNLD|NORWD}</td><td>Default=RWD</td></tr>
<tr><td>OUTFIL</td><td>BLKSIZE=n</td><td>Default=l_3 value in RECORD statement</td></tr>
<tr><td></td><td>EXIT</td><td></td></tr>
<tr><td></td><td>NOTPMK</td><td></td></tr>
<tr><td></td><td>OPEN={RWD|NORWD}</td><td>Default=RWD</td></tr>
<tr><td></td><td>CLOSE={RWD|UNLD|NORWD}</td><td>Default=RWD</td></tr>
<tr><td>OPTION</td><td>PRINT={ALL|NONE|CRITICAL}</td><td>Default=ALL</td></tr>
<tr><td></td><td>STORAGE=n</td><td>Default=Partition size minus sort/merge program origin</td></tr>
<tr><td></td><td>LABEL=(output,input,...,$input_n$,work)
S=standard labels
N=nonstandard labels
U=unlabeled tapes</td><td>Default=Standard labels</td></tr>
<tr><td></td><td>VERIFY
KEYLEN=n
ADDROUT={A|D}
CALCAREA
ALTWORK</td><td></td></tr>
<tr><td>END</td><td>No operands</td><td></td></tr>
</table>

APPENDIX C

FILE-TO-FILE UTILITY MESSAGES

The following lists contain the messages which are printed from the DOS File-To-File Utility Programs.

RESPECTIVE ORDER OF DIAGNOSTIC MESSAGES FOR THE FILE-TO-FILE PROGRAMS

Note: Whenever xxx precedes a message, it indicates in which field definition the error occurred, e.g. cards 1 and 2 each have 5 field definitions: for a format error in the third definition,xxx would be printed as 003; for a format error on the fifth definition of card 2, xxx would be printed as a cumulative 010.

MESSAGE	REASON	ACTION
END CARD MISSING	No END statement supplied (// END),or noncontrol statement read before END.	Job is cancelled.
x INVALID FORMAT. UTILITY MODIFIER CARD	Format specifications for utility modifier statement were not followed or all required parameters were not supplied as follows: x: Decoded message A: Error in input format specifications (A parameter). B: Error in output format specifications (B parameter). E: Error in device type specification (E parameter). a. Invalid format. b. Not consistent with program type in utility modifier card. F: Error in record format specifications (F parameter). I: Invalid input option (I parameter). J: Invalid type of job (J parameter). M: Missing required parameter (F,A,B parameters must be present). N: Invalid type of program (U identifier // U not found, or xx representing the program type is not valid). O: Invalid output option (O parameter). P: Invalid page number option (P parameter). Q: Error in sequence checking specifications (Q parameter). R: Error in starting record specifications (R parameter). S: Invalid spacing option (S parameter). T: Type of job parameter missing (T parameter). U: Undefined parameters (parameter identifier not valid).	

MESSAGE	REASON	ACTION
FIELD SELECT CARD MISSING	Field select was indicated on utility modifier statement, but no field select statement was supplied.	Job is cancelled.
xxx INVALID FORMAT FIELD SELECT CARD	Format specifications for field select statement were not followed. (000 indicates no fields for field select were indicated but CV was present.)	
FIELD SELECT CARD NOT EXPECTED	Field select was not indicated on utility modifier statement, but field select statement was supplied.	
INVALID CONTROL CARD	A control statement (with //b in the first 3 columns) was read which was not a utility modifier, field select, print header, or END statement.	
INVALID INPUT DEVICE AT SYS004	The device assigned to SYS004 is not valid for this program.	
INVALID OUTPUT DEVICE AT SYS005	The device assigned to SYS005 is not valid for this program.	
UNDEFINED FORMAT CAN ONLY DISPLAY	Data display is the only mode that can be indicated for undefined records in printer output programs.	
xxx CANNOT FIELD SELECT INTO 1st 4 CHARACTERS	The indicated field cannot be selected into the record length field of a variable-length record.	
INVALID OUTPUT DEVICE AT SYS006	The device assigned to SYS006 is not valid for this program.	
UNDEFINED FORMAT CAN ONLY COPY	Copy is the only format that can be indicated for undefined records in non-printer program.	
INCORRECT PROGRAM	Utility modifier statement punched with the wrong program initials, such as DT for a disk to card program.	Job is cancelled. Note that all succeeding messages may not have a valid meaning.
x INVALID FORMAT UTIL MOD CARD	x: Utility modifier statement error	Job is cancelled.
	A: For non DASD input a key field was used. B: For nonprinter output, a printer B format was used; for non DASD output a key field was used. K: For non-DASD input or output a key field was used.	
FIXED LENGTH RECORD FORMAT REQUIRED	Card input or card output was not fixed length.	

MESSAGE	REASON			ACTION
INVALID JOB FOR THIS PROGRAM	Program	Valid Types	Invalid Types	Job is cancelled.
	Undefined records			
	a. TP,DP and MP	D*	C,B,BF,F,L, LF,R,RF	
	b. DD,DM,DT, MD,MM,MT, TD,TM, and TT	C	B,BF,D,F,L, LF,R,RF	
	Fixed-length records without key fields			
	a. CP	B,BF,C,D*, F,L,LF	R,RF	
	b. MP,TP, and DP	D,L,LF	B,BF,C,F,R, RF	
	c. CD,CT,DC, DD,DM,DT, MD,MM,MT, TC,TD,TM and TT	C,F,R, RF	B,BF,D,L,LF	
	Fixed-length records with key fields			
	a. CD,DC	F	B,BF,C,D,L, LF,R,RF	
	b. DT,MT,TM, and TD	F,RF	B,BF,C,D,L, LF,R,	
	c. DD,MM,DM, and MD	C,F	B,BF,D,L,LF, R,RF	
	d. DP and MP	D*,L,LF	B,BF,C,F,R,RF	
	Variable-length records without key fields			
	a. MP,TP, and DP	D*,L,LF	B,BF,C,F,R RF	
	b. DD,DM,DT, MD,MM,MT, TD and TT	C,F,R, RF	B,BF,D,L,LF	
	* If first character forms control is specified (S parameter), data display is invalid.			

MESSAGE	REASON	ACTION
INVALID INPUT RECORD LENGTH	a. Card input. Record length was greater than 80 (EBCDIC) or 160 (binary). b. Tape input. Record length was greater than 8192. c. DASD input without key. Block length was not a multiple of the record length. d. DASD record length exceeds 3625 for 2311, 7294 for 2314, or 2000 for data cell.	Job is cancelled.
NONSTANDARD LABEL INVALID INPUT	DASD programs do not allow nonstandard labels.	
NONSTANDARD LABEL INVALID OUTPUT		
INVALID INPUT OPTION	Option is incorrect for the program. No option for DASD input.	
INVALID OUTPUT OPTION	Option is incorrect for the program.	
INVALID CARD SEQUENCE	Card programs. The length parameter specified is over 10 characters or the starting position plus the length exceeds 80 characters.	
I/O AREA CANNOT BE ASSIGNED	Not enough main storage to assign the specified input/output areas.	
FIELD SELECT MUST BE SPECIFIED	When the output record length differs from the input record length, field select must be used. For printer programs (list function) the input record length cannot exceed the size of the print line. For DASD programs with key fields (except DASD to printer or DASD to DASD) field select must be specified.	
xxx INVALID UNPACK OUTPUT LENGTH	The parameter values specified are invalid.	
xxx INVALID PACK OUTPUT LENGTH		
xxx RECORD CAPACITY EXCEEDED BY PACK	The xxxth field select parameter specifies a field not entirely contained within the input or output record.	
xxx RECORD CAPACITY EXCEEDED BY UNPK		
xxx RECORD CAPACITY EXCEEDED BY FS		
xxx RECORD CAPACITY EXCEEDED BY HEX		

MESSAGE	REASON	ACTION
xxx FIELD SELECT PARAMETER FOR NONEXISTENT KEY	A key field was specified in the field select statement, but no key was indicated in the utility modifier statement.	Job is cancelled.
INVALID OUTPUT RECORD LENGTH	a. Card output. Record length was greater than 80 (EBCDIC) or 160 (binary). b. Tape output. Record length was greater than 8192. c. Printer output. Record length was greater than 144. d. DASD output. The output record length is greater than 3625 for 2311, 7294 for 2314, or 2000 for data cell.	
INVALID INPUT KEY LENGTH	For a DASD input the key length is greater than 255.	
INVALID OUTPUT KEY LENGTH	For a DASD output the key length is greater than 255.	
INVALID INPUT BLOCK LENGTH	a. Card input. Block and record length are not equal. b. Tape input. For fixed length record processing the input block length was not a multiple of the record length; otherwise, the block length was not 4 greater than the fixed portion. c. DASD input. Input block length is greater than 3625 for 2311, 7294 for 2314, or 2000 for data cell.	
INVALID OUTPUT BLOCK LENGTH	a. Block length is not a multiple of the record length. b. For DASD the output block length is greater than 3625 for 2311, 7294 for 2314, or 2000 for data cell. c. For the copy function the block lengths must be equal.	
INVALID INPUT DATA LENGTH	DASD input programs with key require data length plus key length to be less than or equal to 3605 for 2311, 7249 for 2314, or 1984 for data cell.	
INVALID OUTPUT DATA LENGTH	DASD output programs with key require data length plus key length to be less than or equal to 3605 for 2311, 7249 for 2314, or 1984 for data cell.	
xxx FS INPUT LENGTH EQUALS ZERO	Input field length has been specified as zero.	
xxx PACK INPUT LENGTH EQUALS ZERO		
xxx UNPK INPUT LENGTH EQUALS ZERO		

MESSAGE	REASON	ACTION
xxx HEX INPUT LENGTH EQUALS ZERO	Input field length has been specified as zero.	Job is cancelled.
xxx CANNOT PROCESS HEX PARAMETER	Hexadecimal indicator valid only for print output programs.	
xxx CANNOT PROCESS PACK PARAMETER	Cannot pack a field for print output programs.	
USER ROUTINE NOT PRESENT	User label checking is specified on the UPSI statement, but a user label routine is not present.	

RESPECTIVE ORDER OF FILE-TO-FILE PROCESSING MESSAGES

Messages (on SYSLST)	Format	Function	Primary Condition	Associated Conditions	Processing
BLOCK NO. xxxxxx, INPUT AREA OVERFLOW	F, V, or U	Copy	Input block length is longer than that specified in the utility modifier statement.	None	The specified input block size is copied and the remainder is truncated. If the records are variable length, the count field is not corrected.
BLOCK NO. xxxxxx, INPUT AREA UNDERFLOW	F	Copy	Input block length is shorter than that specified in the utility modifier statement.		Only the actual block size is copied (no padding).
BLOCK NO. xxxxxx, INPUT AREA UNDERFLOW	F	R, F, RF, L, or LF		The actual block size is a multiple of the specified record size but less than the specified block size.	Processing is performed as specified for the short block. This message is not issued if the starting record number in the record-skipping parameter has not been encountered.
BLOCK NO. xxxxxx, INPUT AREA UNDERFLOW BLOCK NO. xxxxxx, RCD. NO. xx RECORD AND REMAINDER OF BLOCK DROPPED	F	R, F, RF, L, or LF		The last logical record of the input block is less than the specified record size.	Processing is normal up to the short record. The record is dropped and processing continues. This message is not issued if the starting record number in the record-skipping parameter has not been encountered. The short record is counted as one.

MESSAGE			REASON		ACTION
Message (on SYSLST)	Format	Function	Primary Condition	Associated Conditions	Processing
BLOCK NO. xxxxxx, INPUT AREA OVERFLOW	V	R, F, RF, L, or LF	Input block length is longer than that specified in the utility modifier statement.	The last position of the specified block is the last position of a logical record.	The overflow records from the input block are truncated. This message is issued even if the first record to be processed has not been reached. The truncated records are not counted.
BLOCK NO. xxxxxx, INPUT AREA OVERFLOW BLOCK NO. xxxxxx, RCD. NO. xx RECORD AND REMAINDER OF BLOCK DROPPED	V	R, F, RF, L, LF		The last logical record in the specified block size is not complete within the block.	The input block (and the last logical record) are truncated. The truncated record is dropped. The second message is not issued if the starting record number in the record skipping parameter has not been encountered. The dropped part of the block is counted as one.
BLOCK NO. xxxxxx, RCD. NO. xx RECORD AND REMAINDER OF BLOCK DROPPED	V	R, F, RF, L, or LF	An input logical record contains an invalid length field. A record length field is invalid if it is less than 5 or is not equal to the number of bytes read.		Processing of the current block cannot proceed and the block is dropped. This message is issued even if the record-skipping parameter number has not been reached. The part of the block is counted as one.
BLOCK NO. xxxxxx, RCD. NO. xx, SHORT VARIABLE LENGTH RECORD DROPPED	V	F, RF, or LF	The length of a logical input record is less than that specified as the fixed portion of the variable-length records.		The record is dropped and processing continues with the next record, if present. This message is not issued if the record-skipping parameter has not been encountered. The dropped record is counted as one.

MESSAGE			REASON	ACTION	
Messages (on SYSLST)	Format	Function	Primary Condition	Associated Conditions	Processing
BLOCK NO. xxxxxx, OUTPUT AREA OVERFLOW	V	R, F, RF, L, or LF	A generated output record exceeds the block size specified in the utility modifier statement.		The generated block is truncated. The block count and record count are corrected and the block written out.
BLOCK NO. xxxxxx, KEY LENGTH IS xxx	F or V	C,R,F, RF,L, or LF	The key length for this block is invalid, or it differs from the key length specified in the utility modifier statement.	a. For undefined records, the message should not occur. b. For fixed-length records with no key fields specified, or variable length records, only the data portions are processed. c. For fixed-length records with key fields specified, the actual and specified key length differ. Both key and data fields are processed as specified (i.e., if the actual key is less than that specified, the difference is made up with data bytes; if greater, the excess is treated as data bytes.)	Processing continues, with the output record formatted as specified in the utility modifier statement (if valid specification).

RESPECTIVE ORDER OF FILE-TO-FILE INFORMATIONAL MESSAGES	
Control parameter diagnostics are followed by logging messages in this order.	
MESSAGE	**ACTION**
CARD TO DISK CARD TO PRINTER/PUNCH CARD TO TAPE DATA CELL TO DATA CELL DATA CELL TO DISK DATA CELL TO PRINTER DATA CELL TO TAPE DISK TO CARD DISK TO DATA CELL DISK TO DISK DISK TO PRINTER DISK TO TAPE TAPE TO CARD TAPE TO DATA CELL TAPE TO DISK TAPE TO PRINT TAPE TO TAPE } UTILITY	Identifies the particular utility program. The program continues processing.
INPUT {FIXED PORTION xxxx KEY LENGTH xxxx DATA LENGTH xxxxx RECORD LENGTH xxxx BLOCK LENGTH xxxx}	Processing continues. (x represents a digit.)
OUTPUT {FIXED PORTION xxxx KEY LENGTH xxxx DATA LENGTH xxxxx RECORD LENGTH xxxx BLOCK LENGTH xxxx}	
INPUT {CARD BCD CARD BINARY NO REWIND, UNLOAD REWIND REWIND, UNLOAD}	
OUTPUT OPTION {BCD, CHARACTER CARD BCD CARD BINARY DISK WRITE CHECK NO DISK WRITE CHECK PRINT CHARACTER PRINT HEX NO REWIND, UNLOAD{WRITE TAPE MARK} REWIND{WRITE TAPE MARK} REWIND, UNLOAD{WRITE TAPE MARK}}	
{x INPUT, x OUTPUT x INPUT/OUTPUT} AREAS ASSIGNED	
RECORD FORMAT {FIXED VARIABLE UNDEFINED}	

MESSAGE	ACTION
TYPE { COPY / DATA DISPLAY / FIELD SELECT / LIST / LIST, FIELD SELECT / PRINT AND PUNCH / PRINT, PUNCH, FIELD SELECT / REBLOCK / REBLOCK, FIELD SELECT }	Processing continues. (x represents a digit.)
STARTING SEQUENCE COLUMN xx	
SEQUENCE LENGTH xx	
STARTING RECORD NUMBER xxxxxxxx	
INPUT DEVICE TYPE xxxx	These messages are printed for DASD devices. xxxx indicates 2311, 2314, or 2321.
OUTPUT DEVICE TYPE xxxx	
REPLY x	This message is printed to indicate the reply given to a diagnostic printed on SYSLOG. The action taken is indicated by the letter x. Processing continues.
1ST CHARACTER FORMS CONTROL TYPE { A / B / C / D }	Processing continues.
xx ERRORS FOUND IN CONTROL CARDS	
CARD SEQUENCE ERROR, CURRENT SEQ xxxxxxxxxx LAST SEQ xxxxxxxxxx	
END OF DATA	END OF DATA will not be printed for first-character forms control.
FILE MARK WRITTEN IN XT. NO. B1 C1 C2 H1 H2 R xxx xxx xxx xxx xxx xxx xxx	For DASD output programs the decimal value of the EXTENT sequence number and the address of the file mark (written at the end of the file) are logged. The headings represent bin (B1), subcell (C1), strip (C2), cylinder (H1), track (H2), and record (R) numbers for data cell. For disk, they represent cylinder (C2), track (H2), and record (R) numbers.
NUMBER OF { INPUT / OUTPUT } BLOCKS PROCESSED xxxxxx	Processing continues.
SPECIFIED STARTING RECORD NO. LARGER THAN TOTAL NO. OF LOGICAL INPUT RECORDS	
END OF JOB	

APPENDIX D

SORT/MERGE MESSAGES

The following listings contain the messages which are printed by the Sort/Merge Program.

MESSAGES

7000I 'control card image'

Explanation: The control cards will be listed by a series of these messages.

System Action: Job continues normal processing.

Programmer Response: None.

Operator Response: None.

7001I PHASE 0 END, NO DETECTED ERRORS

Explanation: Phase 0 has ended with no errors detected.

System Action: Control is passed directly to the next sort/merge phase or back to the calling routine (EXEC or LOAD and BALR routine) if CALCAREA specified.

Programmer Response: None.

Operator Response: None.

7002D OPERATOR CORRECT ERRORS OR CANCEL

Explanation: This message follows A messages and I warning messages in phase 0 when the console is used to write the messages. It requests the operator to correct the errors specified in preceding messages or else cancel the job.

System Action: Control is given to operator to correct errors detected in Phase 0.

Programmer Response: None.

Operator Response: Operator has two options:

1. Correct storage and/or unit assignments as noted by preceding 'A' type messages, place control statements back in input stream and key in console 'RETRY'
2. Cancel the job by keying into the console 'CANCEL'

7003I EXCESS CARDS

Explanation: The maximum number of control cards that can be handled by the sort/merge program is 25.

System Action: Job is terminated after Phase 0 has completed its error checking of control statements and unit assignments.

Programmer Response: Check for too many control statements. Compress information onto fewer statements.

Operator Response: None.

7004I NO xxxxxx CARD

Explanation: An essential control card has been omitted. Essential control cards are SORT or MERGE (not both), RECORD, and END. xxxxxx will be replaced by the statement definer.

System Action: Job is terminated after Phase 0 has completed its error checking of control statements and unit assignments.

Programmer Response: Supply a SORT or MERGE, RECORD, and END control statement.

Operator Response: None.

7005I STMT DEFINER ERR

Explanation: The first field of a non-continuation card must be a valid statement definer, that is, SORT, MERGE, RECORD, MODS, INPFIL, OUTFIL, OPTION, or END.

System Action: Job is terminated after Phase 0 has completed its error checking of control statements and unit assignments.

Programmer Response: Check all statements for incorrect, misplaced, or mispelled operation definers.

Operator Response: None.

7006I DUPLICATE xxxxxx CONTROL CARD

Explanation: A statement definer, represented by xxxxxx, must not be specified more than once.

System Action: Job is terminated after Phase 0 has completed its error checking of control statements and unit assignments.

Programmer Response: Check for duplicate statement types.

Operator Response: None.

7007I COL. 1 OR 1 - 15 NOT BLANK

Explanation: Column 1 of a control card must be blank, or a control card with a nonblank character in column 72 (indicating that a continuation card follows) must be followed by a card which has blanks in columns 1-15 (a valid continuation card).

System Action: Job is terminated after Phase 0 has completed its error checking of control statements and unit assignments.

Programmer Response: Check control statements for nonblank characters in col. 1, and continuation cards for nonblank characters in col. 1-15.

Operator Response: None.

7008I NO CONTIN CARD

Explanation: A continuation card did not appear where required. The last control card must have a blank in column 72.

System Action: Job is terminated after Phase 0 has completed its error checking of control statements and unit assignments.

Programmer Response: Check for keypunching error, or an overflow of parameters into col. 72.

Operator Response: None.

7009I xxxxxx OPTION INVALID.DEFAULT USED

Explanation: Either

1. The xxxxxx parameter, which is not valid for this job, has been specified in the OPTION statement, or
2. NOTPMK has been specified for a direct access device on the OUTFIL statement.

The default values for the OPTION statement are:

STORAGE= (value obtained by subtracting the sort/merge program origin from the end of the current partition address.
PRINT=ALL

All other options are nullified if invalid.

System Action: Job continues with the defaulted values.

Programmer Response: Check default values to see if appropriate for next run of application. If not, make appropriate changes.

Operator Response: None.

7010I INVALID KEY LENGTH

Explanation: The value assigned to KEYLEN in the OPTION statement must be a valid number less than 256 and greater than zero.

This message will also be printed if the restrictions associated with the KEYLEN option are exceeded. These restrictions are discussed under the topic "OPTION Control Statement" in Section 3.

System Action: Job is terminated after Phase 0 has completed its error checking of control statements and unit assignment.

Programmer Response: Check KEYLEN value in OPTION statement.

Operator Response: None.

7011I INVALID xxxxxx KEYWORD

Explanation: A keyword not recognized by the sort/merge program or a duplicate keyword has been detected in the control statement represented by xxxxxx.

System Action: Job is terminated after Phase 0 has completed its error checking of control statements and unit assignment.

Programmer Response: Check appropriate control statement for invalid or duplicate keyword operand.

Operator Response: None.

7012I INVALID FORMAT

Explanation: The value assigned to f in the FIELDS parameters, or the value assigned to FORMAT, must be one of the following -- CH, ZD, PD, BI, FI, or FL.

System Action: Job is terminated after Phase 0 has completed its error checking of control statements and unit assignment.

Programmer Response: Check format values given in FIELDS parameter or FORMAT value of the SORT or MERGE control statement.

Operator Response: None.

7013I CFxx DISPLACEMENT INVALID

Explanation: The value assigned to p in the FIELDS parameter of a SORT or MERGE statement must be a numeral greater than zero. The control field number is represented by xx.

System Action: Job is terminated after Phase 0 has completed its error checking of control statements and unit assignment.

Programmer Response: Check displacement value specified in SORT or MERGE control statement.

Operator Response: None.

7014I CFxx LENGTH INVALID

Explanation: The value assigned to m in the FIELDS parameter of a SORT or MERGE statement must be a numeral greater than zero. The control field number is represented by xx.

System Action: Job is terminated after Phase 0 has completed its error checking of control statements and unit assignment.

Programmer Response: Check length values specified in SORT or MERGE control statement.

Operator Response: None.

7015I CFxx BEYOND 4092

Explanation: A control field must not extend beyond the first 4092 bytes of the record.

System Action: Job is terminated after Phase 0 has completed its error checking of control statements and unit assignment.

Programmer Response: Check length and displacement value specified in SORT or MERGE control statement for field extending beyond 4092 bytes.

Operator Response: None.

7016I CFxx SEQUENCE INVALID

Explanation: The value assigned to s in the FIELDS parameter of a SORT or MERGE statement must be either A or D. The control field number is represented by xx.

System Action: Job is terminated after Phase 0 has completed its error checking of control statements and unit assignment.

Programmer Response: Check sequence value specified in SORT or MERGE control statements for a keypunching error.

Operator Response: None.

7017I BOTH SORT AND MERGE DEFINED

Explanation: Both SORT and MERGE statements must not be specified for the same execution of the sort/merge program.

System Action: Job is terminated after Phase 0 has completed its error checking of control statements and unit assignment.

Programmer Response: Check application and eliminate SORT or MERGE control statement.

Operator Response: None.

7018I xxxxxx yyyyyy KEYWORD MISSING

Explanation: A parameter which must be specified, and for which there is no default, has been omitted. The parameters which fall in this category are: FIELDS and WORK in the SORT statement, FIELDS and FILES in the MERGE statement, and TYPE and LENGTH in the RECORD statement. xxxxxx represents the statement, and yyyyyy the keyword.

System Action: Job is terminated after Phase 0 has completed its error checking of control statements and unit assignment.

Programmer Response: Check appropriate control statement for missing keyword.

Operator Response: None.

7019I MISSING FORMAT OR SEQUENCE CODE

Explanation: The f or s value in the FIELDS parameter of a SORT or MERGE statement has been omitted.

System Action: Job is terminated after Phase 0 has completed its error checking of control statements and unit assignment.

Programmer Response: Check SORT or MERGE control statement for missing code.

Operator Response: None.

7020I GIVEN FILE SIZE INVALID AND IGNORED

Explanation: The value assigned to the SIZE parameter of a SORT statement must be a numeral, otherwise the maximum capacity is assumed by the sort/merge program. This may lead to an inefficient sort and capacity exceeded problems.

System Action: Job continues with default value.

Programmer Response: Check SORT control statement for invalid SIZE operand.

Operator Response: None.

7021I FILES VALUE INVALID

Explanation: The value assigned to the FILES parameter of a SORT or MERGE statement must be in the permitted range. Permissible values are 1-9 for a sort and 1-8 for a merge-only.

System Action: Job is terminated after Phase 0 has completed its error checking of control statements and unit assignments.

Programmer Response: Check SORT or MERGE statement for invalid FILES operand.

Operator Response: None.

7022I yyyyyy KEYWORD IGNORED BY MERGE

Explanation: The CKPT, SIZE and WORK keywords are ignored when included in a MERGE statement. yyyyyy represents the ignored keyword.

System Action: Job continues ignoring option specified.

Programmer Response: Check application to see if it was set up properly before next run.

Operator Response: None.

7023I SORT WORK VALUE INVALID

Explanation: The WORK parameter in a SORT statement has been assigned a value not recognized by the sort/merge program. Permissible values are 1-8 for direct access and 3-9 for tape devices.

System Action: Job is terminated after Phase 0 has completed its error checking of control statements and unit assignments.

Programmer Response: Check SORT statement for invalid WORK operand.

Operator Response: None.

7024I PH 1/2 EXITS IGNORED BY MERGE

Explanation: Phases 1 and 2 of the sort/merge program are not used for a merge-only operation, therefore any phase 1 or 2 exits specified in the MODS statement of a merge-only operation are ignored.

System Action: Job continues, ignoring exits specified.

Programmer Response: Check application to see if it was set up properly before next run.

Operator Response: None.

7025I EXIT E32 or E38 IGNORED BY SORT

Explanation: Exits E32 and E38 are available only for a merge-only operation. They are ignored when specified in the MODS statement of a sort operation.

System Action: Job continues, ignoring exits specified.

Programmer Response: Check application to see if it was set up properly before next run.

Operator Response: None.

7026I INVALID PHx NAME

Explanation: The phase name specified in a MODS statement must be a valid DOS name, that is, 8 alphameric characters, with the first one alphabetic. The x represents the phase number.

System Action: Job is terminated after Phase 0 has completed its error checking of control statements and unit assignments.

Programmer Response: Check MODS statement for invalid phase name.

Operator Response: None.

7027I INVALID MODS ADDRESS/LENGTH FIELD

Explanation: The address or length specified in a MODS statement must be a valid number, possibly preceded by the character L.

System Action: Job is terminated after Phase 0 has completed its error checking of control statements and unit assignments.

Programmer Response: Check MODS statement for invalid address/length.

Operator Response: None.

7028I INVALID PHx EXIT

Explanation: An exit not recognized by the sort/merge program has been specified in a MODS statement. The valid exits are listed in "Program Modification." The x represents the phase number.

System Action: Job is terminated after Phase 0 has completed its error checking of control statements and unit assignments.

Programmer Response: Check MODS statement for keypunching error or other error resulting in specification of invalid program exit number.

Operator Response: None.

7029I ERR IN LENGTH VALUE

Explanation: An inconsistency has been detected either among the length values specified in the RECORD statement, or in the blocksize of the INPFIL or OUTFIL statement. Physical and logical record lengths must be as shown in Table 3.

System Action: Job is terminated after Phase 0 has completed its error checking of control statements and unit assignments.

Programmer Response: Check RECORD statement for invalid length value and/or OUTFIL and/or INPFIL statement for invalid blocksize for device being supported.

Operator Response: None.

7030I Lx VALUE INVALID

Explanation: A length specified in the RECORD statement must be a numeral. The x identifies the length.

System Action: Job is terminated after Phase 0 has completed its error checking of control statements and unit assignments.

Programmer Response: Check RECORD statement for invalid length value.

Operator Response: None.

7031I RECORD TYPE INVALID

Explanation: The TYPE parameter in the RECORD statement must be either F or V.

System Action: Job is terminated after Phase 0 has completed its error checking of control statements and unit assignments.

Programmer Response: Check RECORD statement for keypunching error resulting in TYPE operand value not being F or V.

Operator Response: None.

7032I ALTERED RECORDS REQUIRE EXIT E15/E35

Explanation: If user routines modify record length, exits E15 and/or E35 must be specified in the MODS statement.

System Action: Job is terminated after Phase 0 has completed its error checking of control statements and unit assignments.

Programmer Response: Check RECORD and MODS statements for inconsistency. Lengthening or shortening of records ($l_1 \neq l_2 \neq l_3$) requires user exit E15 and/or E35.

Operator Response: None.

7033I xxxxxx BLOCK SIZE = RECORD LENGTH

Explanation: Block size has not been specified for either the INPFIL or OUTFIL statement, so it is assumed to be equal to record length for fixed-length records, or record length plus four for variable length records. xxxxxx will be replaced by INPFIL or OUTFIL.

With variable length records, when BLKSIZE is not specified it defaults to the maximum record length (l_3 if specified or else l_1). All records of lesser size will then be blocked to this size whenever possible.

Warning: If this assumption is not reasonably valid, performance reduction or job termination may result.

System Action: Job continues with the defaulted value.

Programmer Response: Check blocksize parameter to see if appropriate for next run of application. If not, make appropriate change on INPFIL/OUTFIL statement.

Operator Response: None.

7034I RECORD CONFLICTS WITH xxxxxx BLKSIZE

Explanation: The block size specified in the INPFIL or OUTFIL statement must be consistent with the record length specified in l_1 or l_3. That is, for variable length records, block size must be at least record length +4, and, for fixed length records, block size must be an exact multiple of record length. xxxxxx will be replaced by INPFIL or OUTFIL. If ADDROUT is specified with variable-length records, the rules for fixed-length records apply for l_3.

System Action: Job is terminated after Phase 0 has completed its error checking of control statements and unit assignments.

Programmer Response: Check RECORD statement and INPFIL or OUTFIL statement for inconsistency in specifying lengths.

Operator Response: None.

7035I VOLUME VALUE(S) INVALID

Explanation: A value assigned to the VOLUME parameter of the INPFIL statement must be a numeral (1 to 9 for a sort, 1 to 8 for a merge only).

System Action: Job is terminated after Phase 0 has completed its error checking of control statements and unit assignments.

Programmer Response: Check INPFIL statement for invalid VOLUME operand.

Operator Response: None.

7036I BYPASS IGNORED FOR OUTFIL

Explanation: If the BYPASS parameter is specified in the OUTFIL statement, it is ignored by the sort/merge program.

System Action: Job continues normal processing.

Programmer Response: Remove option from OUTFIL statement for next run of application.

Operator Response: None.

7037I SYNTAX ERROR - xxxxxx

Explanation: A SYNTAX error has been detected in the control statement represented by xxxxxx. Common syntax errors are:

- Unbalanced parentheses
- Missing commas
- Embedded blanks
- Redundant operands
- Missing Parameters

System Action: Job is terminated after Phase 0 has completed its error checking of control statements and unit assignments.

Programmer Response: Check specified control statement for an error in syntax.

Operator Response: None.

7038I L3 INVALID FOR ADDROUT

Explanation: If ADDROUT=A is specified in the OPTION statement, the value assigned to l_3 in the RECORD statement must be 10. If ADDROUT=Dis specified, then l_3 must be 10 + length of control word (sum of m values in SORT statement) $\pm$ any modification made to control word length at exit E35.

System Action: Job is terminated after Phase 0 has completed its error checking of control statements and unit assignments.

Programmer Response: Check RECORD statements for invalid l_3 for ADDROUT option or OPTION statement for undesired ADDROUT option.

Operator Response: None.

7039I INVALID xxxxxx DELIMITER

Explanation: A punctuation error has been detected in the control statement represented by xxxxxx.

System Action: Job is terminated after Phase 0 has completed its error checking of control statements and unit assignments.

Programmer Response: Check for operands that are incorrectly split between control and continuation cards.

Operator Response: None.

7040I FLD OR VALUE GT 8 CHAR -- xxxxxx

Explanation: A field or value has been detected in the statement represented by xxxxxx which is greater than 8 characters -- the longest valid length.

System Action: Job is terminated after Phase 0 has completed its error checking of control statements and unit assignments.

Programmer Response: Check specified statement for field or value greater than eight characters.

Operator Response: None.

7041I L4 GREATER THAN L1 OR L5

Explanation: The minimum length specified for input records must not be greater than the specified maximum or modal lengths.

System Action: Job is terminated after Phase 0 has completed its error checking of control statements and unit assignments.

Programmer Response: Check RECORD statement for invalid l_1, l_4, and/or l_5.

Operator Response: None.

7042I MULTIPLE DEFINED EXIT Enn

Explanation: An exit number (represented by nn) must not be defined more than once in the MODS statement.

System Action: Job is terminated after Phase 0 has completed its error checking of control statements and unit assignments.

Programmer Response: Check MODS statement for multiply defined exits.

Operator Response: None.

7043I INVALID INTERNAL LISTS

Explanation: An error has been detected in the parameter list provided by the user when he dynamically invokes the sort/merge program.

System Action: Job is terminated after Phase 0 has completed its error checking of control statements and unit assignments.

Programmer Response: Check all control statements and make appropriate corrections in the parameter list passed to sort.

Operator Response: None.

7044I TOO MANY xxxxxx KEYWORDS

Explanation: The maximum number of keywords that can be specified in the statement represented by xxxxxx has been exceeded.

System Action: Job is terminated after Phase 0 has completed its error checking of control statements and unit assignments.

Programmer Responses: Check specified statements for too many keyword operands.

Operator Response: None.

7045I CFxx BEYOND RECORD

Explanation: A control field specified in the FIELDS parameter of the SORT or MERGE statement extends beyond the length of the minimum record. The number of the control field replaces the xx in the printed message.

Operator Response: None. extend beyond the end of the minimum record. Control field number is represented by xx.

System Action: Job is terminated after Phase 0 has completed its error checking of control statements and unit assignments.

Programmer Response: Check SORT or MERGE statements for incorrectly specified control field displacement or length. Check RECORD statements for incorrectly specified record length.

Operator Response: None.

7046I CFxx TOO LONG FOR TYPE

Explanation: A control field with packed or zoned decimal format exceeds 16 or 18 bytes, respectively. Any other control field must not exceed 256 bytes. Control field number is represented by xx.

System Action: Job is terminated after Phase 0 has completed its error checking of control statements and unit assignments.

Programmer Response: Check length and format of specified control field on SORT or MERGE statement for error.

Operator Response: None.

7047I EXIT Enn NOT GIVEN FOR NONSTANDARD LABELS

Explanation: If nonstandard labels are specified in the OPTION statement, exits E11, E17, E31, and/or E37 must be specified in the MODS statement. The n's represent an exit number.

System Action: Job is terminated after Phase 0 has completed its error checking of control statements and unit assignments.

Programmer Response: Check MODS statement for omitted exit and OPTION statement for required label handling.

Operator Response: None.

7048I MINIMUM SORT WORK AREA nnn TRACKS

Explanation: This message is generated if CALCAREA is specified

Operator Response: None.

System Action: Job terminates at the end of Phase 0 due to the CALCAREA option.

Programmer Response: Value to be used in determining applications work space requirements.

Operator Response: None.

7049I nnn TRACKS FOR BEST PERFORMANCE

Explanation: This message is generated if CALCAREA is specified in the OPTION statement.

System Action: Job terminates at the end of phase 0 due to the CALCAREA option.

Programmer Response: Value indicates optimum work space requirements for given application.

Operator Response: None.

7050I NMAX = nnnn

Explanation: This is an estimate of the maximum number of records that can be sorted within the specifications provided by the user.

System Action: Job continues normal processing.

Programmer Response: Value to be used as an estimate of number of input records that can be handled in the given application.

Operator Response: None.

7051I B = nnnn

Explanation: This is the blocking factor used by the sort/merge program for intermediate work files. It depends on the specifications provided by the user.

System Action: Job continues normal processing.

Programmer Response: None.

Operator Response: None.

7052I G = nnnn

Explanation: This is the number of records that can be contained in the record storage area of the internal sort phase. It depends on the specifications provided by the user.

System Action: Job continues normal processing.

Programmer Response: None.

Operator Response: None.

7053A INCORRECT RESPONSE

Explanation: The operator did not correct the error or cancel the job in response to message 7002D.

System Action: Control is returned to operator to give valid response -- 'RETRY' or 'CANCEL'

Programmer Response: None.

Operator Response: Check your keying in of the response -- only 'RETRY' and 'CANCEL' are acceptable.

7054A INSUFFICIENT CORE

Explanation: Insufficient main storage is available to contain the sort/merge program plus the minimum record storage area and user-routines, if present; or the value the user specified for the STORAGE option is less than the sort/merge program design point.

System Action: Job is terminated after Phase 0 or operator may correct 'A' message errors if message 7002D is given.

Programmer Response: Check STORAGE parameter, and use of options -- e.g., user exits, multi-volume input, etc. Increase core to be used by sort. Refer to the "Environmental Requirements" topic in Section 2.

Operator Response: Sort has a minimum of 10K design point for 2400/2311 sorts and 22K for 2314 sorts. For any given application, minimum core requirements depend on intermediate storage files, record and block sizes, options desired, and user exit sizes. Assign larger partition size, rerun, and note problem on programmer report.

7055I TOO MANY xxxxxx POSITIONAL PARAMETERS

Explanation: The number of positional parameters in the statement represented by xxxxxx must not exceed the maximum allowed, as listed below.

Parameter	Maximum Number of Positional Values
FIELDS (SORT or MERGE statement)	36 if FORMAT keyword is used, 48 otherwise
LENGTH (RECORD statement)	5 if TYPE=V 3 if TYPE=F
VOLUME (INPFIL statement)	Value assigned to FILES keyword in SORT or MERGE statement
LABEL (OPTION statement)	Eleven

System Action: Job is terminated after Phase 0 has completed its error checking of control statements and unit assignments.

Programmer Response: Check specified keyword operand for too many parameters.

Operator Response: None.

7056A MIXED UNIT ASSIGNMENT

Explanation: All sort input units must be the same device type. Similarly, all work units must be the same device type. For example, input files could all be on 2400 9-track devices, and all work files on 2311 devices.

System Action: Job is terminated after Phase 0 or operator may correct 'A' type message errors if message 7002D is given.

Programmer Response: Check assign statements -- Input must be of same type -- i.e., 2311, 2314, 2400 -- 9-track, or 2400 -- 7-track. Work must be of same type -- i.e., 2311, 2314, or 2400 -- 7 and/or 9-track.

Operator Response: Assign input so that all units are of the same type. Assign intermediate storage so that all units are of the same type.

7057A RECORD FORMAT NOT SUPPORTED ON 7-TRACK

Explanation: Variable length records are not permitted if the intermediate work files are allocated 7-track tape units.

System Action: Job is terminated after Phase 0 has completed its error checking of control statements and unit assignments.

Programmer Response: Check assignment of intermediate storage files for a 7-track tape with variable length sort application.

Operator Response: None.

7058A INVALID xxxxxx AS WORK UNIT

Explanation: The words INPUT/ OUTPUT or ALTERNATE will replace xxxxxx. The pooling of input, output or alternate work type to an intermediate storage (work) device has been specified incorrectly. Files may share the same device/ extent if the following rules are followed:

1. Output may be assigned to the same device (same disk extent) as the first work file. On disk, both files must have identical starting track addresses for valid pooling.
2. Alternate work file (tape only) may not be pooled.
3. Tape input files may be assigned on the same device only for a sort. For a merge they must be on separate devices.
4. For disk all files for a sort may be on the same physical device. Pooling exists only when extents coincide or overlap. No pooling of input extents is permitted.

System Action: Job is terminated after Phase 0 or operator may correct 'A' type message errors if message 7002D is given.

Programmer Response: Check assignment of input, alternate work, or output files with given pooling rules.

Operator Response: Check assignment of specified file with given pooling rules. If a free drive of same type is available, assign file to free drive. Note problem on programmer report.

7059A UNITS ASGN ERROR -- xxxxxxx

Explanation: Error was detected in Phase 0. The filename of the device assigned in error is indicated by xxxxxxx. The only devices which may be allocated to an input, work, or output file are 2311, 2314, and 2400. Other devices can be used for input or output if the user reads all input at E15 and writes all output at E35. This message is also generated if an expected unit has not been assigned.

System Action: Job is terminated after phase completion, or operator may correct 'A' type message errors if message 7002D is given.

Programmer Response: Check device type of sort files assigned -- only 2400, 2311, and 2314 are supported or omission of a file's assignment.

Operator Response: Check device type of files assigned -- only 2400, 2311, 2314 supported. If device available, assign to file. Note problem on programmer report.

7060I OPTION NOT CHARACTERISTIC OF DEVICE

Explanation: A parameter has been specified in the OPTION statement which is not applicable to the I/O devices being used.

For a merge application with mixed inputs, this message will be printed as many times as there are input devices for which the option is not characteristic of the device.

System Action: Job continues ignoring option specified.

Programmer Response: Check application to see if it was set up properly before next run.

Operator Response: None.

7061I BLANK CARD ENCOUNTERED

Explanation: A completely blank card was found in the control statements. The card is ignored, but is included in the program's count of control cards.

System Action: Job continues normal processing.

Programmer Response: Remove blank card from input stream for next application.

Operator Response: None.

7062A LABEL OPTION HAS INVALID PARAMETER

Explanation: This message is generated if any character except U, N, or S is read in between two successive commas in the label parameter of the Option card.

System Action: Job is terminated after Phase 0, or operator may correct 'A' type message errors if message 7002D is given.

Programmer Response: Check the Option card and correct the label parameter.

Operator Response: Ensure that the Option card contains the correct label parameters - U for unlabeled files, N for nonstandard or user-standard label files, and S (or two consecutive commas) for standard label files.

7101I END SORT PH

Explanation: Phase 1 has ended.

System Action: Job continues normal processing.

Programmer Response: None.

Operator Response: None.

7159A UNITS ASGN ERROR -- xxxxxxx

Explanation: Error was detected in Phase 1. The filename of the device assigned in error is indicated by xxxxxxx. The only devices which may be allocated to an input, work or output file are 2311, 2314, and 2400. Other devices can be used for input or output if the user reads all input at E15 and writes all output at E35. This message is also generated if an expected unit has not been assigned.

System Action: Job is terminated.

Programmer Response: Check device type of sort files assigned -- only 2400, 2311, and 2314 are supported -- or omission of a file's assignment. Verify that SYS NO assignments agree with those required for each file (see Table 7).

Operator Response: Check device type of files assigned -- only 2400, 2311, 2314 supported. If device available, assign to file. Note problem on programmer report.

7201I END MERGE PH

Explanation: Phase 2 has ended.

System Action: Job continues normal processing.

Programmer Response: None.

Operator Response: None.

7302I EOJ

Explanation: End of job has been reached.

System Action: Job returns control to the calling routine (EXEC procedure or LOAD and BALR routine).

Programmer Response: None.

Operator Response: None.

7901A SORT CAPACITY EXCEEDED

Explanation: This message is generated when all intermediate work space has been used and the input file has not been exhausted.

System Action: Job terminates after message 7903I is printed.

Programmer Response: Examine input file size, printed out record count, and work space assignment.

Operator Response: If intermediate storage is on 2400, be sure all reels contain full length tapes. If they do, rerun job with more intermediate storage when available. If direct access, intermediate storage, assign more tracks when available. Note problem on programmer report.

7902A RCD COUNT OFF

Explanation: This message is generated if the number of records leaving a phase does not equal the number of records which entered, discounting any inserted or deleted by user exit routines.

System Action: Job terminates.

Programmer Response: Check for I/O errors during run and handling of these by user routine E18/E38 or BYPASS option. Have hardware checked and job rerun.

Operator Response: Check console sheet for possible I/O errors. Clear drives and rerun job. Note problem on programmer report.

7903I APPROX RCD CNT nnnn

Explanation: This message is generated after 7901A. It indicates the approximate number of records read in (represented by nnnn).

System Action: Job terminates.

Programmer Response: Rerun job with more intermediate storage. This message is always generated with message 7901A.

Operator Response: None.

7904A I/O ERR - xxxxxx

Explanation: This message is generated when a permanent I/O error occurs. xxxxx is replaced by BYPASS if the bypass parameter was specified in the INPFIL statement, or if "skip" was specified at exits E18 or E38. Otherwise xxxxxx is replaced by 24 bytes of information, consisting of the Channel Control Block.

System Action: If no user options are specified, the job terminates. See E18/38 and BYPASS option.

Programmer Response: None.

Operator Response: Examine unit specified and make sure drive clean. Rerun job. If error persists, have the hardware checked.

7905I RCD IN nnnn, OUT nnnn, ESTIMATED nnnn

Explanation: This message is generated at the end of each phase, and also after message 7902A. The first value is the number of records that entered the phase; the second value is the number of records that left the phase, and the third value is the user's estimate of the number of records given in the SIZE parameter of the SORT statement.

System Action: Job continues normal processing.

Programmer Response: None.

Operator Response: None.

7906I RCD INSERT nnnn, DELETE nnnn

Explanation: This message is generated at the end of each phase, and also after message 7902A. The first value is the total number of records inserted at user exits in this and previous phases; and the second value is the total number of records deleted at user exits in this and previous phases.

System Action: Job continues normal processing.

Programmer Response: None.

Operator Response: None.

7907I OUT OF SEQ

Explanation: This message is generated when a record written out by phase 2 or phase 3 is out of collating sequence with the previous record.

System Action: Job terminates.

Programmer Response: If a user routine (35) was modifying the records leaving the final merge phase at the time the message was generated, check the routine thoroughly. If a merge only application, make sure collated properly when created earlier.

Operator Response: Make sure the proper input files were mounted and drives were cleaned. Rerun the job. Note problem on programmer report if error occurs again.

APPENDIX E

TEST DATA

The student assignments at the end of each chapter utilize the following test data.

CARD FORMAT AND DATA

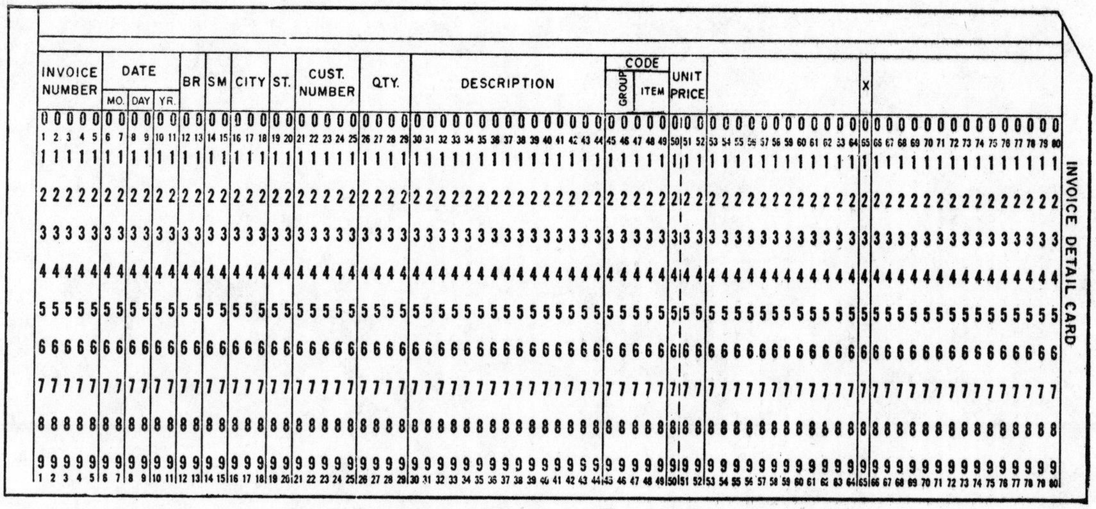

INVOICE NUMBER	DATE MO.	DATE DAY	DATE YR.	BR	SM	CITY	ST.	CUST. NUMBER	QTY.	DESCRIPTION	CODE GROUP	CODE ITEM	UNIT PRICE		X	

INVOICE DETAIL CARD

42401	6/08/67	15	79	323	05	97483	16	BROOMS	65-135	3.10
42401	6/08/67	15	79	323	05	97483	40	CIDER	19-216	1.40
42401	6/08/67	15	79	323	05	97483	200	COFFEE	25-263	2.88
42401	6/08/67	15	79	323	05	97483	150	CONDENSED MILK	76-272	2.60
42402	6/08/67	39	54	512	05	51226	9	AMMONIA	63-015	.93
42402	6/08/67	39	54	512	05	51226	16	APPLE SAUCE	22-032	1.60
42402	6/08/67	39	54	512	05	51226	9	APPLES	22-039	4.70
42402	6/08/67	39	54	512	05	51226	8	APRICOTS	22-048	3.70
42402	6/08/67	39	54	512	05	51226	10	CEYLON TEA	45-169	4.80
42403	6/08/67	39	12	257	05	82224	88	DRIED PEACHES	52-359	.82
42403	6/08/67	39	12	257	05	82224	9	FLOUR	11-383	8.10
42403	6/08/67	39	12	257	05	82224	30	GINGER SNAPS	48-424	3.50
42403	6/08/67	39	12	257	05	82224	15	MATCHES	65-537	.94
42403	6/08/67	39	12	257	05	82224	125	ONIONS	20-583	.17
42404	6/08/67	15	21	105	05	663	80	ANIMAL CRACKERS	48-024	.50
42404	6/08/67	15	21	105	05	663	30	GELATINE	49-408	1.45
42404	6/08/67	15	21	105	05	663	7	GINGER SNAPS	48-424	3.50
42404	6/08/67	15	21	105	05	663	520	LEMON SODA	36-480	.75
42404	6/08/67	15	21	105	05	663	200	PORK AND BEANS	27-697	3.05
42405	6/08/67	39	12	005	05	61520	60	ASPARAGUS	23-056	5.60
42405	6/08/67	39	12	005	05	61520	110	COFFEE	45-263	2.88
42405	6/08/67	39	12	005	05	61520	95	DRIED PEACHES	52-359	.82
42405	6/08/67	39	12	005	05	61520	90	HORSE RADISH	32-456	.42
42405	6/08/67	39	12	005	05	61520	92	ONIONS	20-58	.17
42406	6/09/67	39	12	005	05	77156	176	COCOA	46-257	.83
42406	6/09/67	39	12	005	05	77156	20	CRACKERS	48-312	3.10
42406	6/09/67	39	12	005	05	77156	97	NUTMEG	43-560	.55
42406	6/09/67	39	12	005	05	77156	130	PAPRIKA	43-63	.60
42407	6/09/67	15	21	105	05	93823	20	ANIMAL CRACKERS	48-024	.50
42407	6/09/67	15	21	105	05	93823	17	FLOUR	11-383	8.10
42407	6/09/67	15	21	105	05	93823	60	FLY PAPER	65-393	2.20
42407	6/09/67	15	21	105	05	93823	40	RAISINS	21-744	.54
42407	6/09/67	15	21	105	05	93823	50	STOVE BLACK	66-864	.60
42408	6/09/67	39	54	512	05	36401	28	AMERICAN CHEESE	14-008	.80
42408	6/09/67	39	54	512	05	36401	31	BROOMS	65-135	3.10
42408	6/09/67	39	54	512	05	36401	29	CHOW CHOW	23-207	.34
42408	6/09/67	39	54	512	05	36401	250	CIDER	19-216	1.40
42408	6/09/67	39	54	512	05	36401	160	COFFEE	45-263	2.88
42409	6/09/67	24	81	323	05	51485	360	BUTTER SALT	57-152	.18
42409	6/09/67	24	81	323	05	51485	70	CELERY	20-161	1.10
42409	6/09/67	24	81	323	05	51485	310	COCOA	46-257	.83
42409	6/09/67	24	81	323	05	51485	50	CORN	23-289	2.40
42410	6/09/67	39	12	257	05	12447	70	AMERICAN CHEESE	14-008	.80
42410	6/09/67	39	12	257	05	12447	60	CHICKEN SOUP	74-192	3.78
42410	6/09/67	39	12	257	05	12447	60	CLAM BROTH	74-233	.05

APPENDIX E

TEST DATA

...he student assignments at the end of each chapter utilize the following test data.

CARD FORMAT AND DATA

INDEX